T-LEVELS
THE NEXT LEVEL QUALIFICATION

SCIENCE

CORE

Stephen Hoare

HODDER
EDUCATION
AN HACHETTE UK COMPANY

This edition follows Version 2 of the T Level Technical Qualification in Science, first teaching from September 2023.

This resource has been endorsed by national awarding organisation, NCFE. This means that NCFE has reviewed them and agreed that they meet the necessary endorsement criteria.

Whilst NCFE has exercised reasonable care and skill in endorsing this resource, we make no representation, express or implied, with regard to the continued accuracy of the information contained in this resource. NCFE does not accept any legal responsibility or liability for any errors or omissions from the resource or the consequences thereof.

'T-LEVELS' is a registered trade mark of the Department for Education.

'T Level' is a registered trade mark of the Institute for Apprenticeships and Technical Education. The T Level Technical Qualification is a qualification approved and managed by the Institute for Apprenticeships and Technical Education.

Although every effort has been made to ensure that website addresses are correct at time of going to press, Hodder Education cannot be held responsible for the content of any website mentioned in this book. It is sometimes possible to find a relocated web page by typing in the address of the home page for a website in the URL window of your browser.

Hachette UK's policy is to use papers that are natural, renewable and recyclable products and made from wood grown in well-managed forests and other controlled sources. The logging and manufacturing processes are expected to conform to the environmental regulations of the country of origin.

Orders: please contact Hachette UK Distribution, Hely Hutchinson Centre, Milton Road, Didcot, Oxfordshire, OX11 7HH. Telephone: +44 (0)1235 827827. Email education@hachette.co.uk Lines are open from 9 a.m. to 5 p.m., Monday to Friday.
You can also order through our website: www.hoddereducation.co.uk

ISBN: 978 1 3983 4739 7

© Stephen Hoare 2022

First published in 2022 by
Hodder Education,
An Hachette UK Company
Carmelite House
50 Victoria Embankment
London EC4Y 0DZ

www.hoddereducation.co.uk

Impression number 10 9 8 7 6 5 4 3 2

Year   2026 2025 2024

Cover photo © Gorodenkoff - stock.adobe.com

Typeset in India by Integra Software Services Pvt. Ltd.

Printed by Ashford Colour Press Ltd

A catalogue record for this title is available from the British Library.

Contents

Answers can be found online at: www.hoddereducation.co.uk/subjects/science/products/t-level/science-t-level-core

Acknowledgements

Stephen Hoare

I would like to thank my wife, Janet, for her patience and forbearance during all the hours I was tied to the computer. Also to Rachel and her colleagues at Hodder Education for their support. Finally, to the NCFE reviewers for their helpful and constructive criticisms and suggestions.

Photo credits

Guide to the book

When starting your T Level Technical Qualification in Science course, you should check on the NCFE QualHub website to find out which version of the specification you should be following. This textbook follows Version 2, which is first teaching from September 2023. If you are following a later version, ensure you know how your version has been updated.

The following features can be found in this book.

Learning outcomes

Core knowledge outcomes that you must understand and learn. These are presented at the start of every chapter.

Key term

Definitions to help you understand important terms.

Reflect

Tasks and questions providing an opportunity to reflect on the knowledge learned.

Test yourself

A knowledge consolidation feature containing short questions and tasks to aid understanding and guide you to think about a topic in detail.

Research

Research-based activities – either stretch and challenge activities, enabling you to go beyond the course, or industry placement-based activities encouraging you to discover more about your placement.

Practice points

Helpful tips and guidelines to help develop professional skills during the industry placement.

Case study

Placing knowledge into a fictionalised, real-life context. Useful to introduce problem-solving and dilemmas.

Health and safety

Important points to ensure safety in the workplace.

Project practice

Short scenarios and 1–3 focused activities at the end of each chapter, reflecting one or more of the tasks that you will need to undertake during completion of the ESP. These support the development of the four core skills required.

Assessment practice

Core content containing knowledge-based practice questions at the end of each chapter.

Answers can be found online at: www.hoddereducation.co.uk/subjects/science/products/t-level/science-t-level-core

A1: Working within the health and science sector

Introduction

The health and science sector covers a wide range of organisations and employers as well as a wide range of jobs. Despite this variety, all well-run organisations usually have a common approach based around:

▶ policies and procedures
▶ quality
▶ ethics
▶ professionalism
▶ investment in the development and progression of their employees.

We will cover these aspects in this chapter and will expand on some points in future chapters.

Learning outcomes

The core knowledge outcomes that you must understand and learn:

A1.1 the purpose of organisational policies and procedures in the health and science sector

A1.2 the importance of adhering to quality standards, quality management and audit processes within the health and science sector

A1.3 the key principles of ethical practice in the health and science sector

A1.4 the purpose of following professional codes of conduct

A1.5 the difference between technical, higher technical and professional occupations in health, healthcare science and science, as defined by the Institute for Apprenticeships and Technical Education occupational maps

A1.6 opportunities to support progression within the health and science sector.

A1.1 The purpose of organisational policies and procedures in the health and science sector

In our professional lives we must maintain high standards out of respect for ourselves, our colleagues and those who require our services – customers, patients, etc. It is not enough to have good intentions; we need policies to consult and procedures to follow so that we know we are always working to the highest standards.

Equality, diversity and inclusion policy

Sometimes we can act in a way that is discriminatory without even realising it. If we stop and put ourselves in the other person's place, we might realise the effect our actions would have. Even if we do that, we may still have room to improve. That is why we have policies that cover equality, diversity and inclusion in the workplace which make it clear how to behave (Figure 1.1).

▲ Figure 1.1 Equality, diversity and inclusion should be central to our professional lives

Complying with legislation

One very good reason for having policies that cover equality, diversity and inclusion is to ensure that we comply with the relevant legislation. The main piece of legislation in the UK is the **Equality Act 2010**.

This gives legal protection from discrimination in the workplace and in wider society. Before this **law**

came into force, there were several laws that covered discrimination, including:

- Sex Discrimination Act 1975
- Race Relations Act 1976
- Disability Discrimination Act 1995.

Replacing these and other laws with a single Act made the law easier to understand and gave increased protection in some areas. The Act sets out the different ways in which it is unlawful to treat someone. The Equality Act 2010 is administered by the **Government Equalities Office**, which has produced an easy-to-read publication called 'The Equality Act – making equality real'. You can find this by carrying out an internet search using this title.

Key term

Laws (legislation): passed by Parliament. They state the rights and entitlements of individuals and provide legal rules that have to be followed. The law is upheld through the courts. If an individual or care setting breaks the law by, for example, inappropriately sharing or inaccurately recording information, they can, in certain circumstances, be fined, dismissed or given a prison sentence.

Ensuring equality

The Equality Act places responsibility on employers, providers of goods and services, caregivers, public sector bodies, private clubs and associations, voluntary organisations and many others not to discriminate on the basis of:

- age
- disability
- gender reassignment
- pregnancy and maternity
- race – this includes ethnic or national origins, colour and nationality
- religion or belief
- sex
- sexual orientation.

Eliminating discrimination

These are called **protected characteristics**. By having policies in place to cover these aspects of equality, and promoting diversity and inclusion, organisations can ensure that they comply with the law and also benefit from treating everyone fairly and equally.

We should also be aware of **indirect discrimination**. This is where there is a practice, policy or rule that applies to everyone in the same way but could have a worse effect on some people than others. Here are two examples of indirect discrimination:

- A woman has been on maternity leave. On return to work, she makes a flexible working request so that she can reduce her hours and look after her child instead of using childcare. Her manager refuses her request and says everyone doing that job must work full-time. This could be indirect sex discrimination.
- A Jewish woman works in a large store. She is told that because of a change in shifts, she now must work one Saturday a month. She explains that, as an observant Jew, she cannot work on Saturdays (the Sabbath). Her manager tells her that it would be unfair to everyone else if she were allowed not to work on Saturdays. This could be indirect religious discrimination.

Safeguarding policies

Safeguarding means ensuring individuals are protected from harm. The NHS England website is a useful source of information about safeguarding in the context of healthcare. Its definition of safeguarding is worth consulting:

> 'Safeguarding means protecting a citizen's health, wellbeing and human rights; enabling them to live free from harm, abuse and neglect. It is an integral part of providing high-quality health care. Safeguarding children, young people and adults is a collective responsibility.'
>
> Source: www.england.nhs.uk/safeguarding/about/

Note that the policy specifies 'children, young people and adults' – basically, everyone. We probably think of children and young people as being in greater need of protection. However, adults can also be vulnerable and require protection, such as people with learning difficulties or those with a physical or mental disability.

That is why safeguarding policies are required in all organisations, not just in those dealing with children, young people or the elderly. Organisations in the science sector also require proper safeguarding policies covering employees, customers and others they come into contact with, including visitors.

There are many situations where you might be working with children or vulnerable adults, such as in healthcare, childcare, education or a voluntary organisation such as Scouts or a youth club. In such situations, the employer or organisation is responsible for checking whether you have a criminal record. This is done through the Disclosure and Barring Service (hence the term DBS check – previously known as CRB). Different levels of DBS check are available, depending on how sensitive the job or role is:

- A basic check just shows any unspent convictions and conditional cautions. Convictions become 'spent' (i.e. they no longer appear on your criminal record) after a period of time, depending on age and length of sentence (if any).
- A standard check shows spent *and* unspent convictions, cautions, reprimands and final warnings.
- An enhanced check shows, in addition to the standard check, any information held by local police that is considered relevant to the role.
- An enhanced check with barred lists shows the same as an enhanced check plus whether the applicant is on the list of people barred from doing the role, e.g. someone on the sex offenders register.

You can only request a basic check yourself. For more information, search gov.uk for 'DBS'. Take care, because if you just do an internet search for 'DBS' the top results will be for commercial organisations that want to sell you a DBS check.

Employment contracts

Every employee has an employment contract with their employer. The contract does not have to be written down – in fact, as soon as someone accepts a job offer, they have a contract with their employer. This means that if either side backs out (for example, the employer withdraws the job offer or the employee decides to take a different job), they could risk legal action for compensation. The employment contract is an agreement that sets out:

- employment conditions
- rights
- responsibilities
- duties.

Both employer and employee must stick to the terms of the contract until it ends. That will happen when either side gives notice, i.e. when the employee announces they will be leaving, or the employer decides to end their employment (for example, through redundancy), or an employee is dismissed (they lose their job). The terms of the contract can be changed, usually by agreement between both sides.

Do not confuse an employment contract with a 'contract to provide services', such as when you agree with someone that they will paint your house or mow your lawn. In those circumstances, the decorator or gardener does not become your employee.

The legal parts of a contract are known as the **terms**; these are legally binding on both parties. Contract terms can take different forms:

- a written contract or statement of employment
- a verbal agreement
- in an offer letter from the employer
- in an employee handbook, on a company noticeboard or intranet.

Some terms are required by law, such as the requirement to pay at least the National Minimum Wage to all employees over 18 years of age (and the rate called the National Living Wage for people aged 23 and over), or the right to a minimum of paid holiday.

Practice point

You can find the minimum wage for your age group on the Gov.uk website:
www.gov.uk/national-minimum-wage-rates

Some contracts are based on **collective agreements**. This is where the employer or employers negotiate agreements with trade unions or staff associations which represent a group of employees.

Some terms might be **implied** rather than clearly agreed. Examples include:

- Employees should not steal from their employer.
- Your employer must provide a safe and secure working environment.
- If a job provides a company car, the employee needs a valid driving licence.
- Something that has been done regularly over a long period of time, such as paying an annual bonus or certain days off.

When you start a job, your employer is obliged to give you a **written statement of employment particulars**. This is not an employment contract. There are two statements of employment particulars. The **principal**

statement must be provided on the first day of work and covers things such as:

- the employer's name, the employee's name, job title (or description of work) and start date
- how much and how often you will be paid
- your hours and days of work and how they might change – as well as if you are expected to work Sundays, nights or overtime
- how long the job is expected to last (or, if permanent, that it is indefinite), and the end date if it is a fixed-term contract
- if there is a probation period, how long it will last and what its conditions are, e.g. to achieve satisfactory performance
- other benefits, such as childcare vouchers or free lunches
- any obligatory training.

As well as this, on day one an employer must give the employee information about:

- sick pay and procedures
- other paid leave, such as maternity and paternity leave
- notice periods, both from the employer and the employee (they may be different).

Within two months of starting work, the employer must give a **wider written statement** that covers:

- pensions and pension schemes
- any collective agreements (see above) that might be in place
- any right to other (non-compulsory) training provided by (or on behalf of) the employer
- disciplinary and grievance procedures (see page 5).

Performance reviews

How do you know that you are doing a good job? You might think you are doing well, but does your employer agree? That is why organisations usually have regular performance reviews for staff. However, this is not just a one-way process.

Performance reviews have several objectives:

- Evaluating work performance against standards and expectations: you might have been given targets to achieve or, if you work in a highly regulated sector, you might have formal standards to maintain or strive for.
- Giving feedback: a performance review gives your line manager (the person who manages you directly, i.e. your boss) the opportunity to help you improve your performance. You should expect feedback to be supportive and encouraging.

Providing opportunities to raise concerns or issues: performance reviews are not simply about the organisation evaluating your performance, you can also raise any concerns or issues that you have. Try to be non-confrontational – telling your manager exactly what they do wrong and how you could do it so much better might be a career-limiting move!

Contributing to continuing professional development (CPD): this might mean identifying areas where you need more training or education so that you can develop in your work.

Disciplinary policy

If your employer has concerns about your work, conduct or absence from work (including sickness absence), initially they should raise these concerns in an informal way. However, they can go straight to formal **disciplinary** or even dismissal procedures. A disciplinary procedure is a formal way for an employer to deal with an employee's unacceptable or improper behaviour (this is known as misconduct) or their performance (lack of capability).

Part of this process is that the employer should set and maintain expected standards of work and conduct. You need to know what is expected of you before you can be disciplined for not achieving it!

The disciplinary policy should also ensure consistent and fair treatment of all employees; there should be no favouritism, nor should individual employees feel picked on or bullied.

There should be a process for disciplinary action. This will be part of the disciplinary policy that all employers must have. You should have been given details of this process as part of the wider written statement of employment particulars that you receive within two months of starting work. This should say what performance and behaviour might lead to disciplinary action and what action your employer might take. It should also include the name of someone that you can speak to if you do not agree with your employer's decision.

Your employer's disciplinary procedure should include the following steps:

- A letter setting out the disciplinary issue.
- A meeting to discuss the issue; the employee should have the right to be accompanied by a colleague or trade union representative at this meeting. Some employers may have a policy of allowing a wider range of people to accompany you, such as a friend or relative.

- A decision about the disciplinary issue. This might result in no further action, a first or final written warning, dismissal (i.e. losing your job) or some other sanction.
- A chance to appeal the decision.

Grievance policy

In a well-run organisation, there will be open communication and consultation between managers and their staff. This means that problems and concerns can be raised quickly and settled as part of the normal working relationship.

However, anyone working in an organisation may have problems or concerns about their work or working conditions; they may have problems in their relationships with colleagues. These are all **grievances** that employees want to be addressed and, if possible, resolved. As well as this, the management will want to resolve any problems before they develop into major difficulties.

> ### Key term
>
> *Grievance:* any concern, problem or complaint you may have at work. If you take this up with your employer, it is called 'raising a grievance'.

Issues that may cause grievances include:

- terms and conditions of employment
- health and safety issues and concerns (see Chapter A3)
- relationships with colleagues and management
- bullying and harassment
- working practices, particularly when new practices are introduced
- the working environment
- changes in the organisation
- discrimination – or perceived discrimination.

However, there may be occasions where an employee has a grievance against their line manager and this needs a different approach.

As with disciplinary procedures, all employers should have a written grievance procedure. This should explain what to do if you have a grievance and what happens at each stage in the process. It should provide opportunities for employees to confidentially raise and address grievances. There should be a sequence for raising and resolving grievances. This will usually involve a meeting to discuss the issue. As with disciplinary procedures, you can appeal if you do not agree with your employer's decision.

Research

Acas, the Advisory, Conciliation and Arbitration Service, is an independent public body funded by the government. Acas works with employers and employees to improve relationships in the workplace. It has produced several codes of practice that set out the minimum standards of fairness that employers should follow. These include:
- disciplinary and grievance procedures
- collective bargaining with trade unions
- requests for flexible working.

An **employment tribunal** will use the Acas codes of practice when deciding cases. Employers do not need to follow these codes of practice but if they do not and you take your claim to an employment tribunal, your compensation might be increased, so it is in the employer's interest to follow the Acas codes of practice.

There is more information on the Acas website (www.acas.org.uk/advice). Do you think the information they give is helpful? Does it help you understand your rights as an employee?

Key term

Employment tribunals: responsible for hearing claims from people who think an employer has treated them unlawfully, for example, through unfair dismissal or discrimination.

A1.2 The importance of adhering to quality standards, quality management and audit processes

Adhering to quality standards should be central to any organisation's way of working. Those standards may be national or international standards such as British Standard or ISO (the International Organization for Standardization) or the organisation's own internal quality standards. In the health and science sector, quality standards help improve the quality of care or service provided.

Ensuring consistency

One reason for adhering to quality standards is to ensure **consistency** – always obtaining the same, high-quality outcome.

Quality and consistency are terms you will encounter a lot, both in this book and in your working life. Think about how we should always strive for both quality and consistency. If you go to a restaurant, you want the food to be consistently good. If it is consistently bad, you probably will not want to go. But what about a restaurant that is inconsistent? You might occasionally get a good meal, but is it worth a gamble? An organisation should always strive to achieve consistently high quality.

Maintaining health and safety

You will learn, in subsequent chapters, how adhering to proper procedures can help avoid (or at least reduce) accidents and harm to employees, service or care receivers or the general public.

This is covered in most detail in Chapter A3 Health, safety and environmental regulations in the health and science sector.

Monitoring processes and procedures

It is not good enough to intend to do something properly, you must do it. This applies to doing a favour for a friend but is even more important in the workplace. That is why there will often be a check sheet on the wall of a public toilet showing that it has been cleaned according to the required schedule.

This will be covered in more detail in Chapter A8 Good scientific and clinical practice.

In the summer of 2015, the Smiler roller coaster at the Alton Towers theme park crashed, causing life-altering injuries to four riders (two teenagers had to undergo amputations). The Health & Safety Executive (HSE) report found that there were no mechanical failings in the track, the cars or the system designed to keep the cars separate. The investigation identified a number of human errors that led to the crash. However, the HSE investigators found that Merlin Entertainments (the operator of the theme park) had multiple failings in not performing an adequate risk assessment and not having proper procedures to prevent a series of errors by staff, leading to harm to the public. As a result, Merlin Attractions was fined £5 million.

Do you think 'human error' is ever a valid defence or excuse when harm is caused to employees, patients, care-receivers or members of the public?

Facilitating continuous improvement

Continuous improvement means making many, often small, improvements over time. The success of the GB Olympic cycling team in recent years has been due, in part, to an approach that looks for many tiny performance improvements – in athlete training, equipment or clothing, for example. Each one might shave a hundredth or even a thousandth of a second off a lap time. Cumulatively, they have contributed to many gold medals being won.

We can take the same approach in a science, health or healthcare environment. It starts with adopting quality standards and adhering to them, monitoring performance against those standards and then looking for ways to improve performance.

Facilitating objective, independent review

Audit processes might be a legal requirement – see Chapter A8 for examples. But an audit really means asking the question: 'Did we achieve what we set out to achieve?' We need to have processes that ensure we ask that question in an objective and independent way so that we get useful answers. If we did not achieve our objective, what can we do to achieve it in future? If we did achieve our objective, are there ways we can improve further?

Quality control (QC) means the testing of a product to ensure that it meets required standards. The QC department in an organisation will be responsible for testing products before they are sold. Any product that fails QC tests will have to be reworked or scrapped.

Quality assurance (QA) means having procedures in place that ensure that the product will always meet the required standards.

Which do you think is more important, QC or QA?

A1.3 The key principles of ethical practice in the health and science sector

We are probably all aware of medical ethics – the need for medical professionals to adhere to a set of values or moral principles. This provides a framework for analysing a situation and deciding on the best course of action to take. We will expand upon that in this section. However, aspects of ethical practice are important in all areas of health and science, as we will see.

Beneficence

Put simply, **beneficence** means 'doing good'. All healthcare professionals need to follow the course of action that they believe to be in the best interest of their patient. However, 'doing good' is often too simple in the real world. It is better to think of beneficence as ranking the possible options for a patient, from best to worst, taking account of:
- Will the option resolve the medical problem?
- Is it proportionate to the scale of the problem?
- Is it compatible with the patient's individual circumstances?
- Are the option and its outcomes in line with the patient's expectations?

Several of these points are related to the patient's circumstances or expectations. This forms the basis of patient-centred or person-centred care.

Nonmaleficence

If you have seen the 2014 Disney movie 'Maleficent' you can probably work out that **maleficence** means 'doing harm', so **nonmaleficence** must mean 'not doing harm'. In that sense, beneficence (doing good) and nonmaleficence (not doing harm) go together. In the science and healthcare sector we all have a duty of both beneficence and nonmaleficence to those we are responsible for.

You can think of nonmaleficence as a threshold for treatment. In other words, if a treatment causes more harm than good then we should not consider it. That is different to beneficence, where we consider all the valid treatment options and then rank them in order of preference or benefit to the patient. A treatment could still be the most beneficial and cause more harm than good.

Another difference is that we usually think of beneficence in response to a specific situation – what is the best treatment for a patient? However, nonmaleficence is something that should always be considered in a healthcare setting. If you see someone collapse, you have a duty to provide (or seek) help for that person. Because we must try to prevent harm, it will be better for that person to receive medical attention than to be left there. Even if you are not qualified or able to help, you can at least make sure that help is given or called for (e.g. by calling 999).

We have described beneficence and nonmaleficence in the context of a doctor providing medical treatment. However, the same principles apply to all health workers who are providing care.

Here are some factors to consider in the context of nonmaleficence:
- What are the risks associated with intervening or not intervening?
- Do I have the skills necessary to help this person or carry out this action?
- Are any other factors (staff shortages, lack of resources, etc.) putting the person at risk?
- Is this person being treated with dignity and respect?

Autonomy and informed consent

Autonomy means that everyone has the right to make the final decision about their care or treatment. That means that, as caregivers, we cannot impose care or treatment on any individual, with some limited exceptions (see below).

This has not always been the case – there have been many instances of 'doctor knows best' in the past and some people might still feel the need to defer to what they see as an authority figure.

Informed consent means that before making that final decision, a person receiving care or treatment has the right to be given all the relevant information about the care or treatment. This might include the benefits, the potential risks and what might happen if the care or treatment is not given.

In some cases, the person may not have the **capacity** to give informed consent. To have capacity, the person must be able to:
- understand the information they are given
- retain that information long enough to make a decision
- weigh up or assess the information to make a decision
- communicate their decision.

If the person does not have capacity to give informed consent, the principles of beneficence and nonmaleficence should be applied. In some cases, for example, with children, the parent or guardian would have to give consent.

According to UK law, adults are over 18 years. However, 16- and 17-year-olds are considered able to give informed consent without the need for a parent. Children under 16 can also give informed consent, provided they have sufficient capacity – intelligence, competence and understanding.

In some cases, the beliefs of a parent (e.g. religious beliefs) may lead them to oppose a course of treatment that healthcare staff believe to be in the interests of the child. In such cases it might be necessary to obtain a court order to overrule the parent's wishes. Of course, this might not be possible in an emergency. In such cases, the principles of beneficence and nonmaleficence should be applied. However, this might result in the parent taking legal action. Ethical issues are not always straightforward!

Truthfulness and confidentiality

Confidentiality is central to the relationship between patients, care-receivers or the general public on the one hand and science and healthcare staff on the other. Lack of confidentiality may lead to loss of trust; if a patient feels their confidential information may be disclosed without their consent, they may withhold necessary information or even avoid seeking treatment – either way, they are less likely to receive appropriate treatment.

Truthfulness is an obligation on the part of science and healthcare staff. We have an obligation to be truthful, whether that is answering a patient's questions or reporting the results of experiments or analysis. Being truthful with patients is important, even if it might lead to them deciding against a course of action or treatment that we think will be beneficial for them. This is a consequence of informed consent that healthcare staff must accept.

> How would you apply the principles we have covered to help you deal with the following situations?
>   A colleague has told you that they have a drink problem, but that it does not affect their work. You, however, are not sure because you have noticed that they are not always fully attentive and even show signs of being drunk on duty.
>   A friend has asked if you can access their partner's medical records as they believe the partner is having an affair and they are worried about STIs (sexually transmitted infections).
>   A patient tells you that they have been using illegal drugs.

Justice

Justice can mean fairness, equality and respect for all. Therefore, when we decide whether something is ethical or not, we must think about:
  Is it legal or compatible with the law?
  Is it fair?
  Does it respect the person's right and equality?
  Does it show respect for all concerned?

A1.4 The purpose of following professional codes of conduct

Whatever area of science, health or healthcare we work in, it is likely that we will be expected to follow specific professional **codes of conduct**. It is not enough to have good intentions; we need to achieve good outcomes – codes of conduct are one way to help ensure that.

Professional codes of conduct may be written by professional societies or organisations. Some examples, covering a diverse range of professions, include:
  The Nursing and Midwifery Council (NMC)
  The Royal College of Nursing (RCN)
  The Health Care Compliance Association (HCCA)
  The Royal Society of Chemistry (RSC)
  The Institute of Food Science & Technology (IFST)
  The Science Council
  The Royal Society of Biology (RSB)
  The Society of Radiographers (SoR)
  The Health and Care Professions Council (HCPC)
  The British Association of Sport and Exercise Sciences (BASES)
  The Institute of Biomedical Science (IBMS).

There are many more. Members of these societies or organisations are expected to follow the code of conduct.

In addition, many organisations in the science, health and healthcare sectors have their own codes of conduct:
  Government agencies, such as Public Health England.
  Private companies, such as HCA Healthcare UK.
  Employer-led bodies such as the Sector Skills Councils, including Skills for Care and Skills for Health.

Professional codes of conduct will usually follow the same format:
  They clarify the missions (aims) of the organisation and its values and principles.
  They clarify the standards that everyone must adhere to.
  They outline expected professional behaviours and attitudes.
  They outline rules and responsibilities within organisations.
  They promote confidence in the organisation and profession.

A1.5 The difference between technical, higher technical and professional occupations in health, healthcare science and science, as defined by the IfATE occupational maps

The Institute for Apprenticeships and Technical Education (IfATE) is an employer-led organisation sponsored by the Department for Education. A key element in the work of the Institute is to support employer groups in developing apprenticeships.

The Institute also maintains the **occupational maps** that underpin technical education. These occupational maps show where technical education can lead. They group occupations that have related knowledge, skills and behaviours into **pathways** so that it is easier to see opportunities for career progression within a particular route. Within each pathway, occupations at the same **level** are grouped into clusters to show how skills you have learned can be applied to other related occupations (Figure 1.2).

This is a small selection of the qualifications available at each level:

▷ Level 1 qualifications:
  – GCSE grades 3 to 1 or D to G
  – Level 1 NVQ.
▷ Level 2 qualifications:
  – GCSE grades 9 to 4 or A* to C
  – Intermediate apprenticeship
  – Level 2 award, certificate or diploma.
▷ Level 3 qualifications:
  – AS/A Level
  – T Level
  – Advanced apprenticeship.
▷ Level 4 qualifications:
  – Higher apprenticeship
  – Higher national certificate (HNC).
▷ Level 5 qualifications:
  – Foundation degree
  – Diploma of higher education (DipHE)
  – Higher national diploma (HND).
▷ Level 6 qualifications:
  – Ordinary or honours degree, e.g. BA, BSc.
▷ Level 7 qualifications:
  – Master's degree, e.g. MA, MSc, MChem, Meng.
▷ Level 8 qualifications:
  – Doctorate, e.g. PhD or DPhil.

For a full list, visit **www.gov.uk** and search for 'What qualification levels mean'.

Technical

These are skilled occupations that a college leaver or an apprentice would be entering, typically requiring qualifications at levels 2/3. Examples include:

▷ adult care worker/lead care worker
▷ healthcare support worker

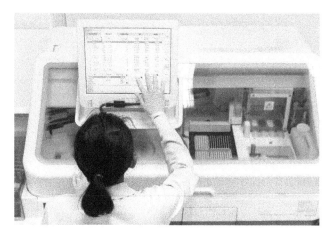

▲ Figure 1.2 Modern laboratory equipment needs qualified and highly trained staff

- dental nurse
- food technologist
- laboratory technician.

Higher technical

These are occupations that require more knowledge and skills. This could be acquired through experience in the workplace or further technical education. They typically require qualifications at levels 4/5. Examples include:
- lead practitioner in adult care
- healthcare assistant practitioner
- nursing associate
- dental technician
- food testing/laboratory manager
- technician scientist.

Professional

These are all occupations where there is a clear career progression from higher technical occupations, as well as occupations where a degree apprenticeship exists (level 6). Examples include:
- social worker
- healthcare science practitioner
- registered nurse or midwife
- biochemist/biologist/chemist/physicist
- research scientist.

Research

You can view the latest occupational maps on the Institute for Apprenticeships & Technical Education website (www.instituteforapprenticeships.org/about/occupational-maps) or search online for 'Institute for Apprenticeships occupational maps'.

Were you able to find relevant information? Will this be a useful resource to help you to plan your career?

A1.6 Opportunities to support progression within the health and science sector

When you were a child, what did you want to be when you grew up? Is that still what you want to do? Some people seem able to plan their careers and then pursue their objectives with single-minded determination. Others may move from job to job without any clear plan. The former group is usually, but not always, more successful than the latter. Whichever category you fall into, the end of your T Level course is just the beginning. It helps if you have a plan as to how you can

progress in your career. Even if you are not sure where you want to go, at the very least you should be aware of the opportunities that are available.

Research

Although it is more relevant to the science sector than health or healthcare sectors, the Royal Society of Chemistry offers a 'careers toolkit' of online resources to its members.

Other professional bodies in your field may offer something similar. You should use all the resources and sources of advice and information available to you. Look at the professional bodies listed in section A1.4. Are any of those relevant to your chosen field of work? If so, their website might have useful resources. Make a list of sources of help and information about how to progress your career.

Undertaking further/higher education programmes

As you come to finish your T Level, it is a good idea to have already planned your next move. You will have achieved a level 3 qualification, so you should normally consider moving on to a level 4 or level 5 qualification, unless you decide to change track – in which case there will be a range of other level 3 qualifications that might be suitable.

If you plan to remain in the science, health or healthcare sector, you will probably consider a level 4 or level 5 qualification appropriate to your chosen field of work, such as Higher Technical Qualifications. In some cases this will mean that you have to become registered with a statutory regulator, such as the Nursing and Midwifery Council or the General Dental Council.

Your T Level will be worth UCAS points, so you can continue into higher education (level 5 or 6) at university or with another education provider if you wish.

Undertaking apprenticeship/degree apprenticeship

An **apprenticeship** is a job with training to industry standards and should involve entry into a recognised occupation. Apprenticeships are employer-led, so employers will:
- set the standards the apprentices need to meet
- create the demand for apprentices to meet their skills needs
- fund the apprenticeship, i.e. pay for training

- employ the apprentice, i.e. pay them and give them work
- be responsible for training the apprentice on the job.

The needs of the apprentice are also important. Apprentices are not meant to be simply a source of cheap labour. The apprentice must be able to achieve competence in a skilled occupation. Not only that, but they should also acquire skills that are transferable and offer the possibility of long-term earnings potential, greater security and the ability to progress in the workplace.

A higher apprenticeship (level 4) might lead on naturally from a level 3 T Level, but entry to a level 6 or level 7 degree apprenticeship is also possible. Degree apprenticeships combine working for an employer with studying at a university. Study periods can be on a day-to-day basis or in blocks, depending on the programme and the needs of the employer.

More information about degree apprenticeships is available on the UCAS website (www.ucas.com) or the Institute for Apprenticeships and Technical Education website (www.instituteforapprenticeships.org).

Undertaking continuing professional development (CPD)

Continuing professional development can take many forms. It is a way in which professionals use different learning activities to maintain, develop and enhance their abilities, skills and knowledge. CPD combines different methods of learning, such as:

- conferences and events
- training workshops
- e-learning programmes
- best practice techniques
- ideas sharing
- shadowing a more experienced professional in the field.

CPD programmes are often run by employers or professional bodies such as those described in section A1.4.

Joining professional bodies

Professional bodies fulfil a number of important functions. As well as being the guardians of professional codes of conduct in their area of expertise, they offer CPD programmes.

In some occupations in the science, health and healthcare sectors you need to be registered with a statutory body, such as one of the professional bodies.

Some professional bodies offer **chartered** status. As well as indicating an in-depth knowledge of the field, chartered status is required in some regulated activities that have to be supervised by a **qualified person**, such as production of pharmaceuticals (see section A8.3 for more information). Examples include:

- Chartered Chemist (CChem) administered by the Royal Society of Chemistry
- Chartered Biologist (CBiol) administered by the Royal Society of Biology
- Chartered Physicist (CPhys) administered by the Institute of Physics
- Chartered Scientist (CSci) administered by the Science Council.

Undertaking an internship

Internships can offer valuable experience in a real work environment – particularly if you have not gained this through an apprenticeship. Internships are usually relatively short and often take place during the summer months, as many are designed for university students. Placements are similar, but generally last longer. Internships and placements are usually offered by large companies, such as GSK (which manufactures pharmaceuticals) or Unilever (consumer products). In some cases, you will be paid at least the UK National Living Wage, but in others it can be much higher than this – though some internships are not paid at all. Bursaries are often available to cover your costs in an unpaid internship. Many of the professional bodies already mentioned will offer help with internships, placements or bursaries. Their websites are the best place to look for advice and information.

Undertaking a scholarship

As well as help with bursaries, many of the professional bodies can offer help with scholarships. These are usually available to help with the costs of obtaining higher qualifications, usually at level 6 or level 7. Educational institutions that offer these qualifications may also offer scholarships or can give guidance on what scholarships and other sources of funding are available.

Project practice

You are working in a science/health/healthcare organisation (choose one according to your own area of work). You have been asked to produce materials to help new apprentices understand the importance of the working practices of the organisation, as well as to inform them about the ways in which their careers might develop.

1 Prepare a summary of the organisation policies that you are aware of in your organisation, or ones that you know should be in place. Give explanations for the relevance and importance of these.

2 Research the professional codes of practice relevant to your area of work. This might require you to use the websites of any relevant professional bodies to gather information.

3 Prepare a list of the types of CPD that are available or recommended in your organisation.

4 Finally, outline the additional ways in which apprentices can progress in their careers.

You should present the information in the form of a poster or short written document, such as an employee handbook.

Assessment practice

1 What piece of legislation covers the requirement for diversity and inclusion for people with certain characteristics?

2 Who is responsible for obtaining a DBS check for work?

3 What is the name for the legal parts of an employment contract?

4 What are collective agreements?

5 Your employer has a disciplinary policy that includes informal and formal written warnings. You have been found stealing and dismissed. You feel that you have been treated unfairly because you were not given any warnings or a notice period. Are you correct?

6 Give **two** reasons why an organisation needs an equality, diversity and inclusion policy.

7 Explain, using an example, what is meant by safeguarding.

8 Give **two** reasons why organisations adhere to quality standards.

9 During the early stages of the COVID-19 pandemic, there were serious concerns that NHS hospitals would be overwhelmed and unable to treat patients. Therefore, hospitals were instructed by the government to discharge any patients who could be transferred back to their care homes. In many cases this led to the introduction of COVID-19 into care homes from hospitals because patients were not tested for COVID-19 or were known to be infected.

Evaluate this instruction, considering the key principles of ethical practice.

Your response should demonstrate:
– reasoned judgements
– informed conclusions.

A2: The science sector

Introduction

The term 'science sector' covers a huge range of organisations and occupations. You can get an idea of its breadth and diversity by thinking first about the different areas of science:
▶ life sciences – this covers all of the various aspects of biology, such as botany, zoology and ecology, as well as areas bordering on medicine, such as anatomy and physiology
▶ chemical sciences, such as physical and organic chemistry
▶ physical sciences, such as physics but also mechanics and engineering.

There are also areas of overlap between these three broad areas, for example:
▶ environmental science
▶ pharmaceutical sciences (incorporating areas of chemistry and biology)
▶ materials science (incorporating areas of chemistry and physics).

There are many others, some of which will be discussed in sections B1 and B2, but the science sector is much too wide to be covered entirely in one textbook.
This chapter aims to provide an overview of the sector and the opportunities that might be open to you.

Learning outcomes

The core knowledge outcomes that you must understand and learn:

A2.1 factors that contribute to the diversity of employers/organisations within the science sector

A2.2 the diversity of work undertaken in different job roles within the science sector

A2.3 possible employers and job roles that require the application of science in non-science sectors

A2.4 the difference between a job description and a person specification

A2.5 how individual roles fit into teams within an organisation

A2.6 the individual's responsibilities in relation to the wider team

A2.7 the principles of good laboratory practice (GLP)

A2.8 the principles of good manufacturing practice (GMP) in ensuring that products meet standards

A2.9 the key principles of continuous improvement in relation to scientific tasks

A2.10 the difference between quality assurance and quality control

A2.11 how organisations in the science sector ensure compliance with internal and external regulations

A2.12 how regulatory controls apply in different working environments within the science sector

A2.13 factors that may have an impact on the commercial activities (for example, pharmaceuticals, cosmetics, manufacturing, services) of science organisations

A2.14 the importance and impact of innovation in the science sector.

A2.1 Factors that contribute to the diversity of employers/organisations within the science sector

In 2019, according to the Office for National Statistics (ONS), 8.5% of UK employees were categorised as 'Professional, Scientific, Technical' whilst 8.2% were categorised as 'Manufacturing'. Of course, not all of these could be described as science jobs. The science sector is so diverse that it can be difficult to get reliable figures for numbers of employees. However, between 2018 and 2019, the second largest increase in employees was in the 'professional, scientific and technical' category, which is encouraging for anyone planning a career in the science sector.

There is a great variety of organisations within the science sector and this is influenced by a number of factors that we shall cover here.

Size of employer/organisation

There is a wide range of sizes of employers/organisations:

▶ The NHS employed about 1.3 million people in the UK in 2021, and is one of the largest employers in the world.

▶ Large pharmaceutical companies employ thousands of people in the UK, for example GSK (13 900 UK employees), AstraZeneca (2600 UK employees) and Novartis (1300 UK employees).

▶ Smaller pharmaceutical companies working in developing new drugs or providing services to the larger pharmaceutical companies may have anything from just a few people to a few hundred employees.

▶ Large chemical manufacturing companies, such as Johnson Matthey employ many people (almost 14 000 employees at the end of 2020).

▶ Newly formed start-up companies may have small workforces to begin with; these tend to be founded by university researchers, often with funding from the university, to commercialise their research.

Case study

BioCity Nottingham

BioCity Nottingham (https://biocity.co.uk) was established in 2003 by the University of Nottingham and Nottingham Trent University, who were each part-owners of the business. The buildings, located on the site where the painkiller ibuprofen was developed many years earlier, were donated by the pharmaceutical company BASF. The function of

BioCity is to act as a 'bioincubator' – a place where new companies can rent lab space to exploit new technologies in the life sciences. These companies are often spin-outs from local universities trying to commercialise their research.

Over the years, many companies have been formed and located in BioCity. Some have not lasted, but most have gone on to grow and attract new investment. Several successful companies have been bought by larger companies and moved on from BioCity, freeing up space for new companies to occupy.

There are a number of similar bioincubators located throughout the UK, including ones in Glasgow and Cheshire also set up by BioCity. There are numerous science parks also, that provide space and facilities for growing companies in all areas of science, not just the life sciences. Many of these are members of the UK Science Park Association (UKSPA). You can get an idea of the range of employment available in the sector on the UKSPA website: www.ukspa.org.uk

Funding streams

Scientific research and development is funded largely through grants provided by:

▶ government bodies (research councils) under the umbrella of UK Research and Innovation (UKRI); those most relevant to the science sector are:

  – the **Biotechnology and Biological Sciences Research Council (BBSRC)**, which supports mostly academic research in the biological sciences in universities

  – the **Engineering and Physical Sciences Research Council (EPSRC)**, which fulfils a similar role in engineering and physical sciences

  – the **Medical Research Council (MRC)**, which supports research in medical sciences, mostly in universities, research institutes and teaching hospitals

  – Innovate UK helps businesses to develop new products, services and processes

▶ charitable organisations, such as:

  – Wellcome, a charitable foundation that funds medical and pharmaceutical research

  – the Wolfson Foundation, which funds buildings and other infrastructure and equipment, mostly in the health sector

  – Cancer Research UK and many other medical charities, large and small

– Lloyd's Register Foundation, which funds research in engineering to help improve the safety of infrastructure in areas such as energy, transport and food.

There is a wide variety of other funding streams for science-based companies, including:
▶ commercial loans from banks and other investors
▶ **venture capital** (organisations that invest in new or growing companies in exchange for a share in the ownership, rather like on the TV programme *Dragons' Den*)
▶ shareholders, who provide funds for companies.

Some science-based companies sell their services, such as research, product development or testing, and use the income generated to fund their activities. Small and medium-sized companies known as **Contract Research Organisations (CROs)** undertake work under contract for large pharmaceutical companies. These contracts might cover parts of the drug discovery process (finding new drug compounds), **toxicology testing** (to see if the drug is safe) or **DMPK studies** (to study how a drug is absorbed and broken down in the body).

Commercial status

From what we have discussed so far in this chapter, you will have noticed that employment in the science sector is spread across the whole range of types of organisation:
▶ national governmental organisations, such as the NHS and the civil service
▶ local government organisations, such as departments responsible for health, environment and standards (for example, Trading Standards)
▶ education, including schools, colleges and universities
▶ research institutes, charities and other not-for-profit organisations
▶ limited companies (see below).

There are two types of limited companies. In both cases they issue shares that investors can purchase, and this provides capital (money) to fund the business. Once established, shares can be bought and traded (that is, sold) and the shareholders usually receive a share in the profits made by the business, known as a **dividend**. **Limited liability companies** are privately owned and usually have 'Limited' or 'Ltd' after the company name. Larger companies are usually **public limited companies** – the shares are traded publicly on the stock market – and have 'plc' after the name.

Working environments

We usually think of scientists and technicians working in laboratories, which is certainly often the case. However, many will be employed in the manufacturing industry, working in manufacturing plants and factories, as well as in laboratories.

Some science-based jobs are 'in the field'. This does not mean you are actually working in a field, but that much of the work will be out and about in relevant locations. Examples include:
▶ performing ecological or geological surveys
▶ environmental monitoring, such as collecting air samples or water samples from rivers, lakes or reservoirs
▶ agricultural and veterinary science technicians.

In other cases, scientists are desk-based rather than laboratory-based. Examples include:
▶ science-related computer work (bioinformaticians, biostatisticians)
▶ science publishing, including educational publishing
▶ commercial roles, such as sales & marketing, purchasing, planning, etc. in science-based industries.

Scientists can also work with patients in hospitals or other healthcare environments.

Geographic location

Employment in the science sector is spread throughout the UK; this is particularly true of jobs in the NHS and local government. However, some sectors are concentrated in particular areas.
▶ Many life science and biotechnology companies are based in the 'golden triangle' of London, Oxford and Cambridge; this is because they were formed as spin-out companies from those universities or have strong research links with them.
▶ Pharmaceutical research is located mostly in the east and south-east of England, although there are some research facilities elsewhere, for example in the north-west of England and Scotland.
▶ Pharmaceutical manufacturing is located largely in the north-west of England.
▶ Chemical manufacturing is concentrated in the north-east of England, particularly on Teesside.

A2.2 The diversity of work undertaken in different job roles within the science sector

Just as there is great diversity in the type and location of employers in the science sector, there is also great diversity in the types of job role available.

Research and development

Whether it is in pharmaceuticals or chemicals, the petroleum industry, electronics or engineering, we tend to associate careers in the science sector with **research and development (R&D)** – scientists working in labs to discover new drugs or develop new products and technologies.

There is usually a high demand for skilled technicians in most R&D labs. A lot of R&D involves sophisticated equipment that must be operated by highly trained staff. Research scientists are usually graduates or have PhDs, but opportunities exist at all levels (as described in section A1.5) and good employers will usually offer the opportunity for progression, as described in section A1.6.

Data analysis

Chapter A5 covers data and information management; you will see from this just how important data is in the science and healthcare sectors. **Data analysts** (people who analyse and process data) are required in many areas. Large organisations, such as the NHS and major pharmaceutical or manufacturing companies, generate huge quantities of data that needs to be processed and analysed as the basis of sound decision making. The government also requires data analysis – just think of the COVID-19 statistics that featured on news reports each evening during the pandemic. These were collected and collated by the **Office for National Statistics (ONS)**, which is responsible for collecting and reporting statistics about the economy, population and society in the UK.

Clinical testing/trials

A study published in 2020 by a group led by the London School of Economics and Political Science (LSE) estimated the average cost of developing a new drug was $1.3 billion. Almost half of that is the cost of clinical trials. These usually occur in three phases:
▶ **phase I:** trialling the drug with small numbers of healthy volunteers to test safety and to establish the dose needed
▶ **phase II:** larger trials to judge how well the drug works
▶ **phase III:** even larger trials carried out in many hospitals, often worldwide, to compare the drug to existing treatments and to look for any potentially harmful side effects.

Although clinicians (doctors and nurses) are heavily involved in such trials, there are numerous opportunities for technical staff who will be involved in many aspects such as administration of the trials, preparation of trial materials, analysis of samples taken from patients and data analysis.

Quality control and quality assurance

Section A2.10 covers the differences between quality control (QC) and quality assurance (QA). In essence, QC involves testing a product to ensure it meets the required specification. This is mostly lab-based analytical work. QA means ensuring that the product is made in a consistent way that will meet the required specification. This involves working more closely with the production department to ensure processes are fit for purpose.

The aim of QA is to prevent quality problems before they arise. In other words, it is proactive. QC aims to find and reject defective products or components once problems have occurred. So, we can say that QC is reactive. The details of QA and QC procedures will vary depending on the type of organisation and the type of processes carried out. The case study illustrates the work done by QA and QC in a small engineering company. Some of these points will apply in many organisations, whilst others will be specific to this type of process in this type of company.

Case study

A small engineering company produces automated chemical synthesisers for use in drug discovery and development. They are developing a new model. The QA process begins at a very early stage in the product development.

These are just a few of the issues that must be addressed in the QA process:
▶ What elements of the design are new? Could these lead to problems during production?
▶ What are the specifications of the product? Can these be achieved?
▶ What has been learned from previous experience in assembling similar equipment? Are there any issues that can be predicted?
▶ What standards, such as ISO, are applicable? How does the production process have to be designed and managed to ensure that these standards are met?
▶ Are all the necessary procedures in place to minimise production problems and ensure that the product meets its specifications?
▶ Are production staff qualified and trained sufficiently to ensure the product will meet the necessary standards and specifications?
▶ Is the production process, including all SOPs, sufficiently well worked out?

QC will involve testing the final product to ensure that it meets the specifications, and so the QC department must prepare SOPs for the various tests that will be required. However, many of the components are bought in from specialist suppliers and some of these need to be tested to ensure that they meet the company's own requirements for quality. Testing of **sub-assemblies** (parts assembled separately to be used within a product) is also required during the production process to ensure they are fully operational before being incorporated into the finished equipment.

QC procedures also include actions required to correct any faults or defects, either during the production process or in the final product.

Think about your own work experience.
▶ Which aspects of QA and QC have been applied in your workplace?
▶ Can you identify which of these are QA (proactive) and which are QC (reactive)?

Product development

Product development covers a wide range of activities across the whole science sector. It involves taking a concept or idea through stages of development to where it can be sold. Typical stages are:
▶ identifying a market need
▶ deciding if people will buy the product
▶ devising how the product will address the need
▶ deciding whether the product will be viable
▶ designing an initial version that has sufficient functionality to be used by customers
▶ releasing this to customers/users and refining the product based on user feedback.

Product development is usually done by multidisciplinary teams, that is, a group of people with different skills and backgrounds working together to bring the product to the point where it can be sold.

Scientific publishing

Books and journals in science cover a wide range of subjects. Their publication requires people with a strong background in science who are able to use their knowledge to produce scientific literature of a high standard. Scientific publishing includes:
▶ research journals covering a range of scientific disciplines, such as *Science, Nature, Proceedings of the National Academy of Sciences of the USA* (*PNAS*)
▶ research journals covering more specific areas of science, such as *Chemical Communications, Journal of Cell Biology, Biochemical Journal, Journal of Immunology*

and the *Journal of Physics*; these are often published by learned societies such as the Royal Society of Chemistry or the Institute of Physics
▶ science magazines such as *New Scientist, Scientific American, The Scientist*
▶ textbooks and reference works.

Although the authors of many of these will be active researchers or experts in their specific field, the process of assembling, editing and publishing these works requires people with a more general scientific background. Some areas of scientific publishing are more like scientific journalism (see section A2.3).

Scientific publishing is moving online increasingly. Like paper-based scientific journals, this is often available only on subscription (so you need a login to access it). However, there are an increasing number of free-to-access online journals.

Manufacturing

Manufacturing processes are becoming increasingly sophisticated and, therefore, there is a requirement for skilled people to operate them. Many manufacturing jobs in the science sector require people with technical skills, which can be gained via apprenticeships or further education. Examples include:
▶ chemical manufacturing
▶ pharmaceutical and biotech manufacturing, including formulated products (medicines, injectable products)
▶ nuclear manufacturing
▶ food manufacturing.

Manufacturing technicians are responsible for operating manufacturing processes in accordance with **standard operating procedures (SOPs)**, either alone or as part of a team. They also need to be able to follow quality procedures, understand the regulatory environment (including health and safety legislation) and follow ethical practice and codes of conduct.

Research

The National Careers Service is a government organisation that helps with career, learning and training course in England.

Visit https://nationalcareers.service.gov.uk/job-categories/science-and-research for information about a selection of careers in the science sector.

Are there any that you were not aware of? Do they sound interesting? Would this resource help you in making decisions about your future?

Reflect

What types of jobs in the science sector do not involve working in a laboratory? Are there more of these than you originally thought? Which type do you think you would find more attractive?

A2.3 Possible employers and job roles that require the application of science in non-science sectors

This covers areas that do not appear to be obviously science-based but where knowledge of science and the scientific method is needed to do the job well.

Communication and outreach

We have covered scientific publishing already, but science journalism is another related area where a good knowledge of science is important. Science journalists may write for national newspapers or for specialist science magazines. Many will work on a freelance basis, meaning that they sell 'stories' to a number of publications rather than being employed by a single publisher.

Public relations (PR) and science communication are roles that you will find in most pharmaceutical and manufacturing companies, as well as in medical charities and patient groups, academic societies, universities and government organisations.

Education

Besides the obvious example of a science teacher, there are other opportunities for those with science and technical qualifications, for example education officers in organisations such as:
- museums or zoos
- charities, such as the RSPB or RSPCA
- universities and research institutes.

Policy

Academic societies (such as the Biochemical Society), trade associations (such as the Association of the British Pharmaceutical Industry) and professional bodies (such as the Royal Society of Chemistry) are all involved in developing science policy. This is concerned with allocation of resources to science in the public interest. This covers aspects such as:
- funding of science
- developing the careers of scientists
- turning scientific discoveries into commercial products
- the role of science in economic growth and development.

These roles involve people with skills and experience in administration as well as science.

Public service

Sir Chris Whitty (Chief Medical Officer for England) and Sir Patrick Vallance (Chief Scientific Adviser to the UK government) featured regularly in government press briefings and on the television news during the COVID-19 pandemic. There are many more science-based jobs in the civil service, although less high profile.

The civil service offers a science and engineering fast track programme for graduates (including degree apprenticeships) who could be working on a wide range of functions:
- understanding and anticipating the implication of science and technology developments
- analysing science and engineering evidence
- assessing risks and opportunities
- designing, developing, testing and evaluating science and engineering services
- monitoring, inspecting, enforcing and advising on regulations that protect society, health and the environment.

This could cover areas of science and technology such as:
- climate change
- health science
- defence and security
- energy technology or innovation in transport.

Individuals would apply to the civil service for such roles.

Many other national and local government departments need scientists with different levels of qualification and experience. Examples include waste management, environmental monitoring, conservation and food hygiene.

A2.4 The difference between a job description and a person specification

Having decided what you want to do as a career, it is helpful to be familiar with some of the terminology you will encounter. One possible confusion is between a **job description** and a **person specification**.

Job description

A job description is a detailed description of the individual roles of a job. It should include the responsibilities, objectives and requirements of the job. In simple terms, it describes what you will do in the job and what you will need to achieve or be in charge of.

As an example, here is part of a job description for a senior science technician (chemistry) in a sixth-form college:

▶ Provide and enable the use of practical resources and facilities for teaching and learning in chemistry-related work and projects.
▶ Prepare solutions; assemble apparatus and equipment for practicals; and set up, test and assist with demonstrations.
▶ Assist in the safe clearing away of used resources and waste in accordance with health and safety regulations following best practice.
▶ Give technical advice and help students and teachers with practical and project work.

Note the last point about giving technical advice and help to teachers (not just students). A good technician in an educational setting will have skills and experience complementary to those of a teacher.

Person specification

A person specification is a profile of the necessary skills and attributes that a person suitable for the job must have. In other words, it describes the type of person who could do the job. The person specification for the senior science technician (chemistry) included the following:

▶ Educated to at least Level 3 in a relevant science subject.
▶ Recent laboratory experience.
▶ Awareness of COSHH (Control of Substances Hazardous to Health) and risk assessments.
▶ Ability to prioritise workload.
▶ Excellent time management and organisational skills.
▶ Ability to work within a team.
▶ Ability to communicate effectively with subject staff, colleagues and students.

A2.5 How individual roles fit into teams within an organisation

Unless you decide on a career as a freelance science journalist, it is quite likely that you will be working in a team within an organisation. In the past, many organisations were very hierarchical: they had many layers of management, with a small group of people at the top of the structure being far removed from those at the bottom, who were doing the work on a daily basis. That type of structure is less common now as it can be inflexible and inefficient. Nevertheless, any organisation will have a structure and it is important that you know where you fit into it.

People you work with

Most companies will have an **organisation chart** or **organigram** (sometimes spelt 'organogram'). Figure 2.1 shows an organisation chart from a typical biotech or small pharmaceutical company.

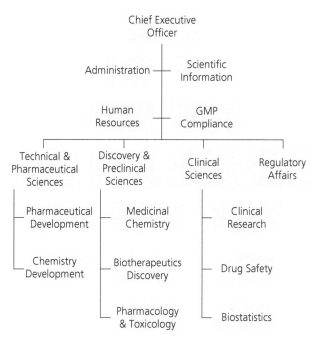

▲ Figure 2.1 A typical organisation chart

As the name suggests, it shows the organisation of the company as well as the reporting relationships. For example, there would be a Head of Clinical Sciences who would be responsible for the three departments: Clinical Research, Drug Safety and Biostatistics. The head of each department would report to the Head of Clinical Sciences who would, in turn, report to the Chief Executive Officer (CEO).

Functions such as Administration and Human Resources operate across the whole organisation, interacting with people from all departments, so these are shown below the CEO. Larger organisations are likely to have other senior roles, such as Chief Operating Officer (COO), Chief Scientific Officer (CSO) and Chief Financial Officer (CFO).

Even larger organisations would be split into **divisions** and each division would have its own structure along these lines. For example, the pharmaceutical company GSK has three main divisions:

- pharmaceuticals
- vaccines
- consumer healthcare (medicines that can be bought without a prescription, as well as vitamins, minerals and supplements).

Similar organisation charts can be used to show individual teams or departments, and might include the names of the people occupying the roles.

People you report to

An organisation chart may show how you fit into the whole organisation, or at least how your particular department fits in. However, you need to know who you **report** to. This will be your manager (sometimes called a **line manager**) or supervisor. When we say 'report to', it does not necessarily mean that you give them a regular report (although that might be the case); it simply means that they are your boss.

People that you manage

As you progress in your career, you may have people who report to you. These might be more junior staff or trainees. You will be responsible for managing them, which is likely to involve providing support and guidance. You will also need to ensure that they carry out their assigned tasks on time and to the required standards, not to mention safely. To do this, you will need to allow them freedom to learn and grow into their roles. This means you have to strike a balance between appearing to be checking up on them too much on the one hand and allowing them to make costly or even dangerous mistakes on the other.

A2.6 The individual's responsibilities in relation to the wider team

Your line manager will have particular requirements of you, such as performing the tasks assigned to you. However, you will also have responsibilities to the wider team and to anyone else you or your work interacts with.

Health and safety

Your actions can have an impact on other colleagues, the whole organisation and even the wider environment. For example, storing, handling and disposing of hazardous substances must be done correctly otherwise colleagues could be harmed and/or the organisation could be in breach of laws and regulations. (This topic is covered in detail in Chapters A3 and A4.)

It is always better to check whether something is being done safely and correctly rather than assuming that it is not your responsibility.

Security

There are numerous reasons why it is important to take security seriously. Many labs will have some form of access control, such as key-card entry or combination locks. These can be annoying, particularly if you forget your key card, but they are there for a reason – to keep unauthorised people out of areas where they could harm themselves or the work being done. They might also be required to contain hazardous materials within a controlled area (see section A4.8 for an example of this). Wedging open doors or letting a stranger through the door with you can have serious consequences. In real life, you might find it difficult to refuse access to someone who claims to have forgotten the combination or lost their key card, but security measures like this will have been put in place for a good reason and rely on everyone playing their part to avoid security breaches.

It is also important to use technology safely and securely. For example, many automated or remote-control systems control dangerous equipment or processes. These systems should be password protected in order to prevent non-authorised personnel from using them and potentially causing damage or harm.

Section A5.10 covers the security measures that are needed to protect data.

Organisational policies and procedures

Organisational policies and procedures, such as SOPs, ensure that work is done in a correct way and that there will be no adverse effect on the quality of the product or service. The importance of following organisational policies and procedures was discussed in Chapter A1, and SOPs are covered in more detail in Chapter A8.

Deadlines

In the health and science sectors it is possible that someone's life may depend on you completing your work to schedule. At the very least, it can mean colleagues are delayed in their work and there could be cost implications for the organisation.

Departmental dependencies

Many organisations in the health and science sectors are highly **integrated**: in other words, the work of one part of the organisation will depend on the work of other departments. That is why meeting deadlines can be so important.

There are other ways in which other departments may depend on you doing your work carefully and accurately. For example, you may be responsible for preparing samples for colleagues to analyse. If you are careless then it is possible that the results of the analysis will be incorrect. This could have serious implications for the work of the organisation and, if nothing else, your colleagues might be blamed for your mistakes.

A2.7 The principles of good laboratory practice

Good laboratory practice (GLP) is important in all types of laboratories, but particularly in those that are subject to additional regulation. GLP helps to ensure that everyone can have confidence in the results of your laboratory work. GLP is a set of principles providing a framework for planning, performing, monitoring, recording, reporting and archiving. It is important not to confuse GLP with standards for laboratory safety.

GLP is particularly important in laboratories involved in testing. According to the Medicines and Healthcare products Regulatory Agency (MHRA), any test facility must comply with GLP when carrying out tests on:
▶ pharmaceuticals
▶ agrochemicals
▶ veterinary medicines
▶ industrial chemicals
▶ cosmetics
▶ additives for human food and animal feed
▶ biocides (chemicals used as disinfectants or to control the growth of micro-organisms).

To ensure compliance, the test facility must belong to the UK GLP compliance monitoring programme that is run by the UK GLP Monitoring Authority.

A key aspect of GLP is following SOPs (see page 122), but these need to have been drawn up to incorporate the principles of GLP.

By following GLP, organisations can ensure:
▶ the quality, reliability and integrity of the studies that they undertake, meaning that the results will be valid and the work carried out to the correct standards
▶ that they are able to report verifiable conclusions that can be accepted by customers or collaborators
▶ that there is full traceability of data, for example by using proper laboratory notebooks or **laboratory information management systems (LIMS)**.

See section A6.3 for more on LIMS.

Case study

Industrial Bio-Test Laboratories (IBT Labs)

IBT Labs was an American company responsible for the safety testing of pharmaceuticals, pesticides and food additives, and supporting registration approval of these products by the US Food and Drug Administration (FDA). The company was founded in 1953 by an American professor and quickly gained a reputation for producing high-quality work at a reasonable price.

In 1966, IBT Labs was bought by another company and rapid expansion of the business followed. In 1976, an investigator from the FDA questioned the work being done by IBT Labs. Further investigation showed irregularities and suspected scientific misconduct in their work. Eventually, the company and three former officials were convicted of falsifying or altering the results of key product safety tests used to obtain approval for two popular pesticides and two commonly used drugs. The company was fined, and senior personnel were imprisoned.

Further investigation by the FDA showed that the problems at IBT Labs were not restricted to just this one company. There were some cases of fraud, but there was a more widespread problem of lack of organisation and poor management.

The FDA decided that the problem could only be dealt with by imposing regulations – the US GLP regulations. Soon after, everyone realised that there was a need for international agreement on GLP and this led the Organisation for Economic Co-operation and Development (OECD) to publish its 'Principles of Good Laboratory Practice' in 1982 based on the US GLP regulations. These became the international standard for GLP.

The main principles of GLP are:

▶ **Resources:** The structure and responsibilities of the organisation should be clearly defined and staffing levels should be sufficient to perform the required tests. Qualifications and training of staff should be documented. Facilities and equipment should also be sufficient and in working order. This is achieved by having a strict programme of qualification (ensuring it is fit for purpose), calibration and maintenance of all equipment.

▶ **Rules:** Research studies will usually have a study plan or protocol. Within this plan, routine procedures will be described in written SOPs.

▶ **Characterisation:** As much as possible should be known about the materials used (identity, purity and so on) and the test systems (usually the animal or plant) to which the materials are administered.

▶ **Documentation:** All raw data must be documented (written down) and a final report must describe the study accurately and interpret the results correctly. Records must be stored (archived) for many years so that they can be consulted in future if necessary.

▶ **Quality assurance (QA):** QA means that there will be a separate team of people who can assure the management that the laboratory has complied with GLP.

A2.8 The principles of good manufacturing practice

GLP is particularly important in research, development and testing. However, once a product has been developed and is ready to be manufactured on a large scale, the principles of **good manufacturing practice (GMP)** are used, ensuring that products:

▶ are of consistent high quality
▶ are appropriate for their intended use
▶ meet the requirements of the product specification.

One way to ensure the product meets the required specification is to analyse or test it (see 'quality control' in section A2.10). However, testing of the final product does not always identify or eliminate possible risks. GMP is designed to minimise the risks involved in any food, cosmetic or pharmaceutical production that cannot be eliminated by testing of the final product.

It is often said that there are ten principles (sometimes described as 'golden rules') of GMP:

1 Write step-by-step operating procedures (SOPs).
2 Ensure that SOPs are followed; over time, operators sometimes find their own ways of working or cut corners in ways that could harm the product or process.
3 Document all work; if it is not written down, assume it is not done! An organisation cannot afford to look back at production records and find blank spaces.
4 Validate all work; does the process or procedure do what it is intended to?
5 Design and construct facilities and equipment that prioritise productivity, quality and safety.
6 Maintain those facilities and equipment so that they go on performing as designed or expected.
7 Define, develop and demonstrate the competence of those employees involved in the process.
8 Protect against contamination, for example by incorporating cleaning, disinfection and decontamination processes into the workflow.
9 Control components and processes; this might mean monitoring storage temperatures to ensure that they are sufficiently low, for example, or verifying that heat-treatment of a product is carried out at the correct temperature – neither too high nor too low.
10 Carry out periodic audits to a planned schedule as a way of ensuring that GMP is being followed correctly. Any areas for improvement can be addressed before they can cause a problem with the product.

Think about the principles of GLP and GMP.

Can you see similarities? What differences are there? Are you confident that you could identify situations where you would be following GLP and when you would follow GMP?

A2.9 The key principles of continuous improvement in relation to scientific tasks

You will see from the previous section that continuous improvement is an important part of GMP. This applies to all aspects of work in the science and health sectors. Rather than ask 'what went wrong?' or 'who was to blame?', it is more important to ask 'how can we make things better?' – ideally before anything does go wrong.

Reviewing costs

Continuous improvement is not just about avoiding problems with a product or process (although that is very important), it also has financial implications.

Continuous improvement might mean using new **reagents** (substances used in chemical processes) or products to reduce the cost of a process. This can have widespread benefits if more money is available to spend elsewhere.

Standardising and optimising procedures

Standardising procedures means establishing a series of rules that are followed when carrying out a procedure, task or series of tasks. Standardisation is a key part of GLP and GMP, particularly in the use of SOPs. It can also make operations more efficient if the same or similar processes are carried out in the same way across all operations. Optimising procedures might mean using new technologies, or it might mean outsourcing to a specialist company that is better able to carry out some of the procedures – perhaps because it has specialist skills, already owns appropriate equipment or can offer economies of scale.

Using the evaluation cycle

Continuous improvement usually follows a pattern – the evaluation cycle.

▶ **Plan:** Identify potential problems and plan required improvements.
▶ **Do:** Implement potential solutions.
▶ **Check:** Analyse the results.
▶ **Act:** Review the solution and retest if necessary.

Capturing data at each stage of production

Remember the importance of capturing and recording data when following the principles of GLP? It is also an important part of continuous improvement. By measuring how effective a process is, we can identify opportunities for improvement. Of course, you have to be careful not just to focus on the aspects that are easiest to measure and miss the opportunity for more significant improvements elsewhere.

A2.10 The difference between quality assurance and quality control

We mentioned QA and QC as aspects of work undertaken in section A2.2. You will have seen QA mentioned as one of the main principles of GLP. QA is also central to the principles of GMP.

Quality assurance procedures

QA procedures are designed to prevent errors and defects in products or processes. It is nearly always cheaper and more effective to make something right in the first place rather than to have to rework (for example, purify) or remanufacture a product that does not meet specifications.

Quality control

QC focuses on the identification of errors and defects in completed products or processes. A product will 'pass' QC if it is tested and shown to meet the required specifications.

It does not matter whether it is a food or a medicine – you want to be sure that it will not contain any harmful contaminants or impurities before you consume or use it.

To summarise: QC will identify faults or problems after they happen, whereas QA focuses on consistency in the process to ensure that faults or problems do not happen in the first place.

> Which do you think is more important, QA or QC? Could you have one without the other?

A2.11 How organisations in the science sector ensure compliance with internal and external regulations

All organisations in the science and health sectors are subject to regulation. Some of these will be external regulations that relate to health, safety and protection of the environment, as described in Chapter A3. Others will affect organisations that have to work to GLP or GMP and these could be both internal and external regulations.

In either case, organisations must ensure that they and their employees comply with all the applicable regulations – failure to do so can have serious consequences. The case study in section A2.7 gives an example of just how serious the consequences can be.

Ensuring that all individuals follow standard operating procedures

You have probably realised the importance of SOPs by now. The second principle of GMP (see section A2.8) is to ensure that SOPs are followed. One potential problem is that operators of manufacturing processes often find short cuts or easier ways of doing things. This can be a good thing if it is part of a process of continuous improvement. However, if operators deviate from an SOP, it can have serious consequences for the product or the process. It might even affect the safety of those carrying out the process.

Health and safety

Bear this in mind in your placement and throughout your working life: SOPs exist for a reason, and deviating from an SOP could have unforeseen and potentially dangerous consequences. Do not be tempted to cut corners, even if you see colleagues doing that.

Complying with requirements for internal and external audits

The tenth principle of GMP is the need to carry out regular audits of the process. This can help to ensure that SOPs are actually being followed, but it can also show areas where there is a need or opportunity for improvement. This type of audit will be **internal**, that is, carried out by and within the organisation.

Many organisations in the science and health sectors are subject to regulation by government bodies, such as the Medicines and Healthcare products Regulatory Agency (MHRA). These bodies carry out audits of the organisations that they are responsible for, and many of these are **statutory** – in other words, they are required by law. These are examples of **external** audits.

In highly regulated industries, such as chemicals and pharmaceuticals, there is often a legal requirement to report the results of internal audits to regulators such as the MHRA or the Health & Safety Executive.

Making sure that staff are adequately trained

Ignorance is not considered a defence against breaking the law. This is true of driving at 60 m.p.h. on a road with a 30 m.p.h. limit, but also of the laws and regulations that apply to organisations within the science and health sectors. Therefore, all good employers will take staff training seriously. Employees need to be familiar with how relevant legislation and regulations apply to their specific occupation.

There are also some occupations or functions within an organisation that must be licensed. An example of this is the **qualified person (QP)** in the pharmaceutical industry. The QP must be an experienced professional who certifies the release of every batch of product that is made. A QP must meet certain eligibility criteria and be registered with either the Royal Society of Chemistry, Royal Pharmaceutical Society or the Royal Society of Biology. Each organisation publishes a list of its members who are registered as QPs.

A2.12 How regulatory controls apply in different working environments within the science sector

Think about a large pharmaceutical company and its different working environments. Here are some examples:
- a data scientist collating results from clinical trials
- a research technician working with highly infectious pathogens
- a chemical plant operator working with large volumes of hazardous chemicals
- a technician working in a facility that fills vials of vaccine under sterile conditions.

How would each of these individuals be exposed to different types of risk? How would this influence the regulatory controls that would apply to each of them?

These regulations, and their application, are covered in more detail in Chapters A3 and A4.

Type and level of required personal protective equipment

Section A3.2 makes clear that risks and hazards in the workplace must be minimised. An important principle is that all potential sources of risk or harm must be removed as much as possible. Only then should personal protective equipment (PPE) be used to protect against any remaining risk (if using PPE is appropriate). The nature of the risk and the extent to which it can be removed or minimised will vary according to the working environment. This is covered in more detail in Chapter A4.

Standards of health and safety and housekeeping

While a data scientist may be able to work safely, if not quite so efficiently, with a cluttered desk, that is not the case with someone working in a lab or production environment. In those circumstances, clutter can be dangerous. It could mean that hazardous materials may not be stored correctly or may not be visible. Other hazards and risks may result from lack of attention to health and safety and good housekeeping. These aspects are covered in more detail in Chapter A3 (particularly section A3.3) and Chapter A4. Clearly, the standards will vary according to the working environment. These principles still apply in office settings. For example, clutter that spills from the desk onto the office floor can create a trip hazard, so there is still a need to maintain high standards in all working environments.

Requirements for mandatory training to comply with guidance or legislation, refreshed as required

The role of a QP was mentioned in section A2.11. Before a QP can become registered, they must meet minimum standards of education or training, as well as experience. They are also encouraged to undertake continual professional development (CPD); although this is not compulsory, it helps them to maintain their skills and knowledge.

The same applies in many roles in the science sector, particularly in highly regulated industries such as pharmaceuticals or chemicals.

Requirements for disposal of waste

If you live in an area where the local authority encourages recycling, you may be familiar with the need to separate paper, card, glass, plastics and so on, to reduce the amount of general waste that goes to landfill or incineration. Many organisations have similar practices. However, other types of waste require special handling, depending on the organisation. Examples include:
- clinical waste, such as used dressings, swabs and other items contaminated with bodily fluids
- human or animal tissue
- syringes, needles and other items classed as 'sharps' (for example, lancets, pipettes and scalpels)
- hazardous chemicals (flammable, corrosive or toxic)
- microbiology waste, such as culture media and Petri dishes, that might be contaminated with pathogens

- radioactive waste
- waste that can harm the environment.

All of these are covered by specific laws and regulations that must be followed. There is more detail about this in Chapter A4.

Requirements for health screening and inoculation

Some occupations may expose you to the risk of infection, for example by working with animals, pathogens, tissue or blood samples, or patients. In such cases you may be required to be vaccinated/inoculated to protect you against contracting a disease from workplace exposure.

Other health risks in the workplace might be due to exposure to harmful chemicals, irritant dusts or ionising radiation. Because of this, many organisations will have a policy of health screening where the health of at-risk employees is regularly monitored so that early action can be taken.

Controls specified within standard operating procedures

SOPs should always take account of the regulations that are applicable in a specific industry or occupation. Therefore, particular regulatory controls might be specified in an SOP. This might include the need for record keeping, for example, or testing of raw materials and intermediate or final products.

A2.13 Factors that may have an impact on the commercial activities of science organisations

We have covered some of the regulations that many companies must observe, such as operating to GLP and/or GMP. Chapters A3 and A4 cover laws around health and safety in more detail and most of these apply to all organisations.

Science-based organisations of all types, not just companies, may be involved in commercial activities (buying and selling products and services). These could be in areas such as:
- pharmaceuticals
- cosmetics
- manufacturing
- providing services to other organisations.

We will consider some of the factors that could have an impact on these commercial activities here.

Government priorities/policies

Besides legal obligations, government priorities and policies can have a significant impact on commercial activities. We will consider the examples of food labelling and environmental policies.

Food labelling

The labelling of foodstuffs is governed by various regulations, mostly retained from EU regulations following Brexit. Some information must be included on the label, such as the weight or volume of the product, the ingredients list including any allergens, as well as the nutritional analysis (Figure 2.2). Some products must be labelled with a 'use by' date. Other types of information might be optional. However, whether the information on the label is there to meet a legal obligation or not, it must be accurate.

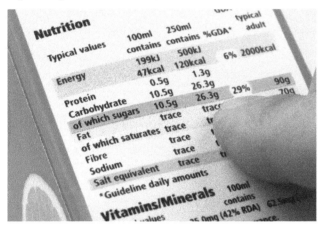

▲ Figure 2.2 Food label showing the nutritional analysis; this is a legal requirement for most packaged foods

Food labelling is regulated in the UK by the Food Standards Agency (FSA). The FSA is a government department that has been set up as an independent watchdog working to protect public health and consumers' wider interests in relation to food in England, Wales and Northern Ireland.

Some government policy is based on encouragement and other non-regulatory means of influencing food manufacturers and the general public. An example of this is the voluntary 'traffic light' system for highlighting nutrition information, usually on the front or side of food packaging. It is used by many supermarkets and food manufacturers to show whether a food contains high, medium or low amounts

of fat, saturated fat, sugars and salt. It also tells you the number of calories and kilojoules in that particular product. Figure 2.3 shows an example of this.

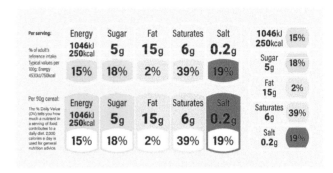

▲ Figure 2.3 Traffic light labelling system on a cereal packet

Environmental policies

The Environmental Protection Act 1990 (EPA) controls emissions and management of waste to reduce air, water and land pollution. There is more detail on the EPA in section A3.1.

In addition to this and related laws and regulations, the government has environmental policies encouraging companies and other organisations to reduce harm to the environment. Examples include:

▶ encouraging recycling
▶ encouraging reductions in energy use
▶ encouraging companies to reduce carbon emissions or reduce the carbon footprint of their products and services
▶ encouraging organisations to draw up an environmental policy.

Public perception and media influence

The COVID-19 pandemic provides many examples of how public perception and media influence have had an impact on companies, universities and the NHS.

Here are some examples, taken from the book *Vaxxers* by Professor Sarah Gilbert and Dr Catherine Green:

▶ Messages circulated on social media that the COVID-19 vaccines contained substances derived from pork or beef, making them unsuitable for Muslims or Hindus. This is not true. However, we tend to trust information from friends and often pass it on. It took social media campaigns to convince people that there was no basis for these claims.
▶ Many pregnant individuals were put off having the vaccine because it had not been tested during

pregnancy. New drugs, including vaccines, are never tested on pregnant individuals until there is a great deal of evidence of safety in the non-pregnant population. Thus, in the early stages of vaccine roll-out, people who were pregnant were advised not to have the vaccine. Now that we have abundant evidence of the safety of COVID-19 vaccines in pregnancy, we can be confident that the very low risk of the vaccine is much less than the real risk of harm, particularly in the last three months. Nevertheless, the message took a long time to get through and many pregnant people continued to be reluctant to be vaccinated.

▶ A German newspaper published a story that erroneously stated that the AstraZeneca vaccine would not be licensed in the EU for use in over-65s because it was apparently 'only 8% effective' for this age group. That was incorrect and it is not clear where the figure came from, however it was repeated in other newspapers, and even the French President said that the vaccine 'seems quasi-ineffective on people older than 65'. Nevertheless, the vaccine was approved the same day for use in all adults. Probably as a result of this publicity, there was a reluctance in some countries to give the AstraZeneca vaccine to older people and some older people refused it.

> If you were involved in the development of a COVID-19 vaccine, how would negative messages on social media affect your plans? What actions could you take?
>
> If you were a member of management in the company mentioned, how would you expect these events to influence your actions?

Funding streams

Funding streams were covered in section A2.1. In that section we focused on grant funding of research or commercial funding (shares, loans and so on) of companies.

However, changes to private or public funding streams can have direct and indirect effects on companies and other science-based organisations. Governments, charities, investors and large companies can change the focus of their funding activities. Many great scientific advances have come from basic research that has no obvious application – sometimes called 'curiosity-led' or 'blue skies' research. However, in the last 30 years there has been a shift away from this type of basic research towards research with a more immediate commercial application. This has benefited some organisations, particularly companies providing services closer to the marketing of a product or service, at the expense of others (mostly universities and research institutes) that have traditionally been involved in more basic research. So, even if the overall level of funding is maintained, there may be winners and losers.

Availability of materials

According to the British Chambers of Commerce, 87% of production and manufacturing companies reported raw material price increases in the first quarter of 2021 because of COVID-19 and post-Brexit disruption.

Shipping containers also became hard to obtain in 2019–20, partly because they were not returned to Asia from other parts of the world during the coronavirus pandemic. As a result, supply lines were disrupted leading to shortages of raw materials, many of which are shipped on container ships from Asia to Europe.

Case study

Economic collapse caused by the global lockdown in 2020 meant that there was a huge fall in demand for crude oil. This led to crude oil prices plummeting, even becoming negative at one point (oil producers had to pay people to take their product). During 2021, demand remained low leading to cuts in production. Oil refineries throughout the world reduced or even stopped production. However, crude oil is not just a source of petrol and diesel and other fuel oils; it is the source of many starting materials or **feedstocks** for the chemical industry. Basic chemicals from crude oil are used in about 95% of all manufactured goods.

As a result of the reduced refinery output, there was a reduction in output of ethylene and propylene. These feedstocks are converted to ethylene oxide and propylene oxide, which are used in production of key products such as glycol and polyester. These are essential raw materials for many industries including the plastics, clothing, furnishing, automotive, agrochemical and pharmaceutical industries.

The weather can also affect the availability of raw materials. In February 2021, extremely low temperatures in the southern United States caused widespread power cuts. This led to the shutdown of many petrochemical plants, which meant that 80–85% of the US production of vital feedstocks such as polyethylene and polypropylene (the two most widely used plastics in the world) was temporarily lost. In addition, about 57% of US production capacity of PVC was taken out of action. These plastics are used in a wide range of industries, including medical devices, electronics, packaging and automotive, so the disruption was widespread.

Market demand

The demands of the market (what people want to buy) can have a significant effect on the commercial activities of companies and other organisations in the science sector. There are obvious examples of this in the food sector, but there are also many ways in which the demands of the market influence development in science-based industries.

Case study

Until the early 1970s, most eggs sold in the UK had white shells. By the end of that decade almost all the eggs sold in supermarkets were brown. Consumers had begun to view brown eggs as being healthier than white eggs, in the way that brown bread was seen as healthier than white bread. In fact, the nutritional value of eggs is not related to shell colour.

In January 2021, the non-profit organisation Veganuary (veganuary.com) reported that more than 500 000 people had pledged to eat only vegan food during the month of January. There has been a significant increase in the availability and sales of vegan foods in UK supermarkets. In January 2019, Greggs launched their vegan sausage roll following an online petition signed by more than 20 000 people and saw an immediate increase in overall sales.

Essentially, **market demand** (what people want to buy) can lead to new or improved products becoming available. Here are some more examples:
▶ Demand for increased battery life in products as diverse as smartphones and electric vehicles has led to advances in battery technology.
▶ Increased demand for organic produce has led to an increase in the amount of organic produce grown, as well as to developments in intensive production of organic fruit and vegetables, particularly in the Netherlands.
▶ Demand for more environmentally friendly products has led to an increase in the use of 'green chemistry' (using processes that are less harmful to the

environment) and the replacement of plastics or man-made fibres with natural products such as bamboo.

Cost-effectiveness

The **National Institute for Health and Care Excellence (NICE)** is responsible for providing evidence-based guidance and advice for health, public health and social care practitioners in the UK. However, you are most likely to have heard about NICE in news items about new drugs, because the appraisal of a new drug or treatment by NICE will determine whether or not it is available on the NHS. One of the criteria that NICE uses is cost-effectiveness. This takes account of the cost of the drug or treatment, the cost of medical staff involved in administering it, and the likelihood and cost of managing any side effects. This is then balanced against the benefits to the length and quality of life as a result of the drug or treatment. If NICE does not give its approval to a drug or treatment, it may still be available through a private healthcare practitioner, but it has to be paid for rather than being available on the NHS.

Do you think that cost-effectiveness should be taken into account when approving a new drug? It might be possible to work out all the costs of administering a new drug, but how would you work out the benefits? NICE uses a measurement called the quality-adjusted life year (QALY) to work out the cost of giving an extra year of healthy life. NICE aims for a cost of less than £20 000–£30 000 per QALY when approving a new drug, although this might be doubled for end-of-life drugs; the view is that, at the end of life, a small amount of extra time can be very precious for patients and families.

Do you think a price can be put on life in this way? What are the alternatives?

Companies and other organisations must also consider cost-effectiveness, for example when planning new products or services. They have to account for the cost of research, development and production, and balance this against the value of the product, for example in increased sales.

Environmental concerns

We looked at how government policies on the environment can have an impact on commercial activities. However, environmental concerns can have a wider impact on companies. This might be driven by the demands of customers or public opinion, but it is leading companies to address environmental concerns as part of their daily operations.

Examples include:

▶ **Reducing waste:** This usually has the added benefit of reducing costs, particularly as the cost of waste disposal is increasing. Wasteful processes mean that companies pay twice: once for the materials that are not used, and again to dispose of them. Other ways in which companies are reducing waste include reducing packaging or moving to reusable packaging, drinks bottles and cups, and so on.

▶ **Reducing the carbon footprint of the business:** The carbon footprint of a product or activity is the sum of all the emissions of carbon dioxide associated with that product or activity, usually expressed in kg or tonnes of carbon dioxide. To calculate the carbon footprint, you have to take account of any carbon dioxide emissions in the production process (direct carbon emissions) as well as the emissions associated with obtaining and transporting the raw materials and finished product (indirect carbon emissions).

▶ **Ensuring sustainable use of resources:** This is leading to increased use of renewable energy sources, such as wind power. It is also driving the switch towards increased use of biofuels, such as E10 petrol, which contains up to 10% bioethanol (produced from plant sources).

Reflect

Demand for electric vehicles (to reduce carbon emissions) has increased the demand for minerals such as cobalt, which is used in lithium-ion batteries. However, the main source of cobalt is the Democratic Republic of Congo (DRC), where a number of ethical issues such as child worker exploitation and other human rights concerns have been reported.

Think about the move towards renewable and sustainable forms of energy and raw materials – are they always good, or are there some downsides?

A2.14 The importance and impact of innovation in the science sector

Innovation is central to the science and health sectors. It is an area that depends on continually finding new products, services, processes and technologies.

Fosters economic development

Innovation can increase the size of the economy. For example, development of genetically modified crops can increase yields or improve the nutritional value of the food. It can also reduce the need for expensive herbicides and pesticides.

Solves large-scale problems

Global warming is described as an existential threat to humanity – meaning that it could lead to the extinction of humans if it is not addressed. That is about as large-scale a problem as you can get!

One way of tackling global warming is to switch away from burning fossil fuels as an energy source towards alternative energy sources, such as wave, wind or solar power. Technological improvements have reduced the cost of wind power so much that the cost of power generation by onshore wind farms is now lower per unit of electricity than power generation by burning coal or gas. However, wind power has one main disadvantage – the wind does not always blow, which means that it is not always a reliable source of power. This has stimulated the need for improved batteries and other novel forms of energy storage; this is an area where there are good opportunities for innovation.

Improves healthcare

The single most cost-effective contribution to reducing death and disease in the last two hundred years is probably the widespread availability of vaccines. The impact of innovation was clear during the COVID-19 pandemic when several vaccines, using different vaccine technologies, were developed and licensed for use within a year of the SARS-CoV-2 gene sequence becoming available. Almost as important as the scientific innovation was the fact that those involved in the development, testing and licensing of the vaccine took an innovative approach that achieved in one year what normally takes ten years, without any compromise in safety.

Another area of innovation has been in diagnosis. The efficiency of diagnosis has been improved through the use of innovative techniques such as:

▶ artificial intelligence (AI) and machine learning

▶ improvements in genomic sequencing, allowing the development of new diagnostics and treatments, for example the RT-PCR test for diagnosing COVID-19

▶ genetic tests to personalise treatments, such as those used in cancer therapy; these tests can look for particular alleles that are associated with certain diseases or show whether a patient will respond to a particular drug or not.

> See section A5.5 for more on AI and machine learning.
>
> See section B1.18 for more on alleles.

Develops new products

Scientific innovation has led to the development of many new drugs for the treatment of a wide range of diseases.

Developing a new drug is an extremely expensive business and can take many years. Partly because of this, it requires the huge resources that only governments or large pharmaceutical companies ('big pharma') have available. Pharmaceutical companies tend to be relatively conservative – the risk of failure can be very high and they have shareholders to account to, so they do not invest their money in everything. For this reason, big pharma tends to focus on diseases of the developed world, such as cancer, cardiovascular disease and respiratory diseases, where there is a large and predictable demand. Many new drugs are also 'me too' drugs – slight improvements on existing drugs – rather than ground-breaking innovation.

Another consequence is that big pharma tends to focus on treatment (so that a drug will continue to be sold) rather than cure (where a drug is only sold once).

Do you think that governments should play a larger role in development of new drugs? Does big pharma need to be more regulated, or would that stifle innovation?

There are many examples of innovation outside of the pharmaceutical industry. One of these is the development of nanomaterials, such as graphene. Graphene is made of a single layer of carbon atoms where each atom is covalently bonded to three other carbons. This leaves one electron for each 'left over' and these electrons become delocalised, giving graphene interesting electrical properties. Graphene is a transparent and flexible electrical conductor. This makes it suitable for use in solar cells, LEDs, touch panels, smart windows (that can regulate the amount of light entering a room) and touchscreens on mobile phones. Fullerenes are spherical or cylindrical forms of carbon (think of graphene formed into spheres or

rolled into tubes). 'Buckyballs' such as the 60-carbon buckminsterfullerene can be used in medicine for drug delivery by trapping a drug inside the sphere until it is delivered to the site of action. Cylindrical fullerenes are known as carbon nanotubes and are used in high performance composite materials (see section B1.33) such as carbon fibre. They are also electrical conductors, making them useful in electronic applications.

> See section B1.33 for more on composite materials.

Enables new scientific discoveries

Innovation is the basis of new scientific discoveries, some recent examples being:

- ▶ Genome editing, where systems such as CRISPR-Cas9 can be used as a form of molecular 'cut and paste', removing and replacing bases in DNA. This technique is being used to develop treatments for inherited diseases such as cystic fibrosis.
- ▶ Bioinformatics, which is the use of huge databases of protein and DNA sequences to investigate the molecular basis of disease, help design new drugs and understand how organisms are related and how they evolved.
- ▶ Computational biology, which takes bioinformatics as a starting point and then uses the principles of AI and machine learning to develop algorithms or models that help in our understanding of highly complex biological systems.
- ▶ Nanoscience, the science of very small particles the size of a few hundred atoms, has enabled the development of novel catalysts, new composite materials (such as carbon fibre), new methods of drug delivery and novel electronic materials. Coating metal nanoparticles with DNA allows them to self-assemble into larger structures that have novel properties.
- ▶ The Large Hadron Collider (LHC) is the world's largest and highest-energy particle collider. In 2012 it observed a new particle that was later confirmed to be the Higgs boson. The Higgs boson is the particle that gives mass to all other particles. This forms the basis of the Standard Model, which describes the origin of the weak nuclear force and the electromagnetic force.

Project practice

You have been asked to contribute to a local careers fair for students aged 16+ years. Choose **one** of the following topics, ideally one that is relevant to your placement:

▶ types of jobs in the science sector
▶ individuals and teams working in the science sector
▶ employers in the science sector – sources of funding and how they spend it
▶ innovation and regulation in the science sector.

For your chosen topic you should:

▶ research the background
▶ think about the wider implications of the topic – try to bring in related areas you have covered in this chapter.

Once you have done that, prepare a presentation (using slide presentation software, a poster or infographic) that will:

▶ convey information about the topic
▶ engage and enthuse the reader, and make them want to consider a career in the science sector.

Present your work to the group and lead a group discussion.

Reflect on your work and write an evaluation of it.

Assessment practice

1 For each of the following organisations, name a different source of funding that would be available:

 a university research department
 b contract research company
 c newly formed start-up company.

2 A job advert stated that the applicant must be:

 – qualified to Level 3 or higher in at least one science subject
 – experienced in testing foodstuffs.

 Explain whether this is **most likely** a person specification or job description.

3 You are leaving the lab at the end of the day. As you do, someone you have not seen before asks if you could let them in as they have forgotten their key card and need to go back to get it. What should you do?

4 A new company is developing a product incorporating a novel technology for use in home testing for allergies.

 Discuss the factors that could affect the commercial success of the company.

5 A chemical company produces a product that is used in the manufacture of several pharmaceutical compounds. The SOP requires written records to be kept of the temperature in the chemical reactor making the product.
 Explain the importance of these records.

6 For each of the following, state whether it is an example of either quality control (QC) or quality assurance (QA):

 a Testing of raw materials to ensure that they meet the required specification.
 b Testing of final products to ensure that they meet the required specification.
 c Following an SOP so that the production process is the same every time.

7 State **two** reasons why employers need to ensure that their employees receive regular training.

8 Discuss ways in which innovation has improved healthcare.

9 Describe the stages of the evaluation cycle and explain its importance in continuous improvement.

A3: Health, safety and environmental regulations in the health and science sector

Introduction

When you work in the health and science sector you should feel safe. Your employer must make sure that you have a safe working environment and are not exposed to any unnecessary risks – we say that they owe you a **duty of care**. However, it is also the duty of every employee to play their part. You will need to be aware of the legislation and regulations that help to keep you and your colleagues safe, and to understand your rights, duties and responsibilities.

Laboratory and industrial processes can cause harm to the wider public and to the environment. So, you also need to be aware of your responsibility to help in eliminating (as far as possible) such harm.

This chapter covers the relevant legislation and regulations while Chapter A4 will cover how these are applied in the workplace.

Learning outcomes

The core knowledge outcomes that you must understand and learn:

A3.1   the purpose of legislation and regulations in the health and science sector

A3.2   how to assess and minimise potential hazards and risks, including specific levels of risk, by using the Health and Safety Executive's 5 Steps to Risk Assessment

A3.3   how health and safety at work is promoted

A3.4   how to deal with situations that can occur in a health or science environment that could cause harm to self or others (for example, spillage of hazardous material).

A3.1 The purpose of legislation and regulations in the health and science sector

Health and Safety at Work etc. Act 1974

The Health and Safety at Work etc. Act 1974 (sometimes referred to as HSWA 1974 or HASAWA) sets out the duties of employers and employees to ensure the health of anyone at work or who may be affected by work activities. Under HSWA 1974, the employer must ensure, as far as is reasonably possible, the health, safety and welfare of all employees while they are at work.

The employer's specific duties include the following:
- providing and maintaining **plant** and systems of work that are safe and without risks to health
- arrangements for ensuring safety and absence of risks to health in the use, handling, storage and transport of articles and substances
- giving the information, instruction, training and supervision necessary to ensure the health and safety at work of employees
- maintaining any place of work in a condition that is safe and without risks to health
- providing and maintaining all means of entry and exit (i.e. doorways, corridors, walkways, etc.) that are safe and without such risks
- providing and maintaining a working environment for employees that is safe, without risks to health, and adequate as regards facilities and arrangements for their welfare at work.

The legislation recognises that it is impossible to avoid all risk and so all of the above duties are to be carried out 'so far as is reasonably practicable'.

Key term

Plant: any equipment used in the workplace, e.g. laboratory equipment.

Health and safety is not the sole responsibility of the employer. Employees must play their part. HSWA 1974 states that it is the duty of every employee:
- to take reasonable care for the health and safety of themselves and of others who may be affected by what they do, or don't do, at work
- to cooperate with the employer to enable the employer to perform their duty under the legislation.

Reflect

HSWA 1974 means that employees have a duty to themselves and to their co-workers. You should not do anything that could cause harm. But not doing something you should have done can also affect health and safety in the workplace. The act says that employees must co-operate with the employer in ensuring safe working practices are carried out. Think about how health and safety in the workplace is a partnership between the employer and the employee – how can you play your part in how you work? Is there anything that you are doing that you should not be doing? Is there anything that you are not doing that you should be doing?

HSWA 1974 is an enabling act – that means that the government can introduce regulations at any time to modify the act or to establish up-to-date standards. Many regulations have been introduced under HSWA 1974. The ones that are most relevant to work within the health and science sectors are covered in the following sections.

Research

The Health and Safety Executive (HSE) is responsible for enforcement of HSWA 1974 and associated regulations. It has a website that is an excellent source of information about all matters related to health and safety in the workplace: www.hse.gov.uk

The COVID-19 pandemic meant there was an increase in the number of people working alone (**lone working**). This includes people working from home as well as those more likely to be alone in the workplace because of COVID-safe practices. Find out what the HSE recommends that lone workers should do to protect their health and safety.

Management of Health and Safety at Work Regulations 1999

These regulations aim to reduce the number and severity of accidents in the workplace, through assessing and managing risks. They make good health and safety management a legal requirement for, and impose duties on, both the employer and the employee. The main duty of the employer is to assess risks to health and safety in the workplace as they affect both employees and any others present (e.g. members of the public, visitors, etc.). The employer must then ensure effective planning, organisation, control, monitoring and review of measures taken to prevent or protect against risk. The employer must also appoint one or more 'competent persons' to assist in carrying out these legal responsibilities – this could be the employer themselves or employees appointed to the role.

The duties of employees under the regulations are to use machinery, equipment, dangerous substances, transport equipment and safety devices in accordance with their training and instructions. Also, employees have a duty to inform their employer and other employees of any dangers or shortcomings in the arrangements.

Control of Substances Hazardous to Health (COSHH) Regulations 1994 and subsequent amendments 2004

COSHH is designed to protect people against risks to health arising from work-related exposure to hazardous substances. It requires assessment and control of risks before any work takes place.

Anyone in charge of, or working in, a laboratory should be familiar with COSHH and actively involved in implementing it. The steps involved are shown in the table.

Health and safety

The use of material safety data sheets is covered in sections A4.18 and A4.19. For now, you should be familiar with the labelling of hazardous substances using the GHS hazard pictograms (Figure 3.1).

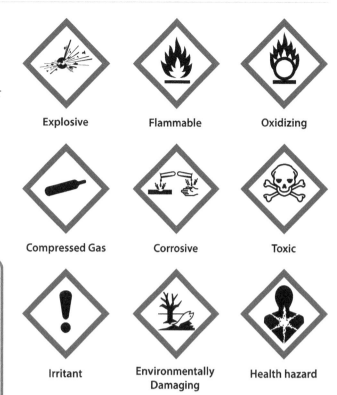

Explosive	Flammable	Oxidizing
Compressed Gas	Corrosive	Toxic
Irritant	Environmentally Damaging	Health hazard

▲ Figure 3.1 The GHS hazard pictograms

Carry out an assessment of the tasks in the laboratory:
- Record the scope of assessment (who the tasks are carried out by, what is being assessed, when the tasks are carried out).
- List significant laboratory tasks.
- List substances involved.

Assess the factors that decide the appropriate control approach:
- What are the hazard categories associated with each task?
- What degree of exposure is likely? (This depends on the quantity handled as well as the nature of the substance.)

Determine the control approach:
- None: open bench working (i.e. not in a fume cupboard or other containment) with general ventilation.
- Intermediate: fume cupboard or other exhaust ventilation.
- High: glove box or similar containment.
- Special: purpose-designed facility.
- Use of personal protective equipment (PPE) for eyes and skin where there is a risk from skin contact or for inhalational protection.

Implement and review:
- Assess other tasks and related risks.
- Planning (how it will be implemented and resources needed).
- Consider safety and environmental risks.
- Consider other aspects of COSHH – monitoring, health surveillance and training.
- Use and maintenance of control measures (including regular checks and reporting defects).
- Set up proper record-keeping and review procedures.

Personal Protective Equipment at Work (Amendment) Regulations 2022

Managing and reducing risk is always the priority. If a factory roof leaks, it is better to repair it than to issue every member of staff with their own umbrella. Use of PPE should be considered only after everything else has been done to reduce risk. However, in most workplaces in the health and science sectors, some level of PPE will always be necessary. These regulations define employers' responsibilities to provide appropriate PPE to reduce harm to employees, visitors and clients. This can include providing safety helmets, masks, goggles and gloves.

Health and safety

PPE is not always convenient or comfortable. Healthcare staff have spoken about the discomfort of wearing full PPE throughout shifts lasting eight hours or longer during the COVID-19 pandemic. That is one reason why all other steps to remove or reduce risk should be taken before considering the use of PPE. However, the pandemic also illustrated the importance of appropriate PPE when faced with an invisible and potentially life-threatening risk.

Reporting of Injuries, Diseases and Dangerous Occurrences Regulations 2013 (RIDDOR)

The purpose of RIDDOR is to define employers' duties to report serious workplace accidents, occupational diseases and specified dangerous occurrences ('near misses').

RIDDOR puts duties on employers, the self-employed and people in control of work premises (the **Responsible Person**) to report certain serious workplace **accidents**, occupational diseases and specified dangerous occurrences as well as **reportable injuries**.

Reportable accidents are those that result in death, major injury, or being absent from work or unable to do normal work for more than seven days. Accidents that result in absence from work for more than three days but less than seven days must be recorded but do not need to be reported unless they are reportable injuries.

RIDDOR requires employers and self-employed workers to report cases of occupational cancer and any disease or acute illness caused by **work-related** exposure to a biological agent (e.g. bacteria, viruses or toxins). This may take place because of identifiable events or unidentified events.

Key terms

The following terms have a specific meaning in terms of RIDDOR:

Accident: a separate, identifiable, unintended incident, which causes physical injury. This specifically includes acts of violence to people at work.

Reportable injuries: the following injuries are reportable under RIDDOR when they result from a work-related accident:
- the death of any person
- specified injuries to workers (see the HSE website for more information)
- injuries to workers which result in them being unable to work for more than seven days
- injuries to non-workers which result in them being taken directly to hospital for treatment, or specified injuries to non-workers which occur on the premises.

Work-related: an accident in the workplace does not always mean that the accident is work-related – the work activity itself must contribute to the accident. An accident is 'work-related' if any of the following played a significant role: the way the work was carried out; any machinery, plant, substances or equipment used for the work; the condition of the site or premises where the accident happened.

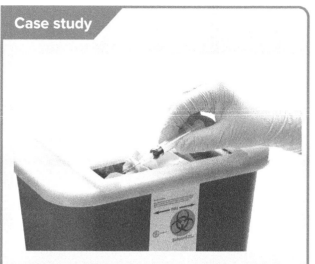

▲ Figure 3.2 Safe disposal of contaminated syringe needles, 'sharps'

Samah made a list of identifiable events in her workplace, including accidental breakage of a laboratory flask, accidental injury with a contaminated syringe needle (Figure 3.2) or an animal bite. She found it harder to think of unidentified events.

▶ Samah read that an unidentified event could be when a worker is exposed to legionella bacteria while conducting routine maintenance on a hot water service system. What other unidentified events could be present in your workplace?

▶ Do you think laboratory workers are more likely to be exposed to identifiable or unidentified events?

Environmental Protection Act (EPA) 1990

▲ Figure 3.3 Controlled waste can be harmful to the environment

The EPA aims to improve control of pollution to the air, water and land by regulating the management of waste and the control of emissions.

▶ The EPA enables the Secretary of State to make any process or substance subject to strict controls and to set limits on its emissions into the environment. Carrying out such processes requires approval and there are criminal sanctions against offenders.

▶ The EPA also covers regulation and licensing of the disposal of controlled waste – this has a very broad meaning and includes any household, industrial or commercial waste.

▶ The way in which the EPA is enforced is quite complex. The Environment Agency (EA) or, in Scotland, the Scottish Environment Protection Agency (SEPA), together with local authorities, are responsible for control of processes specified by the EPA.

▶ Local authorities are responsible for the collection of some controlled waste, such as household waste.

▶ Local authorities also regulate and license the disposal of controlled waste from industrial or commercial premises such as shops, offices, laboratories, hospitals, GP surgeries, care homes, etc.

Thus, the EPA 1990 applies equally to the contents of your wheelie bin and to waste or emissions in the workplace.

Other legislation

The table lists some of the other legislation relevant to the health and science sectors and explains the relevance of each.

Special Waste Regulations 1996
Measures relating to: • the regulation and control of the transit, import and export of waste (including recyclable materials) • the prevention, reduction and elimination of pollution caused by waste • the requirement for assessing the impact on the environment of projects likely to have significant effects on the environment.
Hazardous Waste Regulations 2005
Controls the storage, transport and disposal of hazardous waste to ensure it is appropriately managed and any risks are minimised.
Waste Electrical and Electronic Equipment Regulations (WEEE) 2013
Aims to reduce the amount of electronic and electrical equipment incinerated or sent to landfill sites. Places onus on all businesses to correctly store and transport electrical waste.

Regulatory Reform (Fire Safety) Order (RRO) 2005

Aims to reduce death, damage and injury caused by fire, by placing legal responsibilities on employers to carry out a fire risk assessment. Because of this, all organisations are required to have procedures for evacuation in the event of a fire.

Manual Handling Operations Regulations 1992 (as amended)

Requires employers to assess and minimise the risk to health of employees involved in the manual handling, moving and positioning of an object, person or animal. It also covers workplace ergonomics, i.e. ensuring furniture is suitable and adaptable to different body sizes and types.

Health and Safety (Display Screen Equipment) Regulations 1992

Defines employers' responsibilities in carrying out risk assessments of workstations used by employees, including the use of display screen equipment, to minimise identified risks, such as:
- tiredness caused by poorly designed or adjusted workstations
- repetitive strain injury (RSI) and carpal tunnel syndrome
- eye strain leading to headaches, fatigue and sore eyes.

Test yourself

1. Which act (rather than regulations) listed above controls the disposal of waste in the workplace?
2. Name two sets of regulations that cover accidents in the workplace.
3. Name three sets of regulations that cover hazardous substances.
4. How does RIDDOR define a work-related accident?
5. What is a reportable injury under RIDDOR?
6. Who is responsible for safety in the workplace – the employer or the employee?

A3.2 How to assess and minimise potential hazards and risks

Minimising risk is extremely important in reducing harm in the workplace. Before we can minimise risk, we must first identify the risks. This is generally done by carrying out a risk assessment.

Risk assessment is a part of the risk management process and involves identifying a **hazard** and then deciding on the likelihood of exposure to that hazard (the **risk**). You need to decide:
- how likely a hazard is to actually cause harm
- the severity of that harm – just how serious the consequences could be.

This means that if a highly dangerous substance (i.e. the hazard) is contained so effectively that there is almost no possibility of coming into contact with it, the risk will be low. However, the probability of exposure to a less hazardous substance might be much higher and could pose a greater risk.

The Health and Safety Executive's 5 Steps to Risk Assessment are outlined in the following sections.

Key terms

Hazard: something that has the potential to cause harm.

Risk: how likely a hazard is to cause that harm.

Step 1: Identifying the hazards

Look around your workplace and think about what hazards there are. It is easy to overlook some hazards when you work in a place every day, so it is useful to follow some guidelines:
- Check manufacturers' instructions, particularly data sheets for chemicals and equipment.
- Look at accident or sickness records to see what has caused issues before.
- Think about non-routine activities such as changes in procedures or maintenance and cleaning, for example, when equipment is taken out of service.
- Consider long-term hazards to health, such as high levels of noise or exposure to harmful substances.

Step 2: Deciding who might be harmed and how

Once you have identified the hazards, you can then think about how employees (and others, such as visitors and contractors) might be harmed. That means you identify groups of people that might be at risk, including those with particular requirements (e.g. new employees, young workers, people with disabilities, visitors or temporary workers).

Step 3: Evaluating the risks and deciding on precautions

Risk is a part of our everyday lives and it is impossible to eliminate all risks. What you must do is identify the main risks and what action you need to take to manage those risks. An employer must do everything reasonable to protect people from harm. This means there is a balance between the level of risk and the measures needed to control the real risk.

You need to ask:
- Can the hazard be eliminated?
- If not, what can be done to control the risks so that they are unlikely to cause harm?

Once you have evaluated the risks, you can consider some suitable precautions:
- Try a safer way of doing things.
- Restrict access to the hazards.
- Organise the work so that it reduces exposure to the hazard.
- Finally, once all other precautions have been taken, use PPE.

It is important that all employees are involved in this stage, to make sure that any course of action will work in practice.

Step 4: Recording findings and implementing them, including completing risk assessment documentation

This is important: remember, to be effective, documentation should be simple and focus on the controls needed. The key points that should be recorded are:
- A proper check was carried out.
- Everyone who might be affected has been consulted.
- All the obvious significant hazards were dealt with.
- Reasonable precautions were put in place and any remaining risk was low.
- Employees, or their representatives, were involved in the process.

Step 5: Reviewing your assessment and updating if necessary

Workplaces change and evolve; new equipment is brought in or new substances and procedures are used. This means that a risk assessment has to be reviewed regularly to answer some key questions:
- Have there been any significant changes?
- Are there any improvements that could be made?
- Have any employees noticed any problems?
- Have there been any accidents or near misses that you could learn from?

Test yourself

1 What is meant by a hazard?
2 What are the five steps in a risk assessment?

A3.3 How health and safety at work is promoted

So far, we have seen how health and safety in the workplace depends on everyone playing their part. This is made clear in HSWA 1974 and other legislation, but it should be obvious: an employer must provide a safe working environment, but employees must do all they can to take care of their own safety and that of their colleagues.

Therefore, a good approach to health and safety must become engrained in the workplace. It must become part of the culture, not just something added on or viewed reluctantly or simply as rules to be followed.

For this reason, health and safety must be promoted by providing a framework of good practice and getting everyone to pull together in ensuring a safe working environment. This section will look at ways that can be achieved. Some of this involves following organisational policies and standard operating procedures (SOPs).

Encouraging individuals to take reasonable care of their own and others' safety

Everyone in an organisation must play their part – to increase the safety of themselves and their colleagues. This starts with the need for individuals to be given the knowledge and information about how to keep safe.

Modelling good practice

Management must lead by example, such as washing hands and wearing appropriate PPE. This helps to create a culture of good practice.

This may not be enough in small organisations or where staff often work alone – they may not often be able to see good practice.

Following organisational policies and SOPs, including site-specific emergency procedures

As well as complying with regulations, following policies and SOPs should enhance safety. However, it does mean that these policies and/or SOPs need to be high quality, well designed and with health and safety addressed at every stage. Everyone has a part to play here. If you see any issues or weaknesses, you should raise these with your supervisor or line manager.

Ensuring there is clearly visible information and guidance

The HSE produces a health and safety law poster (Figure 3.4) and it is a legal requirement for any employer to display a copy where it can be easily read. As well as outlining health and safety laws, the poster lists clearly and simply what employers and employees must do to ensure safety in the workplace.

It is human nature that after a while we no longer notice the things we see every day. It can therefore be helpful to use a range of eye-catching and informative posters and other available materials that keep changing to continually promote good practice.

Following processes for recording and reporting issues and concerns

We can all learn from our mistakes – but it is better, where safety is concerned, if we learn not to make those mistakes. Employees must be free to raise issues and report their concerns. But that will be effective only if there are procedures in place to record and, most importantly, act on those concerns.

▲ Figure 3.4 The HSE safety law poster

Maintaining equipment and removing faulty equipment

Badly maintained equipment increases the likelihood of failure and can become a serious hazard. Faulty equipment can certainly represent a hazard and should be removed, or at least taken out of use, before it can cause harm.

Following correct manual handling techniques

Manual handling covers a wide variety of activities such as lifting, lowering, pushing, pulling and carrying – these are the cause of over a third of all workplace injuries. When these activities cannot be avoided, the risks of the task should be assessed and sensible measures should be put in place to prevent and avoid injury.

You should take account of:
- the strength and capability of the individual
- the type of load (box, crate, container for liquids, bulky item, etc.)

- environmental conditions, for example:
  - is the floor or the item being lifted wet?
  - could strong wind cause problems with light but bulky items?
- whether people have received adequate training
- how the work is organised – can it be reorganised to minimise handling or reduce the risk?
  - try to store materials in smaller quantities
  - try to store materials closer to where they are used.

Some points to remember when lifting by hand:

- Avoid, as much as possible, any twisting, stooping and reaching.
- Try not to lift from floor level or above shoulder height – particularly heavy loads.
- Arrange storage areas so as to minimise the need to lift and/or carry.
- Minimise carrying distances.
- Assess whether the load is too heavy – can it be broken down into smaller components, or can a colleague help?

If you have to use lifting equipment:

- use an aid such as a forklift or pallet truck, hoist (electric or hand-powered) or conveyor belt
- consider storage as part of the delivery process, i.e. heavy items could be delivered directly to the storage area.

> **Case study**
>
> Jack works as a technician in a hospital laboratory. Several times a day he collects items from the stores two floors below in a separate building. These include non-hazardous chemicals (such as 2.5, 10 and 25 litre bottles of solvent or 1–25kg bottles of powder), large cartons of lightweight plasticware and heavy gas cylinders (compressed nitrogen and helium).
> - What risks or hazards would Jack face in handling these items?
> - What precautions should Jack take to minimise the risks?
> - What changes could Jack's employer make to the way the work is organised?

Ensuring working environments are clean, tidy and hazard free

You may have seen signs in various workplaces saying, 'A tidy area is a safer area', and they are displayed for a good reason. Even when obvious hazards are removed, clutter can itself be a hazard:

- Items on the floor can be a trip hazard.
- Unnecessary items can conceal other hazards.

- The need to move items to get at other things can be a hazard.
- Dirty work areas can be a source of chemical or bacterial contamination.

Regular cleaning and tidying are important, and it may not be appropriate to leave the job to regular cleaning staff, for example, if there is specialist equipment or specific hazards. Making cleaning and tidying part of the regular laboratory routine will make for a much safer workplace.

Appropriately storing equipment and materials

The saying 'a place for everything and everything in its place' is an important way of ensuring safety in the workplace. It is the responsibility of the employer to ensure that appropriate storage facilities are provided – but it is everyone's responsibility to make sure that they are used properly.

Storage of hazardous substances is covered by regulations including COSHH Regulations 1994, subsequent amendments 2002, and the Hazardous Waste Regulations 2005.

We also need to think about moving equipment and materials in and out of storage and so the Manual Handling Operations Regulations 1992 (as amended) are also relevant. See pages 35, 37 and 38 for more on these three sets of regulations.

Completing statutory training

Question: who needs training in health and safety? Answer: everyone!

HSWA 1974 requires employers to provide all information, instruction, training and supervision needed to ensure (as far as reasonably practicable) the health and safety of employees.

The Management of Health and Safety at Work Regulations 1999 identify situations where training is particularly important, such as:

- when new people start work
- when there is exposure to new or increased risks
- when existing knowledge or skills have become rusty or out of date.

There are other regulations that include specific health and safety training requirements, for specific industries or for exposure to specific hazards, e.g. asbestos.

If you work with contractors or other self-employed people, they may still be classed as employees for health and safety purposes. This means that they need the same level of protection and appropriate training as regular employees – and they have the same responsibility to ensure their own health and safety and that of co-workers.

Research

The HSE publishes a leaflet, 'Health and safety training: a brief guide'. This is aimed at employers but is also very useful for employees. You can download a copy from www.hse.gov.uk – search for 'health and safety training'.

Use the leaflet to identify types of employees who may have particular training needs. Think about whether this would include you.

Test yourself

1 Identify three ways that management can promote health and safety at work.
2 Identify three ways that employees can promote health and safety at work.
3 Why is it important that equipment is properly maintained?
4 Why is it important that the workplace is kept clean and tidy?
5 A delivery of boxes has just arrived. You have been given the task of moving them to the storeroom and putting them onto shelves. Name three things you would have to consider to ensure safe handling.
6 Give three situations where the Management of Health and Safety at Work Regulations 1999 says that training is particularly important.

A3.4 How to deal with situations that can occur in a health or science environment that could cause harm to self or others

This is where we can put into practice what we have learned in the previous sections. Health and safety at work is governed by law and regulation (see section A3.1) that requires us to assess and minimise any potential hazards (section A3.2) and do all we can to ensure a safe working environment (section A3.3).

In any workplace, the response to a situation that could cause harm should follow a similar pattern.

Following organisational health and safety procedures

As well as helping to minimise the risk of harm, all organisations should have health and safety procedures that everyone can follow if something does go wrong. You need to be familiar with these and think about how they could prepare you to deal with any situation.

Keeping oneself and others safe, including evacuation as appropriate

In dealing with any situation, the priority is not to make things worse. For example, cleaning up a chemical spillage is important to prevent harm, but it must not be done in a way that exposes you or your colleagues to greater risk. Think about what actions and precautions you would need to take:

▶ The use of equipment such as a fire extinguisher or chemical spillage kit.
▶ The use of appropriate PPE.
▶ Do you have the training and capability to deal with the situation?
▶ Are you aware of procedures to be followed?

You should also realise that some situations are too hazardous for you to deal with yourself. In such cases, it might be necessary to evacuate the area. That could mean evacuating a room, a building or even a whole site. Are you familiar with the evacuation procedures?

Securing the area

It might be necessary to evacuate a room or building – but you do not want people wandering back in. Even if evacuation is not necessary, it is important to prevent unnecessary access while clean-up is taking place. Assuming that everyone is safe, the next step is to make sure that the situation does not get worse, for example, that a chemical spill is not continuing.

Reporting and/or escalating as appropriate

Even if the situation is relatively minor and fully controlled, you will still need to inform your supervisor, line manager, safety officer or other responsible person. If the situation is more serious than you can control by yourself, you need to know what action should be taken to escalate the response. This might involve bringing in more senior or experienced members of staff, a specialist response team or the emergency services, as appropriate.

Debriefing and reflecting on the root causes, to prevent the situation from recurring

Once the situation has been dealt with, it is important to learn lessons. This means that those involved need to review what happened and why. This will usually mean that management, safety officers or others will debrief those involved in the incident. The purpose is not to assign blame but to learn important safety lessons. It may be that procedures or working practices need to be changed to reduce the risk of future harm.

Test yourself

1   There is a chemical spillage in a school laboratory. Describe **three** actions that you would take to deal with it.
2   Under what circumstances should you evacuate an area?
3   Following a small fire in a chemical store, a debriefing is held for all employees involved. What is the main purpose of this?
    a   To ensure there is no adverse publicity for the company.
    b   To investigate the cause and assign blame.
    c   To investigate the cause and learn safety lessons.
    d   To remind employees that existing procedures must be followed without question.

Project practice

You are working in a small start-up company, G-Chem Pharma Ltd., based in a new building on a science park where other high-tech companies are located. The company was formed to take a 'green chemistry' process, invented in the local university, and develop it for use in the pharmaceutical industry. The company's founders are its Chief Executive Officer (CEO) and Chief Scientific Officer (CSO). The CSO is responsible for research and development, including managing laboratory and scientific staff.

G-Chem Pharma Ltd. employs 15 people, including PhD chemists, lab technicians and admin and support staff. The company has its own chemical laboratory as well as facilities for storage of chemicals and laboratory consumables. Some of the chemicals stored on-site are hazardous, including some flammable and corrosive substances. At present they use solvents in quantities up to 25 litres, but this is likely to increase in future.

You report to the Chief Operating Officer (COO), who has some experience of the industry. The COO is responsible for administration and human resources. You have been asked by the COO to prepare a strategy for ensuring the company meets all requirements of current relevant health and safety legislation and regulations. Future work is likely to include developing a strategy to implement your

recommendations. For now, you need to prepare a report for the COO to present to the CEO and CSO.

You need to complete the following steps:

1   Research a strategy.
    –   Carry out a review of the relevant legislation and regulations.
    –   Select the legislation and regulations that are applicable.
    –   Justify why you have selected some legislation/regulations and not others.
2   Plan a project based on the legislation and/or regulations you have selected.
    –   Summarise each piece of legislation and/or regulations.
    –   Describe the steps you would need to take to perform risk assessments, including identifying hazards.
    –   Identify the recording and reporting requirements.
3   Present your findings in a written report for the management team. You should cover the following areas:
    –   obligations of the company and employees
    –   roles and responsibilities
    –   specific risk areas that need to be considered
    –   how to promote the strategy within the company.

4 Discuss the following questions:
   - Does the company have the necessary information and expertise to prepare an adequate strategy for health and safety?
   - What further steps should be taken to draw up and implement the strategy within the company?
   - Who should be responsible for promoting the strategy within the company?
   - How should the roles and responsibilities be assigned? Think particularly about who should draw up the risk assessments and SOPs.

5 Reflection – write a reflective evaluation of your work.

Assessment practice

1 Who does the Health and Safety at Work etc. Act, 1974 make responsible for the health of everyone at work?

2 To what extent does the Health and Safety at Work etc. Act 1974 compel employers to remove all risk from the workplace?

3 Which should come first, writing the SOPs or preparing the risk assessment?

4 Which of the regulations, **W** to **Z**, shown in the table below, are relevant in each of the situations **A** and **B**? More than one of the regulations might apply to each situation.

Regulations		Situation in the workplace	
W	Management of Health and Safety at Work Regulations 1999	A	A company that does not handle hazardous materials needs to update its health and safety policies.
X	Control of Substances Hazardous to Health (COSHH) Regulations 2002		
Y	Environmental Protection Act (EPA) 1990	B	A technician in a school laboratory needs to dispose of old chemicals.
Z	Special Waste Regulations 1996		

5 A company is reviewing its training policy. All new employees undergo a half-day induction process that includes health and safety training. Half-day refresher courses are offered to all staff every five years. Evaluate this policy and suggest how it could be improved and why.

6 A new company has developed an approach to health and safety that includes the following:
   - The human resources department has the responsibility for carrying out all risk assessments.
   - The production department has the responsibility for preparing all SOPs.
   - Risk assessments and SOPs are made available on the company intranet.

   Assess what might be missing from the company's approach to risk assessment and SOPs. Your answer should include reasoned judgements.

7 Explain how the use of personal protective equipment fits into a risk management strategy.

8 You have been asked to dispose of large drums of solid and liquid chemical waste. Describe what the following legislation would require the process to include:
   - Environmental Protection Act (EPA) 1990
   - Control of Substances Hazardous to Health (COSHH) Regulations 2002
   - Manual Handling Operations Regulations 1992.

9 A company kept an accident book that was used to record all accidents that led to an employee being off work for more than seven days. Describe what the Reporting of Injuries, Diseases and Dangerous Occurrences Regulations 2013 (RIDDOR) requires in terms of recording and reporting accidents.

10 A technician has been given the task of producing SOPs for maintenance and disposal of scientific equipment. Outline the legislation and regulations that they would need to consider in producing the SOPs.

A4: Application of safety, health and environmental practices in the workplace

Introduction

In the previous chapter we looked at the regulations that cover health, safety and the environment in the health and science sectors. We are now going to look in more detail at the application of those regulations in the workplace.

Learning outcomes

The core knowledge outcomes that you must understand and learn:

A4.1 the purposes of Registration, Evaluation, Authorisation and Restriction of Chemicals (REACH) guidelines in relation to the use of chemicals in the science sector

A4.2 how the Environmental Protection Act 1990 relates to practices in scientific workplaces

A4.3 the consequences of breaching environmental legislation

A4.4 the purpose of the Control of Major Accident Hazards Regulations 2015 (COMAH)

A4.5 the COSHH definition of a biohazard (biological agent)

A4.6 the four hazard groups in relation to biohazards (biological agents)

A4.7 the potential implications of not adhering to COSHH regulations when dealing with biohazards (biological agents)

A4.8 containment measures that are used in relation to the four hazard groups

A4.9 the procedures to be followed when working with regulated substances and controlled drugs

A4.10 the purpose of pressurised clean rooms and localised extraction and ventilation

A4.11 the purpose of the Control of Noise at Work Regulations 2005

A4.12 how employers can protect employees from noise

A4.13 employers' responsibilities in relation to the Dangerous Substances and Explosive Atmospheres Regulations 2002 (DSEAR)

A4.14 how to work safely in high-risk environments or with substances that can cause harm to health

A4.15 the purpose of the Control of Electromagnetic Fields at Work Regulations 2016

A4.16 the consequences of using devices such as radios and mobile phones in the proximity of specific equipment and instrumentation

A4.17 how to decontaminate a range of common scientific equipment and substances

A4.18 the purpose of material safety data sheets and associated hazard and precautionary codes

A4.19 the importance of ensuring that material data sheets are kept up to date, in line with relevant legislation.

A4.1 The purposes of Registration, Evaluation, Authorisation and Restriction of Chemicals (REACH) guidelines in relation to the use of chemicals in the science sector

Registration, Evaluation, Authorisation and Restriction of Chemicals (REACH) is an EU regulation that came into force in 2006. Following the withdrawal of the UK from the EU, this regulation has been brought into UK law and is now known as **UK REACH**. If the UK regulations diverge from the original EU REACH in the future, it is likely that companies in the UK will still have to follow both sets of regulations.

The aims of UK REACH include:

▶ providing a high level of protection of human health and the environment from the use of chemicals

▶ making the people who place chemicals on the market (manufacturers and importers) responsible for understanding and managing the risks associated with their use

▶ promoting alternative methods for the assessment of the hazardous properties of substances, for example **quantitative structure–activity relationships (QSARs)** and **read across**.

The last point needs some further explanation. If you look ahead to Chapter B1 Core science concepts: Biology, you will see in sections B1.7, B1.8 and B1.9 a discussion of the relationship between the structure and activity of three of the main classes of biological molecules: proteins, carbohydrates and lipids. This theme is expanded on in Chapter B2 Further science concepts: Biology, where we will look in more detail at these relationships between structure and activity, particularly in enzymes (section B2.4). These **structure–activity relationships (SARs)** are **qualitative**, meaning that they show how there is a relationship between a particular molecular structure and the potential for that molecule to have a biological effect. SARs are very important in areas such as the development of new pharmaceuticals.

Key terms

Quantitative structure–activity relationship (QSAR): the use of statistical methods to refine the approach of correlating chemical structure with biological activity. In the context of REACH, the QSAR is used to predict hazards, such as toxicity, associated with a particular substance without having to test the substance, for example on animals.

Read across: a technique that attempts to fill gaps in experimental data about the properties, toxicity or potential environmental harm of substances. It does this by looking for patterns or chemical similarities in substances to predict the properties of substances that have not been fully tested. For example, compounds containing the nitrile (cyanide) group are highly toxic, so it is reasonable to suppose that a new compound containing the nitrile group is also likely to be highly toxic.

When REACH was first introduced, there was criticism that it would require a huge increase in the numbers of animals used for safety testing of substances. The use of QSARs and read across aims to avoid this by providing alternatives to animal testing when evaluating the risk associated with substances.

REACH puts the obligation on the manufacturer or importer of any substance to register the hazardous properties of that substance. This means that the risk associated with the substance can be evaluated. The use of some high-risk substances (for example, ones that are toxic or cause cancer) needs to be authorised and, in some cases, restricted. This is meant to encourage replacement of these substances with other, less hazardous alternatives.

Some substances are exempt from REACH, for example:

▶ those that are known to be harmless, such as water

▶ polymers, because the size of the molecules is thought to make them low risk

▶ biocides and active ingredients of medicines, as these are regulated separately.

Following the departure of the UK from the EU, Northern Ireland will remain part of the EU regulatory systems (for the duration of the Northern Ireland Protocol). This means that organisations in Northern Ireland will continue to follow EU REACH rather than UK REACH. This should not have any practical implications as long as UK REACH does not diverge from the standards of EU REACH.

Test yourself

1 Who is responsible for registering a substance under REACH?
2 What methods are used to reduce the use of animals in safety testing for REACH?

A4.2 How the Environmental Protection Act 1990 relates to practices in scientific workplaces

We saw, in section A3.1, how the Environmental Protection Act (EPA) aims to control pollution of the air, water and land by regulating waste and the control of emissions. We will now look at how that applies specifically in the science sector.

Waste management collection, treatment and disposal

You should know by now that you must never flush a chemical down a sink in the laboratory without considering the impact that it might have. In fact, you cannot dispose of anything in the laboratory without considering the environmental impact.

In practice, some substances can be disposed of down a drain, as long as they are washed away with a very large volume (an excess) of water. These include:
▶ acids and alkalis
▶ harmless soluble inorganic salts, including drying agents such as calcium chloride and magnesium sulfate
▶ solutions produced when hazardous compounds such as sodium or cyanides are destroyed chemically.

Some laboratory waste can and should be recycled, such as clean paper and glass. The rest will be categorised as:

▶ controlled waste, which is like household waste (dirty paper, plastic, wood, rubber, and so on) and can be collected by local authority waste collection services
▶ solvent waste, which is collected in approved containers (for example 2.5 litre brown glass bottles known as Winchesters, or heavy-duty plastic bottles), transferred to a central point and then disposed of by specialist incineration; the exact procedures may vary but it is generally necessary to keep chlorinated solvents separate from other solvents
▶ sharps, such as syringes and needles contaminated with a biohazard (see section A4.5), which are collected in special containers for incineration; Pasteur pipettes and other glass items, including broken glass, should be disposed of as controlled waste, unless they are contaminated with a biohazard. Glass items that are disposed of in controlled waste could lead to a risk of injury for anyone handling the waste if the glass is broken, so it can be a sensible precaution to wrap glass before disposal, for example in several layers of paper.
▶ waste for special disposal, including highly toxic substances and cancer-causing solids (for example asbestos). This involves special handling and record keeping and is highly regulated. This can mean it is troublesome and expensive to dispose of, so should be kept to a minimum.

Containment and uses of genetically modified organisms

Genetically modified organisms (GMOs) are animals, plants or micro-organisms whose DNA has been altered using genetic engineering techniques, for example transferring a gene or genes from one organism into another, or by modifying the DNA and reinserting it into the same organism.

Use of GMOs in the workplace is covered by the Genetically Modified Organisms (Contained Use) Regulations 2014. Contained use of GMOs means any activity where control measures (physical, chemical or biological barriers) limit the contact between GMOs and humans or the environment.

Most work with GMOs in contained use is very safe because most GMOs are incapable of growing outside of the laboratory environment. Nevertheless, it is necessary to carry out a risk assessment before starting work and to ensure that controls are put in place to protect people and the environment. The risk assessment must be submitted to the regulatory authorities for approval before work can start. Laboratories handling GMOs are subject to inspection by inspectors from the Health and Safety Executive (HSE).

The release of GMOs into the environment is regulated by the EPA and follows a similar process of risk assessment, notification, approval and inspection.

A4.3 The consequences of breaching environmental legislation

The consequences of breaching the EPA and other environmental legislation can be severe. Penalties include:

▶ Enforcement notices that can be served on a business requiring it to rectify the breach.
▶ Business closure can be ordered, alongside an enforcement notice, until the breach is rectified.
▶ Clean-up orders, which require a business to clean up any contamination (including water pollution) that they have caused.
▶ Fines and prison sentences; most breaches of environmental law are criminal offences. The penalties depend on where the case is tried:
  – in a magistrates' court the maximum penalty is an unlimited fine and/or six months' imprisonment
  – in the Crown Court the maximum penalty is an unlimited fine and/or two years' imprisonment.

Conviction for an environmental offence is also likely to lead to adverse publicity and damage to reputation, and could lead to an increase in business insurance premiums. Even if a trial ends with a not-guilty verdict, the damage to the company's reputation could still be significant.

A4.4 The purpose of the Control of Major Accident Hazards Regulations 2015 (COMAH)

The COMAH regulations aim to:
▶ prevent or limit the consequences of major accidents involving dangerous substances
▶ mitigate the effects on people and the environment of those that do occur.

The regulations apply to any establishment that stores or handles large quantities of hazardous chemicals. They came about as a result of several major accidents around the world that led to loss of life and major environmental damage.

The regulations operate on two levels – lower and upper tier – depending on the amount of hazardous material kept.
▶ Lower-tier establishments must document a Major Accident Prevention Policy that is signed off by the managing director or chief executive.
▶ Upper-tier establishments must produce a full safety report showing that all necessary measures have been taken to minimise risks to the local population and the environment.

The penalties for unauthorised storage of hazardous materials are severe. Prosecutions are rare because the first step is usually an enforcement notice; only the most serious cases result in prosecution. In 2016 a company was fined £600 000 for a breach of the regulations.

A4.5 The COSHH definition of a biohazard (biological agent)

The COSHH regulations were covered in section A3.1. In this and subsequent sections we are going to look at how COSHH applies particularly to **biohazards** or **biological agents**.

The biohazard symbol (Figure 4.1) is used for the labelling of:

▶ containers that contain or might contain an infectious biological agent or other biohazard

▶ cupboards and cabinets where biohazard materials are stored or handled

▶ the doors of laboratories where biohazards may be encountered

▶ waste containers, including those for the disposal of sharps, that may be contaminated with infectious agents or other biohazards.

▲ Figure 4.1 The internationally recognised biohazard symbol

Human **pathogens** infect only or mostly humans, while animal pathogens infect non-human animals. However, it is quite common for animal viruses to cross over into humans; experts believe the SARS-CoV-2 virus that causes COVID-19 is the most recent example of this, and there are likely to be more examples in future.

Plant pathogens can also infect humans. These are usually opportunistic fungal pathogens that generally infect people with weakened immune systems.

Viruses would usually be classed as biohazards if they are able to cause human, animal or plant disease. Some viruses, known as **bacteriophages**, infect only bacterial cells and cannot infect eukaryotic cells; as such, they are not classed as plant, animal or human pathogens and thus are not considered to be biohazards.

Biohazards are classified into four groups, according to the degree of hazard. These are discussed in the next section.

A4.6 The four hazard groups in relation to biohazards (biological agents)

The four hazard groups are:

▶ **Group 1:** These are unlikely to cause human disease. Examples include tissues and cell lines from animals other than humans or primates (gorillas, chimpanzees, monkeys); human or primate cell lines that have been safely used for many years; disabled, **attenuated** or **non-pathogenic** strains of some bacteria and viruses.

▶ **Group 2:** These can cause human disease and may be a hazard to laboratory workers, but infection is unlikely to spread to the community and there are usually effective vaccines or treatments available. Examples include tissues and primary cell lines of human or primate origin, and micro-organisms such as adenovirus, clostridium and most strains of *E. coli*.

▶ **Group 3:** These can cause severe human disease and may be a serious hazard to laboratory workers. Infection may spread to the community, but there are usually effective vaccines or other preventative measures and treatments available. Examples include HIV, hepatitis B, *E. coli* 0157 (a pathogenic strain of *E. coli* that causes potentially fatal food poisoning), *Salmonella typhi*, SARS-CoV-2 and similar coronaviruses.

▶ **Group 4:** This group causes severe human disease and may be a serious hazard to laboratory workers. Infection is likely to spread to the community and there are no effective vaccines or other preventative measures or treatments available. Examples include the rabies and Ebola viruses. These agents are only permitted in specialised laboratories.

Key terms

Attenuated: a strain of a pathogenic organism that has been modified or weakened so that it does not cause disease.

Non-pathogenic: an infectious organism that does not cause disease.

The HSE publishes an approved list of biological agents and their classification; it can be downloaded from www.hse.gov.uk/pubns/misc208.pdf

Besides infectious agents, other biohazards are either allergens/sensitising agents or toxins. These can be treated in the same way as the chemical hazards described in section A4.17.

1 Explain the difference between an attenuated pathogen and a non-pathogenic organism.
2 Explain why allergens and toxins are classed as biohazards, but can be treated in the same way as chemical hazards.

A4.7 The potential implications of not adhering to COSHH regulations when dealing with biohazards (biological agents)

The main risk associated with working with biological agents is the potential for infection. There are three main potential routes of infection to be aware of:

▶ **inhalation**, for example breathing in a fine aerosol or vapour mist that may contain a viable organism
▶ **ingestion** through poor hygiene practice (handwashing, and so on) or things that should never be done in a laboratory, namely mouth pipetting (drawing liquid into a volumetric pipette by sucking with your mouth) or eating/drinking
▶ **skin penetration** – this could be the result of injury with a contaminated sharp object, contact with the mucous membranes of the eyes/nose/mouth, or entry via an uncovered wound.

Failure to adhere to COSHH regulations when dealing with biohazards can lead to:

▶ risks to employees' health – the short- and long-term effects of infection
▶ risks to the wider population, namely spread of disease

▶ risks to the environment, such as vegetation, water supply or soil.

The HSE publishes advice on the assessment of risk and control measures needed when dealing with biohazards; you can download a copy at **www.hse.gov.uk/biosafety/management-containment-labs.pdf**

A4.8 Containment measures that are used in relation to the four hazard groups

There are four levels of containment corresponding to the four hazard groups of infectious agent. They are summarised in Table 4.1.

We saw in section A3.1 how COSHH regulations control the use of hazardous materials of all kinds in the workplace, including in laboratories. Following on from a risk assessment, we must put in place appropriate control measures. Some features of biohazards require particular types of control measure, but we should still follow the same type of hierarchical approach to control.

The hierarchy reflects the fact that eliminating and controlling risk by using physical controls and safeguards (keeping pathogens isolated) is more dependable than relying solely on systems of work. To be more precise:

▶ eliminating risks, for example by substituting a hazardous biological agent with something less hazardous or non-hazardous, such as using a less-harmful strain of a biological agent
▶ controlling risks at source, for example using a **biological safety cabinet** (**BSC**; see sections A10.3 and A10.8) when work could create an infectious aerosol, or using needle safety devices to prevent and control needlestick injuries

Risk group	Biosafety level	Laboratory type	Laboratory practices	Safety equipment
1	Basic (Biosafety Level 1)	Basic teaching, research	Good microbiological techniques (GMT)	None; open bench work
2	Basic (Biosafety Level 2)	Primary health services, diagnostic services, research	GMT plus protective clothing, biohazard sign	Open bench plus biological safety cabinet (BSC) for potential aerosols
3	Containment (Biosafety Level 3)	Special diagnostic services, research	As Level 2 plus special clothing, controlled access, directional airflow	BSC and/or other primary devices for all activities
4	Maximum containment (Biosafety Level 4)	Dangerous pathogen units	As Level 3 plus airlock entry, shower exit, special waste disposal	Class III BSC, or positive pressure suits in conjunction with Class II BSCs, double-ended autoclave (through the wall), filtered air

▲ Table 4.1 Containment measures required for each risk group

▶ minimising risks by designing suitable systems of working, for example having an effective hand-hygiene policy in place; this option also includes the use of PPE, but PPE should only be used as a last resort after considering elimination or tackling at source

▶ using good aseptic technique (see section A10.8).

Levels of PPE

PPE and clothing may act as a barrier to minimise the risk of exposure to aerosols, splashes and accidental contamination with infectious materials. Coats should be long-sleeved and back-opening gowns or coveralls will give better protection than a coat. Gloves should be pulled over the wrists of the gown or coat so that there is no skin exposed at the wrist (Figure 4.2). Aprons can be worn over the lab coat or gown to give additional protection.

▲ Figure 4.2 A technician wiping down a Biological Safety Cabinet whilst wearing gloves, coverall and a surgical mask

The type of eye protection required depends on the nature of the risk. Safety glasses (including prescription safety glasses) are the minimum requirement (Figure 4.3). Goggles give some protection against splashes and impacts (Figure 4.4), but face shields made of shatterproof plastic will give the greatest protection.

▲ Figure 4.3 Safety glasses

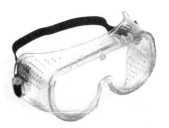

▲ Figure 4.4 Safety goggles

Respirators may be required for high-risk procedures, for example cleaning up a spill of infectious material (see Figure 4.5). Different filters are available for the particular application and/or hazard. To give the best protection, the respirator should be individually fitted to the operator's face and tested. Surgical face masks are designed to protect the patient from the surgeon, so do not offer any respiratory protection to lab workers.

▲ Figure 4.5 A respirator being worn together with goggles, coverall and long gloves

Disposable gloves are available in latex (usually white), vinyl or nitrile (usually blue or purple) – nitrile gloves give greater chemical resistance if required. Used gloves should be discarded with infected laboratory waste. Powdered latex gloves have been associated with the development of dermatitis and immediate hypersensitivity and thus are not recommended.

For even greater protection against chemical hazards, heavy-duty gloves are available in materials such as neoprene or butyl rubber.

Whatever the level of protection, all PPE should be removed before leaving the laboratory.

When we put control measures in place, PPE is generally the last topic that we consider. Why are other measures put in place before PPE?

Laboratory location, access and controls

The location of the laboratory is influenced by several factors. One of the most important is space, because overcrowding will always increase risk.

There must be an absolute ban on eating and drinking in labs, but it is also good practice to have separate areas adjacent to the laboratory or nearby for writing so that the laboratory itself is used only for laboratory work.

Only authorised persons should be allowed access to the laboratory and certainly no children or animals (except laboratory animals used in the work of the lab). Additional security depends on the risk group/biosafety level (see Table 4.1), but there should be a minimum level of security, such as:
▶ vision panels in the entrance door(s)
▶ the biohazard symbol (see Figure 4.1) should be displayed on the door of rooms at Biosafety Level 2 and above. This should carry the warning 'BIOHAZARD' and the text 'Admittance to Authorised Personnel Only', as well as the following information:
  – the biosafety level
  – the responsible person (laboratory manager, and so on)
  – contact to call in case of emergency (daytime and home phone).

A Biosafety Level 3 laboratory needs extra security and containment. Ideally it should be located in a part of the building that has additional access control, such as swipe cards. Access should be via an anteroom (double-door entry), ideally interlocked and self-closing, so only one door can be open at a time.

Required laboratory facilities

There needs to be adequate storage in the lab, with larger quantities of materials stored elsewhere – ideally nearby. Handwashing facilities should be provided near the exit door as well as space for storage of lab coats and supplies of disposable gloves – remember that these should not be worn outside of the lab.

Surfaces (floor, ceiling and walls) need to be:
▶ smooth and easy to clean
▶ impermeable to liquids
▶ resistant to the chemicals and disinfectants that will be used.

Work surfaces (benchtops) should follow the same rules, but also be resistant to acids, alkalis, organic solvents and (moderate) heat. Open spaces between and under benches need to be accessible for cleaning. Under-bench cupboards and drawers should provide sufficient space for storage of consumables to avoid clutter on benchtops.

A Biosafety Level 3 laboratory needs an anteroom that should:
▶ have facilities for separating clean and dirty clothing, and possibly also a shower
▶ be sealable (this also applies to all service ducts and openings, ventilation systems, and so on) so that gas can be used for decontamination if necessary
▶ have windows that are sealed and resistant to breakage
▶ have hands-free handwashing facilities located near the exit door.

Air management is more stringent in Biosafety Level 3 laboratories, where the ventilation system must direct airflow into the laboratory. This must be monitored so that laboratory staff can see that proper airflow is being maintained at all times. In addition, the ventilation system in the whole building must be designed so that air from the containment laboratory is not recirculated to other areas in the building. The air can be filtered through **high-efficiency particulate air (HEPA) filters**, reconditioned and recirculated into the containment lab. Alternatively, exhaust air from the laboratory may be discharged through HEPA filters to the outside, well away from occupied buildings or air intakes.

Complying with specific waste disposal regulations

As well as complying with waste disposal regulations that apply to all laboratories, there are specific requirements for laboratories handling biohazards.

There should be a system for identification and separation of infectious materials (and their containers). Categories should include:
▶ non-contaminated (non-infectious) waste that can be reused or recycled, or treated as controlled waste
▶ contaminated (infectious) sharps – hypodermic needles, scalpels, knives and broken glass – which should always be collected in biohazard-labelled, puncture-proof containers fitted with covers and treated as infectious
▶ contaminated material that can be decontaminated by chemical treatment or by autoclaving and then washed for reuse or recycling
▶ contaminated material for autoclaving and disposal
▶ contaminated material for direct incineration.

All infectious materials should be decontaminated, autoclaved or incinerated within the laboratory wherever possible so as to remove the need for transport of biohazardous waste. Steam autoclaving is the preferred method for all decontamination processes unless an incinerator is available on site.

Sharps disposal containers should never be filled more than three-quarters full. They should then be placed in infectious waste containers. Other contaminated materials should be incinerated. If there is no incinerator on site, all contaminated materials should be autoclaved in leak-proof containers before transfer to the incinerator.

Once decontaminated, all materials should be incinerated following local regulations.

Test yourself

1   Describe **two** differences in the facilities needed when working with Biosafety Level 2 and Level 3.
2   Describe **one** additional feature of a Level 4 containment facility.
3   Describe the relative advantages of different types of eye protection.
4   Describe how biohazardous waste should be disposed of.

A4.9 The procedures to be followed when working with regulated substances and controlled drugs

There are two sets of regulations that you need to know about, covering two types of substance:
▶ regulated substances as defined by the Control of Poisons and Explosive Precursors Regulations 2015
▶ controlled drugs as defined by the Misuse of Drugs Act 1971 and the Misuse of Drugs Regulations 2001.

These regulations aim to restrict the supply of substances that could cause harm (poisons), be used to make explosives (explosive precursors) or be abused (drugs of abuse). In doing so, they put obligations on those working in the science sector who have a legitimate use for any of these substances. As a result, we must take care to follow the regulations as the penalties for failing to do so can be severe and, as always, ignorance of the law is not a valid defence.

Organisations that hold or use substances listed under the Misuse of Drugs Act 1971 must first obtain a licence – this will be the responsibility of your employer. Organisations that hold licences should check whether the regulations apply to any work done and should incorporate the following into their SOPs:

▶ orders and arrangements for receipt of controlled drugs
▶ acceptance of deliveries and procedures upon receipt
▶ QC and QA procedures (QA and QC were covered in detail in section A2.2).
▶ production and packing runs
▶ procedures for accepting orders and controlled drug dispatch
▶ record keeping and cross-checking processes
▶ controlled drug store access, operative and management responsibilities
▶ controlled drug destructions
▶ theft, loss or adverse incident reporting and handling.

You can see from this list that there may be many areas of your work that will be affected if your organisation uses controlled substances.

Undertaking health and safety training

Anyone who works with any of the substances listed in these regulations must undergo health and safety training to ensure that they are familiar with the risks associated with their use and the precautions that must be taken to minimise those risks. If you are unsure about anything or think you need further or refresher training, you should always raise this with your employer.

Ensuring safe and secure storage

Any substances kept in the laboratory must be stored in a way that minimises risk – this is a basic principle of the COSHH regulations. Poisons, like any toxic substances, require special storage. However, many of the substances covered by these regulations may not be hazardous. For example, explosive precursors are substances that can be used to make explosives but may not themselves be explosive or hazardous in other ways.

These regulations focus on keeping these substances safe and secure, in other words, preventing unauthorised use or access. This means that these substances must be kept in locked cupboards that can only be accessed by authorised personnel. In addition, the following precautions are recommended:
▶ external doors and windows should be fitted with secure locks
▶ the premises and area where substances are kept should be protected by an intruder detection system (burglar alarm); this should be externally monitored and an appropriate level of police response arranged
▶ electronic access control systems, such as swipe cards or fobs, should be used by all staff

▶ the substances should be stored in a locked cabinet, safe or strong room, depending on the quantities stored and category of substance.

Undertaking inventory record-keeping

Full records must be kept of all regulated substances and controlled drugs so that they can always be accounted for. This should form part of any SOP that you follow.

Organisations should use a form to record any significant disappearance or theft of scheduled explosives precursors or poisons, as these must be reported to the police.

The use of controlled drugs must also be logged in a **controlled drugs register** with records kept of:
▶ who is handling the substances
▶ what they are being used for
▶ any destruction or disposal – this must be done in the presence of an authorised person, such as a police officer.

Following sign-in/sign-out protocols

Part of the record-keeping process is to have sign-in and sign-out protocols for issuing controlled substances. This aids tracking of stock (inventory) and record-keeping. For example:
▶ how much has been issued, how much has been used and how much has been returned to stock
▶ who has used controlled substances in the past; this might be necessary to comply with internal controls or audit procedures
▶ who is in possession of or responsible for a controlled substance at any time.

Research

The list of substances covered by the two sets of regulations is much too long to reproduce here. However, it is worth taking some time to find out which, if any, of the substances that you are likely to encounter in the workplace are covered.

If you visit www.gov.uk and search for 'controlled drugs list' you will find the most up-to-date information.

On the same website, search for 'supplying explosive precursors and poisons' for a list of explosive precursors and poisons.

Are there any substances on these lists that surprise you? Are there any substances that you use that you were not aware of being subject to these regulations?

Test yourself

1 How should both regulated substances and controlled drugs be stored?
2 What action would you take if you found that a quantity of a regulated substance had gone missing?
3 What information should be kept in the controlled drugs register?

A4.10 The purpose of pressurised clean rooms and localised extraction and ventilation

We touched on this subject in section A4.8 when thinking about the facilities needed for working with different classes of biohazard. The principles are:
▶ to protect individuals and materials against contamination; this is to avoid, or at least reduce, harm to operatives and also to reduce the risk of contamination of samples or cultures that could lead to invalid results
▶ to protect the external environment from contamination.

Pressurised rooms can be either **positive pressure** (this maintains an outflow of air) or **negative pressure** (this maintains an inflow of air), as shown in Figure 4.6.

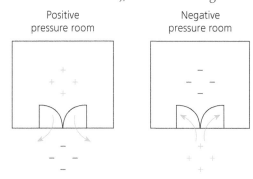

▲ Figure 4.6 Positive and negative pressure rooms

Protection of individuals and materials against contamination

A **positive** pressurised clean room, with a higher pressure in the room compared to outside, will maintain a flow of air **out** of the room. This helps to prevent entry of airborne contaminants such as dirt and microorganisms which provides protection against contamination of individuals and materials inside the clean room. This is illustrated in Figure 4.7.

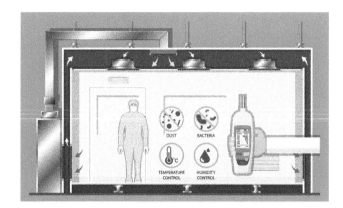

▲ Figure 4.7 A positive pressure clean room

A simple positive pressure clean room is not suitable for use with hazardous substances, unless the air leaving the room is filtered or directed to the outside via a fume handling system.

Protection of the environment against contamination

In this context environment can mean the surrounding rooms in the building or the wider environment. A **negative** pressure room will prevent contamination of the environment as airflow will be directed **into** the room. A negative pressure room can be a clean room as long as the airflow into the room is filtered to remove airborne contaminants.

Localised extraction

This includes fume cupboards and glove boxes (see section A10.3) that are used for handling hazardous substances. Other methods of localised extraction include spot extractor systems. These involve inverted funnels attached to a main extraction system that might also take the exhaust from a fume hood (Figure 4.8).

▲ Figure 4.8 Spot fume extraction positioned over individual pieces of equipment

A4.11 The purpose of the Control of Noise at Work Regulations 2005

The aim of the Noise Regulations is to ensure that workers' hearing is protected from excessive noise at their place of work, which could cause them to lose their hearing and/or to suffer from tinnitus (permanent ringing in the ears). The regulations also apply to members of the public that might be exposed to work-related noise, such as visitors to a factory or construction site. They do not apply to members of the public who are exposed to noise because of non-work activities, however, such as at a rock concert or listening to loud music on headphones.

The regulations state that if noise levels in the workplace exceed 80 dB(A) (80 decibels), the employer must assess the risk to workers' health and provide them with information and training.

If employees are exposed to noise of 85 dB(A) on a daily or weekly basis then employers must provide hearing protection. Also, employees must not be exposed to noise above 87 dB(A), taking account of hearing protection measures such as those described in section A4.12.

Practice points

The decibel (dB) is a measure of the loudness of sound. It is a **logarithmic** scale, which means that 80 dB is ten times louder than 70 dB.

You will also see the unit dBA or dB(A). This is a weighted scale that takes account of the fact that the human ear is more sensitive to some frequencies of sound; our ears are less sensitive to low frequency and very high frequency sounds than they are to sounds in the mid-range. dB(A) is more useful when measuring perceived loudness and stress-inducing capacity.

A4.12 How employers can protect employees from noise

Generating and ensuring compliance with risk assessments

As with all sources of potential harm at work, employers must carry out a risk assessment. This should cover the following points:

- identifying where there may be a risk from noise and who is likely to be affected
- making a reliable estimate of the employees' exposures, and comparing the exposure with the exposure action values (85 dB(A)) and exposure limit values (87 dB(A))
- identifying what must be done to comply with the law, for example whether noise-control measures or hearing protection are needed and, if so, where and what type
- identifying any employees who need to be provided with health surveillance and whether any are at particular risk.

As with all aspects of health and safety at work, if the risk assessment shows that there is a risk of harm to employees, the first step is to look for alternative processes, equipment or working methods that would allow a quieter working environment.

Also, employees have a duty to protect themselves and their co-workers by complying with all necessary methods of harm reduction.

Providing PPE

Once all steps have been taken to reduce noise, if noise levels are still above the action value, employees should be provided with suitable PPE, such as ear defenders (Figure 4.9).

Appropriate PPE should also be provided to visitors and contractors who might be exposed to noise.

▲ Figure 4.9 Ear defenders being used in a noisy environment

Providing regular health checks for employees

Health surveillance is an important part of maintaining the health of employees. This means that, if the risk assessment shows that there is a likely exposure to noise, free hearing checks must be offered to all members of staff who might be affected. Again,

it is the employee's duty to co-operate by attending for hearing checks.

If the results of the hearing checks show that workers' hearing is being affected, then employers must review the controls that have been put in place and make changes, if necessary, to ensure that everyone is protected.

Test yourself

1 What is the decibel (dB)?
2 What is the difference between dB and dB(A)?
3 An employer carries out a risk assessment and finds that there is a risk of harm to employees from high levels of noise. What **three** actions should they take?

A4.13 Employers' responsibilities in relation to the Dangerous Substances and Explosive Atmospheres Regulations 2002 (DSEAR)

Dangerous substances can increase the risk of damage from fire, explosion or corrosion of metal. DSEAR makes it a duty for employers (and the self-employed) to protect the people they work with from these risks. This means protecting employees and visitors to the workplace as well as members of the public who might be put at risk by a work activity. After all, an explosion may not be contained within the workplace.

DSEAR requires employers to:
- keep a record of which dangerous substances are in their workplace and what the risks are
- put control measures in place to either remove those risks or, where this is not possible, control them
- put controls in place to reduce the effects of any potential incidents involving dangerous substances
- prepare plans and procedures to deal with accidents, incidents and emergencies involving dangerous substances
- make sure employees are properly informed about and trained to control or deal with the risks from the dangerous substances
- identify and classify areas of the workplace where explosive atmospheres may occur and avoid ignition sources (from unprotected equipment, for example) in those areas.

You will see that this follows the general pattern of how other potential sources of harm should be controlled.

A4.14 How to work safely in high-risk environments or with substances that can cause harm to health

Substances that can cause harm to health include gases, explosive environments, lasers and ionising radiation (see section B1.57).

The approach to working safely in high-risk environments follows the usual principles:
▶ carrying out risk assessments
▶ following SOPs
▶ adhering to regulations
▶ undertaking appropriate training
▶ wearing appropriate PPE
▶ reporting all accidents, however minor.

A4.15 The purpose of the Control of Electromagnetic Fields at Work Regulations 2016

Electromagnetic fields (EMFs) are produced wherever a piece of electrical or electronic equipment is used. You will see in Chapter B1 Core science concepts: Physics how magnetic fields can be created when a current flows through a wire; this is how EMFs are produced. EMFs are present in almost all workplaces and, if they are of a high enough intensity, an employer may need to take action to ensure employees are protected from any adverse effects.

Employers have a duty to assess an employee's potential exposure to EMFs. As with noise, this exposure has to be judged in relation to action levels and exposure limit values.

Effects from EMFs that are covered by the regulations include:
▶ sensory effects, including nausea, vertigo and flickering sensations in peripheral vision
▶ health effects, including shocks, nerve stimulation and heating effects
▶ indirect effects, including:
  – interference with implanted or body-worn medical devices, such as pacemakers
  – uncontrolled attraction of ferromagnetic objects, that is, the risk of injury from objects in a large

static magnetic field being attracted to magnets in the workplace and hitting anyone in the way. See sections B1.51 and B1.52 for more about magnetic fields.

There have been suggestions that long-term exposure to EMFs might lead to certain types of cancer, such as leukaemia, as well as neurological disorders. However, the evidence for a link is weak or only from animal research. At present there is no well-established scientific evidence, so the regulations do not address these suggested long-term effects of exposure to EMFs.

Sources of high-strength EMFs include:
▶ magnetic resonance imaging (MRI) scanners
▶ **nuclear magnetic resonance (NMR)** machines used in chemical analysis
▶ electrical generators
▶ high-voltage electric power cables.

Employees have a duty to inform their employer if they are in a group that is at particular risk, for example:
▶ those who are pregnant
▶ those who have implanted or body-worn medical devices.

Action to reduce harm from EMFs includes:
▶ ensuring employees keep well away from areas of high-strength EMF
▶ installing screening.

A4.16 The consequences of using devices such as radios and mobile phones in the proximity of specific equipment and instrumentation

EMFs can cause **electromagnetic interference (EMI)** with sensitive electronic equipment.

Early concerns about the use of mobile phones in the vicinity of medical equipment led to their use in hospitals being forbidden. This has been relaxed in recent years, and a 2016 review (**https://doi. org/10.1016/j.jare.2016.04.004**) concluded that almost all of the more recent (up to 4G) mobile phones have little or no interference with medical devices. What is more, current medical devices are designed to operate safely under all conditions of use, and manufacturers of medical devices are required to minimise the risk that their devices can cause or be affected by EMI.

Nevertheless, we should be aware of any effects that electronic devices might have in a laboratory situation.

One effect that is often overlooked is the possibility that a mobile phone might cause a distraction at a critical moment, and this could lead to accidents. For this reason, mobile phones are prohibited or restricted in many laboratories, just as they are while driving. On the other hand, there may be occasions when a person is working alone. Lone working is usually discouraged on safety grounds but, if it is necessary, a mobile phone might be a control measure allowing the individual to summon help if necessary.

Another potential source of EMI are portable two-way radios. These may be necessary on a large site where workers might not be close to regular (wired) telephones or where mobile phones cannot be used, for example if there is no mobile signal.

Care should be taken to ensure that EMI does not:
▶ interfere with the normal operation of sensitive equipment
▶ affect the reliability of results
▶ cause damage to the equipment (both the scientific instrumentation and the devices).

It may not always be obvious that EMI is having these effects, which is why radios and mobile phones are often prohibited. Nevertheless, we should always be aware of the possible effects and look out for any anomalous readings or results that might indicate interference.

Health and safety

When you are working in a laboratory or workshop, you need to be aware of the sounds around you. For this reason, you should never wear headphones as they usually cover the ears and block external sounds. You should never use wired earphones because of the risk of the wire becoming entangled in moving parts or machinery. Wireless ear buds may be worn in some circumstances, but you should only wear one so that you can still hear sounds around you, and you should never touch or adjust earbuds with contaminated hands or gloves.

A4.17 How to decontaminate a range of common scientific equipment and substances

Sterilisation

When handling biohazards, sterilisation is an important part of decontamination prior to waste disposal (see section A4.8). Autoclaving is the preferred method, but chemical antiseptics or disinfectants can also be used, as well as sterilising with ultraviolet light. There is more detail about this in section A10.8.

Disinfection

A basic knowledge of disinfection and sterilisation is essential to ensure safety in a laboratory handling biohazards. Heavily soiled items cannot be disinfected effectively or sterilised – they must be cleaned first, for example by washing with detergent and hot water, as dirt, soil and organic matter can interfere with the action of a disinfectant.

It is important to appreciate the following differences:
▶ a bacteriostat or antimicrobial will stop the growth of micro-organisms but not necessarily kill them
▶ a disinfectant or germicide will kill micro-organisms but not necessarily their spores
▶ a sporicide will also kill spores.

Hydrogen peroxide is widely used as a disinfectant as it is a strong oxidising agent and, therefore, a broad-spectrum disinfectant that is effective against a wide range of micro-organisms. It is also safer, both to humans and the environment, than chlorine-based disinfectants such as bleach.

Incineration

See section A4.8.

Dissolution

Solid contaminants that are soluble, either in water or in an organic solvent, can be removed by rinsing. If the contaminant is non-hazardous and the solvent is water, then the waste can be flushed, well-diluted, down the drain. However, if the contaminant is itself hazardous or an organic solvent is used, this must be disposed of according to the relevant regulations (see section A4.2).

Neutralisation

We referred to dilution in section A4.2 as a way of disposing of acids or alkalis. However, that is not always a practical solution, for example if there is a spillage. Various spillage kits are available for use with acids or alkalis. These contain a powder that is spread on the spilled liquid. As well as absorbing the liquid, the powder neutralises the acid or alkali – there is usually an indicator incorporated to show when neutralisation has occurred; once neutralised, the powder can be swept up and disposed of in controlled waste.

A4.18 The purpose of material safety data sheets and associated hazard and precautionary codes

We have covered the need for risk assessment as required by the COSHH regulations in section A3.1. When handling certain chemicals, we need to know the hazards associated with those chemicals so that we can assess the risk. This information is contained in the **material safety data sheet (MSDS)**.

The MSDS will include:
▶ product information, including a scientific name and/or common name, as well as the name of the manufacturer and supplier, addresses and emergency phone numbers
▶ hazards identification
▶ composition/information on ingredients
▶ first-aid measures
▶ fire-fighting measures
▶ accidental release measures
▶ information on recommended handling and storage conditions
▶ exposure controls/PPE: this section often contains the exposure limits per country, as well as the PPE recommended or required by law
▶ physical and chemical properties
▶ stability and reactivity data
▶ toxicological information
▶ ecological information (ecotoxicity) of the chemical product
▶ disposal considerations
▶ transport information
▶ regulatory information, for example if it is a controlled substance.

MSDSs will often include **Risk phrases** and **Safety phrases (R and S phrases)** but these are being phased out in favour of the **Globally Harmonized System of Classification and Labelling of Chemicals (GHS)**. The GHS labelling system uses a series of pictograms (Figure 4.10) to indicate hazards. These are used on labels as well as on MSDSs.

Explosive **Flammable** **Oxidizing**

Compressed Gas **Corrosive** **Toxic**

Irritant **Environmentally Damaging** **Health hazard**

▲ Figure 4.10 The GHS hazard pictograms

The labelling of hazardous substances is controlled under an EU regulation that covers the classification, labelling and packaging of substances and mixtures, known as the CLP Regulation. This adopts the United Nations' Globally Harmonized System (GHS) for classification and labelling and gives it a legal basis. The EU CLP Regulation has been adopted into GB law, so the provisions still apply to Great Britain and are known as the GB CLP Regulation. (Note that Northern Ireland, which remains within the EU single market, is still covered by the EU CLP Regulation.) As long as GB CLP remains aligned with the EU Regulation, this difference between GB and Northern Ireland should not have any effect on most people working in the science sector.

Test yourself

1   Match the following GHS pictograms to the relevant hazard taken from the following list:

Harmful	Environmentally damaging	Flammable	Corrosive	Toxic	Oxidising
(flame symbol)			(exclamation mark symbol)		
(flame over circle symbol)			(environment/tree symbol)		
(skull and crossbones symbol)			(corrosion symbol)		

A4.19 The importance of ensuring that material data sheets are kept up to date, in line with relevant legislation

Just as it is important to keep SOPs updated (this will be covered in Chapter A8) and only use the most up-to-date version, the same is true of MSDSs. Updates may occur when:

▶ new hazard information, or information that may affect risk-management measures, becomes available
▶ a substance or mixture is classified or reclassified according to the classification, labelling and packaging of substances and mixtures regulation (EU or GB CLP)
▶ an authorisation under REACH is granted or refused
▶ a restriction under REACH has been imposed.

As with SOPs, it is important that you and your workplace have a system in place to ensure that MSDSs are kept up to date and only the most recent versions are used.

Project practice

You are working as a technician in a small, newly formed company on a university science park. The company will be handling a wide range of chemicals, some of which might be controlled substances, including:

▶ sodium nitrate
▶ formaldehyde
▶ phosphoric acid
▶ sodium hydroxide
▶ ammonium sulfate
▶ stanozolol
▶ 19-norandrostenedione.

The company will also be undertaking research into novel vaccines using a range of pathogens (bacteria and viruses) including the following:

▶ *Lactobacillus* species
▶ *Yersinia pestis*
▶ SARS-CoV and MERS-CoV.

The company is also considering working with Ebola and Marburg viruses.

Research the background:
▶ Perform a literature search to establish the hazard groups of the relevant pathogens.
▶ Search the **www.gov.uk** website to establish if any of the chemicals are controlled substances.

Prepare a report for the company management that covers:
▶ the legislation and regulations that will apply to the company's activities
▶ the risk assessments and SOPs that will need to be prepared
▶ the procedures, equipment and laboratory facilities that will be required:
  – to handle the proposed pathogens
  – to handle any controlled substances
  – for the disposal of general, chemical and biohazard waste
▶ the safety implications of working with Ebola and Marburg viruses.

Present your findings to the group and lead a group discussion.

Prepare a reflective evaluation of your work.

Assessment practice

1 Consider the following disposal method in terms of identifying any potential hazards. Suggest how any risks can be mitigated.

Small amounts of waste sodium metal can be destroyed by immersing them in an organic solvent such as pentane and then slowly adding a large volume of ethanol or propan-2-ol (isopropanol); the reaction produces hydrogen gas. Once all the sodium has reacted, water is added slowly with mixing. Finally, the solution produced is washed down the drain with a large excess of water.

2 Consider the data in the table about different viruses. During 2019 there were 1213 deaths from seasonal flu in the UK. In 2020 there were 72 178 deaths from COVID-19 (caused by SARS-CoV-2).

On the basis of this information, explain why SARS-CoV-2 is classified as biohazard level 3.

Virus	Infectivity	Mortality rate	Vaccine or effective treatment	Biohazard level
Ebola	Very high	Up to 90%	No	4
Marburg	Very high	Up to 90%	No	4
Seasonal flu	Moderate	0.1–0.2%	Yes	2
SARS-CoV-2	High	0.5–3%	Yes	3

3 A research group is undertaking a new project that uses small quantities of a controlled drug. What must the group do before starting work?

   A Nothing – the quantities involved are very small.

   B Keep stocks of the controlled drug in a locked cupboard in the lab.

   C Obtain a licence for working with controlled drugs.

   D Inform the safety officer at the next quarterly meeting.

4 A group has been working with cell cultures for some time, but they now need to study infection of the cells with the SARS-CoV-2 virus and related coronaviruses.

   a What is **one** task that must be done before commencing work with viruses?

   b The laboratory they use has been designed to protect the sensitive cell cultures from contamination by:
      – maintaining it at positive pressure
      – using a laminar flow cabinet for all work involving the cell cultures
      – requiring staff to put on clean lab coats before entry to the lab.

     Assess the suitability of these measures for working with the viruses that they plan to use.

5 A laboratory is working with sensitive cultures and is maintained at positive pressure to prevent entry of contaminants. However, the technicians also need to handle toxic substances. Discuss the practicality of using a fume cupboard, which requires a high flow of air through it, in a positive pressure laboratory.

6 Nuclear magnetic resonance (NMR) is widely used in analytical organic chemistry. Instruments are usually very large and incorporate very powerful superconducting electromagnets that generate very strong magnetic fields. These need to be kept at extremely low temperatures (a few degrees above absolute zero).

   a What regulations would NMR instruments be subject to?

   b What potential risks to personnel are there from NMR instruments and how can they be minimised?

7 The use of mobile phones in the laboratory can be controversial.

   a Give **two** reasons why mobile phones should not be used in the laboratory.

   b Give **one** situation in which a mobile phone might be important in a laboratory.

8 What precautions should be taken if you want to listen to music in the laboratory?

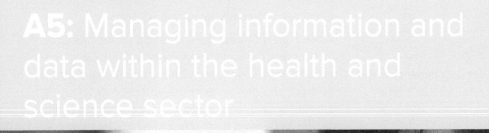

A5: Managing information and data within the health and science sector

Introduction

Before we can think about managing information and data, we have to make sure we know what is meant by the terms **information** and **data**. We often use them interchangeably, as if they have the same meaning – but do they?

Data is a collection of values. These values can be characters or words, numbers or other data types. You may sometimes see the phrase 'units of information' as a description of data.

Information is data that has been processed in a way that we can read it, understand it and use it.

So, managing information and data allows us to use the data to generate useful information that can increase our understanding.

Learning outcomes

The core knowledge outcomes that you must understand and learn:

A5.1 a range of methods used to collect data

A5.2 the considerations to make when selecting a range of ways to collect and record information and data

A5.3 the importance of accuracy, attention to detail and legibility of any written information or data

A5.4 the strengths and limitations of a range of data sources when applied in a range of health and science environments

A5.5 how new technology is applied in the recording and reporting of information and data

A5.6 how personal information is protected by data protection legislation, regulations and local ways of working/organisational policies

A5.7 how to ensure confidentiality when using screens to input or retrieve information or data

A5.8 the positive use of, and restrictions on the use of, social media in health and science sectors

A5.9 the advantages and risks of using IT systems to record, retrieve and store information and data

A5.10 how security measures protect data stored by organisations

A5.11 what to do if information is not stored securely.

The methods that we use to collect data depend on the type of data. Data types will be discussed in a little more detail in A5.2.

Focus groups

Focus groups are used widely in health research, as well as in parts of the science sector, as a way of discovering what individuals feel or believe. They can be very useful in understanding and explaining the factors that influence the feelings and attitudes of individuals as well as how they behave.

A focus group interview is usually highly structured. Participants are usually selected on the basis that they will have something to say on the topic – hence the term 'focus' – rather than being randomly selected. The focus group interview will have a facilitator or moderator whose job is to guide the discussion and manage the interactions between the participants. It is important that participants feel comfortable with expressing their views and opinions.

Focus groups can produce huge quantities of data that need to be processed and analysed to give useful information.

▲ Figure 5.1 Focus groups need a facilitator or moderator to lead the discussion

Surveys

A survey is a way of gathering factual information as well as views and opinions. They can be of two types: **closed-question** or **open-question**.

Closed-question surveys

Closed questions are questions that require a simple answer (see the table below for examples). The advantage of closed-question surveys is that they can be quick and easy to carry out and produce a large amount of data that can be analysed quite easily. This can be made even more efficient when the survey is online or in some other electronic format so that the data is already available in electronic format for further analysis.

However, the questions must be carefully worded to obtain reliable data. This might involve giving people information and asking for their opinion, but there can still be ways in which the results might be influenced by the wording of the questions. Consider the following two questions:

1. People under the age of 40 are less likely to die of COVID-19. Do you think the risks of the COVID-19 vaccine outweigh the benefits in this age group?
2. People under the age of 40 are more likely to suffer with 'long Covid'. Do you think the benefits of the COVID-19 vaccine outweigh the risks in this age group?

Do you think you would have given different answers about the risks and benefits if you were asked these two questions?

Open-question surveys

Open questions are questions that require a longer answer or explanation (see the table). The advantage of open-question surveys is that they allow people to provide more information. This can be important when you are gathering information about complex issues or questions that do not have a simple yes/no answer. However, analysis of the data will take longer and be more complex.

Closed question	Open question
Have you been vaccinated against COVID-19?	What is your opinion of the COVID-19 vaccine?
Have you been admitted to hospital for an overnight stay?	What is your experience of staying in hospital overnight?
Who is your GP?	What is your opinion of the service provided by your GP?
When was the last time you were unwell?	How do you feel about your general state of health?

Interviews

An open-question survey can also be carried out face to face, in which case it can become an interview where the person asking the questions has the opportunity to follow up on answers to questions or adapt the interview to the individual. This can improve the quality of the data collected because it may not necessarily impose ideas or opinions on the individual responding to the survey. However, interviews can be even more complex to analyse and draw conclusions from.

Closed-question surveys are probably better at obtaining large amounts of data from a large number of people. Open-question surveys or interviews are probably better when the number of people being surveyed is relatively small.

Observation

This can be a good way of gathering data about behaviour, which explains its widespread use in healthcare and social science or animal behaviour studies (see case studies). It is an important part of the method known as **qualitative research** (see A5.2).

In clinical and pharmaceutical research, randomised controlled trials are considered to be the most effective and reliable. In these, an experimental treatment or drug is compared either to a **placebo** (dummy treatment or drug) or to an existing treatment or drug. However, it is not always possible or desirable to carry out this type of trial. One example is where it is unethical to give a dummy treatment or drug.

In experimental studies, a researcher will usually have two groups: a test group and a control group. In observational studies, the researcher simply looks at (observes) groups of patients and compares the effectiveness of different types or treatment. An example is the **cohort** study, where a group of people (the cohort) is followed over a period of time. This can allow comparison of different treatment or care pathways to see which is the most effective. Cohort studies can be **prospective** or **retrospective**.

> ### Key terms
>
> **Prospective:** these studies take a group of people and observe them over a period of time. This could involve looking for correlations between factors such as diet or exercise and development of cardiovascular disease. The advantage is that the data collection methods can be tailored to the question being asked. The disadvantage is that these can take many years to complete.
>
> **Retrospective:** these studies look backwards at data from a group of people over many years. This often involves examining published data classifying people according to risk factors or medical outcomes. Although these can give results more quickly than prospective studies, the disadvantage of retrospective studies is that there is little control over data collection. This type of study typically looks at published data from many different sources that might involve different methods of data collection or analysis.

In clinical medicine, observation is very important in making diagnoses. However, the observational method used in qualitative research involves watching people to establish behaviours in a natural setting.

Public databases, journals and articles

You will notice, as you work through this book, that many of the Project practice items at the end of each chapter will start with asking you to research a topic. This is often to search published literature (see page 74), often for background information, but also to make sure that you are not just repeating work that has already been done.

Health Education England and the National Institute for Health and Care Excellence (NICE) provide access to a wide range of journals and other evidence-based resources for health and social care staff in England. Access to these requires an NHS OpenAthens account; you may be eligible for this if your course involves an NHS placement.

The Millennium Cohort Study (MCS)

The MCS is following the lives of about 19000 young people born in the UK in 2000–2002. The study collects data on:

▶ Physical, social, emotional, cognitive and behavioural development (cognitive development is the ability to carry out conscious mental activities such as thinking, understanding, learning and remembering).

▶ Daily life, behaviour and experiences.

▶ Economic circumstances, parenting, relationships and family life.

▶ GCSE exam results.

The MCS has provided important evidence of how circumstances in the very early years of life can influence later health and development.

Research based on the MCS has shown that children born at or just before the weekend are less likely to be breastfed, because breastfeeding support services are less available in hospitals at weekends. Breastfeeding has been shown to have a strong influence on cognitive development.

The MCS has also contributed crucial evidence on two major health issues facing this generation – obesity and the high rates of poor mental health. There is more

information about the MCS on the website https://cls.ucl.ac.uk/cls-studies/millennium-cohort-study.

1 How do prospective studies like the MCS differ from retrospective studies?

2 Do you think a prospective study can produce better quality data?

Observation of clinical practice

Observational research has been used to collect data on errors and potentially harmful events. A UK study on 10 wards in two hospitals showed that almost half of intravenous drug preparations and administrations had at least one error. This research was an example of an ethnographic study – one where the researcher is immersed in the community being observed. In this case, the nursing staff were told that the observer was investigating common problems of preparing and administering intravenous drugs. The word 'error' was avoided so that the study did not appear to threaten the staff.

Source: K. Taxis & N. Barber (2003) BMJ 326, 684 (https://doi.org/10.1136/bmj.326.7391.684)

1 Do you think that observational research into patient care is important?

2 Is it always possible to build a complete and accurate picture of patient care by comparing outcomes in different hospitals or clinics?

The US National Library of Medicine has an online database, PubMed.gov, that has over 32 million biomedical research articles. Many of these have links to full text content, including free-to-access as well as paid-for content. PubMed is a free resource.

As well as literature databases (books, journals, articles, etc.) there are clinical databases that contain data rather than published research. These include:

▷ observational data on patients who meet certain criteria, such as disease type or population

▷ clinical trial data, such as www.clinicaltrials.gov (US based).

The World Health Organization (WHO) publishes databases in many health-related fields, including:

▷ life expectancy

▷ immunisation and vaccination data

▷ mortality.

From September 2021 NHS Digital (the part of the NHS that designs, develops and operates IT and data services in the NHS) began to collect patient data from GP medical records in England. This data includes information about symptoms, test results, diagnoses and medication as well as about physical, mental and sexual health. The intention is to use this data to improve health and care services through better planning, preventing spread of infectious diseases, help with research and monitoring the long-term safety and effectiveness of care. The data will not include people's names or where they live, but there are concerns that it will not be completely anonymous.

Another concern is that this data will be sold to private companies, such as pharmaceutical companies, as well as being made available to research organisations such as universities and charities.

Do you think that making this type of data available could help in the development of new treatments? NHS Digital plans to charge for access to this data. It says that this covers its costs in processing the data and delivering the service, but some people think that this amounts to selling patient data. Do you think this is ethical? Do you think the benefit of using the data to improve treatment outweighs the risk to patient privacy and confidentiality?

These databases are invaluable sources for health and healthcare researchers.

Carrying out practical investigations

When we think of data, we probably think first about the results of experiments in the lab rather than some of the methods just discussed. Practical investigations in the field of health and healthcare can include laboratory experiments, for example, in basic medical science, using knowledge and techniques that build on the subjects covered in B2: Further science concepts.

Practical investigations can form the basis of evidence-based practice. These include:
- clinical trials of pharmaceuticals, usually run by pharmaceutical companies
- investigation of different types of care or treatment, such as comparison of drug treatment with talking therapies (talking to a therapist or counsellor) for treating mental illnesses
- investigation of different types of therapy, such as the RECOVERY trial into treatments for COVID-19.

Case study

The RECOVERY trial is the world's largest clinical trial into treatments for COVID-19. It was funded by the Medical Research Council and the National Institute for Health Research and led by the University of Oxford. The trial found one of the world's first effective drug treatments, dexamethasone. This cheap and widely available steroid medicine was shown to reduce deaths of patients in hospital with COVID-19 by one third.

Later the trial showed that tocilizumab, an anti-inflammatory drug used to treat rheumatoid arthritis, also reduces the risk of death in patients in hospital with severe COVID-19.

Do you think that low-cost treatments like dexamethasone would have been found if the investigations had been left to pharmaceutical companies that need to make profit for their shareholders?

Official statistics

Organisations such as Public Health England (PHE) and the WHO collect and publish statistics on disease, public health, health protection and health improvement. Official statistics available from PHE include those about:
- general public health
- alcohol, tobacco and drug use

- cancer
- cardiovascular disease
- child and maternal health
- chronic disease
- COVID-19
- diet and physical activity
- obesity
- end of life care
- immunisation and infectious diseases
- mental health
- sexual and reproductive health.

The WHO publishes statistics on an even wider range of health-related topics worldwide and by country.

Reflect

Collecting and publishing all these statistics is time consuming and can be expensive. So why do organisations publish official statistics? Think about how such statistics could be used.

Sometimes governments may be embarrassed if official statistics show that policies are not always working. For this reason, it is important that the collection and publication of official statistics is done in a transparent way so that we can have confidence in them.

Test yourself

1 State whether each of these questions is closed or open:
   a How often do you visit your dentist?
   b Have you found it difficult to find an NHS dentist in your area?
   c What type of extra services does your dentist offer?
2 What type of data collection would you use for the following?
   a To discover trends in the incidence of cardiovascular disease in the second half of the twentieth century.
   b To compare the amount of exercise taken by 18–24 year olds and 40–50 year olds in a local authority area.
   c To discover if vigorous exercise increases heart rate more in young people compared to middle-aged people.
   d To discover if young people are better than older people at using unfamiliar equipment.

A5.2 The considerations to make when selecting a range of ways to collect and record information and data

There are many ways we can collect data and turn it into useful information. If we choose the wrong way, we might discover that it is harder to analyse or present and so we cannot get and communicate the best information and understanding from our data.

Data type

Quantitative data includes measurements such as length, height, age, time or mass. Quantitative data can be either:

▷ **discrete** (or discontinuous), meaning something that you can count, such as number of patients, number of visits to the GP, number of cases of flu in a year. Discrete data is usually in whole numbers – you cannot have half a patient or visit your GP on 2.75 occasions in a year

▷ **continuous**, meaning something that can be measured, such as height, weight or blood glucose concentration. Continuous data can have any value within a range and so is usually not in whole numbers – you could have a weight of 94.7 kg or a blood glucose concentration of 5.2 mmol/litre.

Once you process data, for example, by taking an average, it is quite possible for discrete data to no longer be in whole numbers. For example, the average UK household size has been 2.4 for a number of years, but each household will have a whole number of members.

Source: www.ons.gov.uk/peoplepopulationandcommunity/birthsdeathsandmarriages/families/bulletins/familiesandhouseholds/2020

Qualitative data is usually text-based, describing something in a way that may involve numbers (for instance, a rating of how good something is), but will also contain descriptive text. For example, a patient's medical history may contain their age, date of birth and blood pressure (all types of quantitative data) as well as qualitative data such as any diseases or other health conditions they have, operations or medical procedures they have undergone, and other information about their health and wellbeing.

Key terms

Quantitative data: is numerical, for example, the results from a laboratory experiment.

Discrete data: is numerical and can be **counted**. For example, number of patients (you cannot have half a patient). This is sometimes referred to as **integer** (only whole numbers).

Continuous data: is numerical and can be measured. It is possible to have any intermediate value, for example, height, mass, length.

Qualitative data: is descriptive, for example, a patient's medical history.

The most appropriate method of data collection

A laboratory notebook is the traditional method of collecting experimental data in the sciences and medicine, although paper notebooks are being replaced by electronic versions. These offer advantages in terms of security (e.g. backup of data) and data sharing (e.g. collaboration).

Quantitative data can be collected automatically. For example, a data logger can be connected to a piece of electronic equipment to capture the output (data) and transfer it to a computer. This offers several advantages:

▷ Data can be collected without the need for a human operator to be present.

▷ Data can be captured continuously, for long periods if necessary.

▷ Once captured by the computer, the data can be analysed and processed.

Qualitative data is usually based on observation and so cannot normally be collected automatically. However, qualitative data may be collected via questionnaires or surveys, and these can be set up in electronic format, either on computer, tablet or online. This means that much of the data collection can be automated. As with collection of quantitative data, this allows some (if not all) of the data processing and analysis to be automated.

Application of new technology to collection and analysis of data will be covered in more detail in A5.5.

The most appropriate way to present the information or data

Before considering how to present information or data, we need to think about **dependent** and **independent variables**.

> ### Key terms
>
> **Dependent variable:** (often denoted by *y*) a variable whose value depends on that of another variable. In an experiment, we usually count or measure the dependent variable.
>
> **Independent variable:** (often denoted by *x*) a variable whose value does not depend on that of another variable. In an experiment, the independent variable is usually what we change.

Tables

When we collect data, we usually record it in a table, and this can often be a useful way to present the data.

There are rules to follow when presenting data in a table. This ensures uniformity so that anyone reading the table will be familiar with the layout of the data.

- Put the **independent** variable in the first column; the **dependent** variable should be in columns to the right.
- Put any processed data, such as means, rates or statistical calculations, in columns to the far right.
- Do not include calculations in the table, only **calculated values** (the results of your calculations).
- Head each column with the physical quantity and correct units and separate the units with brackets or a slash ('/'), for example, 'mass (g)' or 'body temperature/°C'.
- Do not include units in the body of the table, only in the column headings.
- Use consistent numbers of decimal places or significant figures throughout (see section B1.64 for more information about significant figures), even if it means writing 20.0 rather than 20.
- Calculated values, e.g. mean or other processed data, should be given to the same number of decimal places as the raw data, or one greater.

The table shows an example of how data should be presented. In this case, four 30-year-old men were asked to walk or run at a steady rate on a treadmill. The subject's heart rate (dependent variable) was measured in beats per minute (bpm) after 10 minutes. Following a period of recovery, the speed of the treadmill (independent variable) was increased and the experiment repeated. This was carried out at eight different speeds. The experiment was repeated for each of the four subjects and a mean value at each angle was calculated.

Speed of treadmill (km/hr)	Heart rate after 10 minutes (bpm)				
	Subject 1	Subject 2	Subject 3	Subject 4	Mean
2	75	78	80	76	77
4	80	82	85	88	84
6	88	89	92	90	90
8	95	98	101	105	100
10	110	105	115	112	111
12	125	100	128	130	128
14	155	158	160	165	160
16	170	175	178	185	177

One reason for presenting data like this is that it can help identify anomalous results – ones that stand out as being 'different'. If you look at the highlighted cell in the table (subject 2 at 12km/hr) you will see that the heart rate is lower than the same subject at 10km/hr and is much lower than the other subjects at the same speed. Something is obviously wrong – maybe the heart rate monitor was not working properly – and so we discount this result as anomalous, and we have not included it in the calculation of the mean value (you can check that for yourself).

As well as dependent and independent variables, there are **control** variables. These are things that we must keep constant (control) during the experiment. In this case, it means that anything that might affect heart rate, other than the speed of the treadmill, must be kept constant. One control variable would be the angle (steepness) of the treadmill as this makes walking or running more difficult.

What other control variables can you think of?

Graphs and charts

The type of graph we use will depend on the nature of our data as well as what we hope to get from the graph.

Scatter graph

Scatter graphs are used when investigating the relationship between two variables that can be measured in pairs, for example, the age and height of children in a school. The graph can then be used to establish whether there is a relationship between the variables. This could be a **positive correlation** (as variable *x* increases, variable *y* increases, as in Figure 5.2), a **negative correlation** (as variable *x* increases, variable *y* decreases) or **no correlation** at all. To tell whether the correlation is significant or not you need to carry out a statistical test.

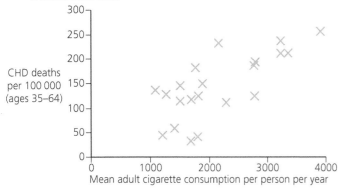

▲ Figure 5.2 Scatter graph showing the correlation between smoking and deaths from coronary heart disease (CHD)

Does correlation imply causation?

This question means: just because there is a positive correlation between two variables, does that mean that one causes the other? For example, there is a strong positive correlation between ice-cream sales and deaths from drowning. Does that mean ice cream is a major cause of drowning? Should we ban ice-cream sales anywhere near open water? Or is there some other factor that is correlated with both? Hot weather, for example.

Line graph

Line graphs are used to show **continuous** data (see Figure 5.3). The independent variable, the one that we change, should be on the *x*-axis (the horizontal axis – think of it like changing a baby's nappy: the thing that is changed goes on the bottom). The dependent variable, the one that changes in response to changes in the independent variable, goes on the *y*-axis (the vertical axis). You would usually join the points with straight lines. You can draw a smooth curve or line of best fit if you think that intermediate values will fall on the curve/line.

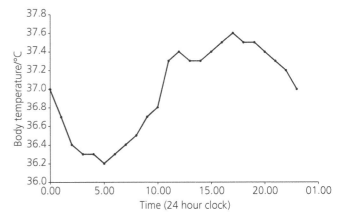

▲ Figure 5.3 Line graph showing how body temperature varies slightly during a 24-hour period

Bar charts and histograms

Bar charts are used to display **categorical data**. If you have an independent variable that is non-numerical (e.g. blood group, ethnic group of patient), then you should use a bar chart. These can be made up of lines, or blocks of equal width, that do not touch (Figure 5.4). The lines or blocks can be arranged in any order, although it can help make comparisons if they are arranged in order of increasing or decreasing size.

Bar charts can be turned through 90 degrees if it is easier to show or read horizontal rather than vertical bars.

Key term

Categorical data: is divided into groups or categories, such as male and female, ethnic group, city or country of residence.

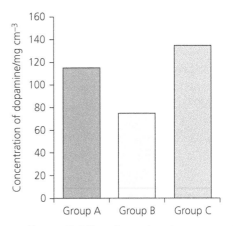

▲ Figure 5.4 Bar chart showing concentration of the neurotransmitter dopamine in three groups of patients

Histograms are sometimes called frequency diagrams. These can be used for either discrete data or continuous data grouped into classes, such as a range of heights or weights. The independent variable is usually on the *x*-axis and is grouped into classes. For example, the height of students in a class could be measured and grouped into 5 cm classes. Height is a continuous variable – students could be any height within a range – and so blocks are drawn touching. The axis is labelled with the class boundaries, e.g. 1600 mm, 1650 mm, 1700 mm, 1750 mm, 1800 mm, etc. and the *y*-axis would show the number or frequency within each class represented by the height of the bar (Figure 5.5).

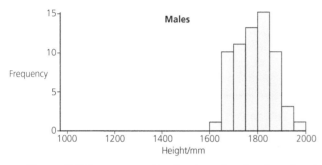

▲ Figure 5.5 Frequency distribution showing height of 18-year-old male students

Pie chart

Pie charts can be used when you need to show proportions or percentages (Figure 5.6). If you are drawing a pie chart by hand, you need to calculate the angle of each sector – divide the percentage by 100 and multiply by 360°, or just multiply the proportion by 360°. However, it is usually much easier to put your data into a spreadsheet and use the chart function to draw the pie chart!

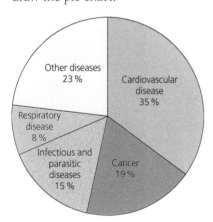

▲ Figure 5.6 Pie chart showing worldwide causes of death from disease, 2000–2019

Source: www.who.int/data/gho/data/themes/mortality-and-global-health-estimates/ghe-leading-causes-of-death

Case study

You have been asked to present data about the incidence (number of cases) of cardiovascular disease in the UK. You have been given the following data:

1 Total incidence of cardiovascular disease by age group (21–30, 31–40, 41–50, 51–60, 61–70, 71 and above).

2 Incidence of cardiovascular disease by the same age groups but separated into males and females.

3 Total incidence of cardiovascular disease (all ages) broken down by ethnic group.

You decide to use a histogram for the first data set. For the second data set, you have two options. You could draw two separate histograms, one for males and one for females. Alternatively, you could use a single histogram and draw two columns for each age group, one for males and one for females.

▶ Which do you think would be better?

▶ Are there advantages and disadvantages of each?

For the third data set you know that you cannot use a histogram. You could use a bar chart or you could use a pie chart.

▶ Think about the two ways of presenting the data. What advantages and disadvantages of each can you see?

Depth of analysis required

Collecting data is the start of the process. To convert raw data into useful information we usually need to carry out some form of analysis. Using graphs and charts can help in analysing the data or they can be the end point, i.e. how we present the data.

The depth of analysis required can determine how we record and present the data. Spreadsheet programs, such as Microsoft Excel or Apple Numbers, are convenient ways of recording data. We could enter the data manually, or even download data collected by a data logger directly into a spreadsheet. Once in a spreadsheet we can use a range of tools to analyse the data:

- simple analysis such as calculating the mean (average)
- more complex analysis such as statistical tests.

Another advantage of using a spreadsheet to record data is that we can use the graphing tools to present our data in different ways, such as the types of graphs and charts described on pages 70–71.

Qualitative or text-based data, e.g. from observations, surveys or focus groups, would normally be recorded and stored in a database, such as Microsoft Access or Apple FileMaker. Databases can contain huge quantities of data. They can also be used to analyse qualitative data as well as to organise it in ways that helps gain insight into connections, trends or groupings within the data.

Databases also allow others, such as collaborators or the general public, to access the data (if permitted), so this can be a way of presenting and sharing information as well as collecting, storing and analysing data.

The intended audience

This is a consideration largely when deciding how to present our data. Simple graphs and charts can make data and information more accessible to a non-specialist audience. This might be the case with an information leaflet or poster aimed at patients or the general public. Meanwhile, a more specialist audience might require more complex graphs that may give greater insight into the finer detail of the data. This would be the case if we wrote a research paper aimed at fellow health professionals for publication in a scientific or healthcare journal.

Collecting and storing data in electronic format (spreadsheet or database) also allows collaborators to get easy access to the data so that they can use it in their own work.

Storage method

How we store data depends in part on how it is collected and how it is used. Paper-based surveys, questionnaires or records of interviews or observations can, of course, simply be kept on paper. The same is true of results of experiments in a lab notebook. However, if they can be converted to digital format, then we can analyse them in the ways described. Data that has been collected in digital form would usually be stored in that form.

A5.3 The importance of accuracy, attention to detail and legibility of any written information or data

There are a few rules we must always follow in keeping written records of our work:

▷ We must be able to understand the record.
▷ It should be legible and contain enough detail for someone else to understand it and be able to repeat the work.
▷ It should be a faithful, honest and accurate record of what we did.

Always write up your work as you go along; do not make notes on scraps of paper and write it all up later – even if you do not lose the scraps of paper, you may not remember all the details. It is always good practice to date all your notes. In some organisations there may even be a legal requirement to do so.

There are numerous reasons why we must pay attention to detail and accuracy in written information or data.

▷ It may be necessary to comply with legal requirements, for example, the General Data Protection Regulation (GDPR) – see section A5.6.

We may need to limit liability, either our own or that of the organisation – for example, by ensuring anonymity and to record informed consent.

We should be able to provide an accurate account of events. That might be obvious when recording the details of an experiment, but it is also important in a healthcare situation. Knowing exactly what happened in the lead-up to a patient's condition improving or deteriorating can be essential in learning from experience. It might also be needed if there is an investigation into the patient's care.

It will help collaboration in integrated working and data sharing. You may be able to remember the details of an experiment, treatment or patient's observations. However, unless you record those details clearly and accurately, colleagues may not have sufficient information without asking you for more detail. Sharing of data is essential in multidisciplinary teams, where healthcare professionals with different specialisms must work together.

It helps to ensure accurate analysis of findings. Accurate analysis begins with good quality data. If the data is poor, then the processed or analysed data will also be poor, and the conclusions may be invalid. Recording a patient's **clinical observations** is an important aspect of monitoring a patient. If these observations are not correctly and accurately recorded, it may not be possible to identify whether they are within the normal range.

It can provide evidence needed in support of audit trails. This could include things as diverse as:
- keeping a record of a patient's clinical observations
- providing evidence that a medication or treatment has been given and when.

To help ensure reproducibility of results. Our work can be reproduced only if it is clear what we did, how we did it and what our results were.

Results are **repeatable** when we carry out an investigation several times in the same place, using the same method under the same conditions, and get the same result.

Results are **reproducible** when investigations are carried out by different people, in different places using different methods or equipment, and get the same result. This means that our findings can be replicated by others.

Case study

A study at a large hospital in England found that nurses sometimes completed documentation retrospectively (after the event) without full knowledge that the recorded care had been completed. One nurse described how a patient had collapsed, but their notes did not contain any information about why they had been admitted. In another case, nurses completed the documentation before they carried out procedures because they were worried about forgetting to complete the paperwork. Accurate record keeping was particularly important for older patients because of the complexity of care that they require and also the problems they may have with communication. However, nurses working with older patients found completing the documentation was very time consuming and took them away from patient care.

The authors of the study recommended that electronic methods of documenting patient care could be used to reduce the amount of unnecessary paperwork as well as to make sharing of information more effective.
- Can you see lessons for your own practice?
- Do you feel that there is sometimes a conflict between accurate record keeping and providing care?
- Should that be the case?

A5.4 The strengths and limitations of a range of data sources when applied in a range of health and science environments

If we are going to be able to draw valid and useful conclusions from data – get good quality information – we need to have an appreciation of the strengths and limitations of the type of data source. Otherwise we risk producing poor quality information based on invalid conclusions.

Results of investigations

Scientific or medical experiments ('investigations') should, if properly designed, give consistent and reliable results. This is because they should be well designed. This includes:
- formulating a clear hypothesis to be tested
- designing an experiment to test that hypothesis
- controlling all variables

- repeating measurements, excluding anomalous results (these are outliers, or results that lie outside of the range of the others) and calculating mean values
- performing a statistical analysis to test the significance of the results.

However, there can be limitations. For example, we might be tempted to apply the results of an investigation done under highly controlled conditions to the more complex situation of the 'real world'. A potential new drug might show promise when tested in test-tube experiments, for instance. However, complex interactions in living systems might mean that the drug does not work as hoped when tested in humans, or it might have unexpected and even harmful side-effects.

Patient history

At its simplest level, this approach involves taking medical histories of patients. This is a skill that many healthcare professionals need to develop. Done well, a medical history will provide detailed information about a patient over a long period. However, the data may be incomplete, particularly if it relies on the patient's memory.

If you consult some of the databases described in A5.1 you might come across many studies looking at patient histories over time. Some of these will look backwards at patient data – these are known as **retrospective** studies. Others will follow a group of normal subjects (a selection of the general population, perhaps of a particular age) or patients over a long period to look for development or progress of disease over a period of time – these are known as **prospective** studies.

In either case, the studies are likely to involve a **cohort**, which is a group of individuals that have something in common. It could be age, sex, state of health or disease, occupation, etc. Examples of this are the two Whitehall studies. Each of these followed a large group of British civil servants over many years and looked at factors that affected the health of the cohort, in particular cardiovascular disease and mortality rates. One finding was that there was a correlation between employment grade and reduced risk of cardiovascular disease – more senior civil servants enjoyed better health.

The strength of this type of study is that it can provide detailed information over time. However, there are cases where the data may not be accurate or complete, particularly with retrospective studies.

Patient test results

Patient test results usually mean the results of biochemical and other laboratory analysis. These should offer a high level of accuracy. Also, testing laboratories and the tests themselves must usually go through an accreditation process to ensure that they operate to approved standards and give reliable results. This means that test results this week should be comparable with results of tests taken weeks or months ago. It should also be possible to compare test results performed in different laboratories.

However, as humans are complex, a patient's test results will often need careful interpretation. This can introduce an element of subjectivity. Two healthcare professionals might look at the same set of patient test results and make different interpretations or draw different conclusions.

Published literature

The medical and scientific literature is a huge and invaluable resource in the health and science sector. This is the reason why most research projects will start with a literature review, to see what work has already been done and how that can inform and influence the research you plan to undertake. It is also an important tool for learning and diagnosis.

As well as the sheer volume available, another strength of published literature is that it goes through a process of **peer review**. This is where, before it can be published, a research paper is examined by other experts in the field. They might find flaws in the experiments or the logic or interpretation of the results and request that any deficiencies are put right before publication. This process improves the validity of the published research.

This process is not perfect. Professional rivalry means that reviewers are not always impartial. Also, the process can be slow and cumbersome, taking weeks, months or even years. This has led to a trend for **preprints**. These are papers that are published rapidly before peer review. This has the advantage of putting information into the scientific community very quickly. This is important during an event such as the COVID-19 pandemic, because information about new treatments can be made available rapidly in order to save lives.

However, not all published literature is of equal high quality. Some research might be based on studies with very small sample sizes, for example, clinical trials with only a few patients. The methods used in the experiment or study might be biased in a way that produces the results the investigators hoped for. The work might even be based on fraudulent data.

Andrew Wakefield and the MMR autism fraud

Andrew Wakefield was an academic physician who published a paper in the medical journal *The Lancet* in 1998. In this paper, Wakefield claimed that there was a link between the measles-mumps-rubella (MMR) vaccine and autism, based on a study involving 12 children. The publication, and particularly news conferences given by Wakefield, led to a loss of confidence in the MMR vaccine.

After the original publication, other researchers were unable to reproduce Wakefield's findings and it was found that Wakefield had a number of undeclared financial conflicts of interest. As a result, he was charged with gross professional misconduct and in 2010 the General Medical Council's Fitness to Practise Panel judged that Wakefield should be struck off the UK medical register, meaning he would no longer be allowed to practise as a doctor. Following this judgement, The Lancet retracted the paper, stating:

> 'It has become clear that several elements of the 1998 paper by Wakefield et al are incorrect, contrary to the findings of an earlier investigation. In particular, the claims in the original paper that children were "consecutively referred" and that investigations were "approved" by the local ethics committee have been proven to be false.'

Source: www.thelancet.com/journals/lancet/article/PIIS0140-6736(10)60175-4/fulltext

As a result of the original publication and the publicity it received, many parents chose not to have their children vaccinated and by 2012/13, cases of measles, mumps and rubella had reached the highest level for 18 years.

Source: www.gov.uk/government/publications/measles-confirmed-cases/confirmed-cases-of-measles-mumps-and-rubella-in-england-and-wales-2012-to-2013

Real-time observation

Some prospective observational studies can last for many years. However, observation does not need to be carried out over such a long period. We can get a great deal of information observing patients or test subjects for a relatively short time. The advantage of this is that we get the data immediately.

Of course, the disadvantage is that our observations might become subjective. **Objective** observations are things that can be measured, such as pulse, temperature or respiration rate. **Subjective** observations are signs that cannot be measured, such as how a person or patient feels. These can be very important, particularly when we observe someone over time and can tell when there is a change in their condition or behaviour. However, we have to take care that we do not misinterpret or over-interpret such observations.

1 Give two advantages and one disadvantage of using results of experiments or investigations in healthcare.
2 Give one advantage of using data or information from published research.
3 What is the difference between **objective** and **subjective** observations?

A5.5 How new technology is applied in the recording and reporting of information and data

In A5.2 we saw the advantages of automated data collection. New technologies are being used to improve both the quality and quantity of data that can be collected, even allowing us to collect new kinds of data. New technology has also transformed the way we can analyse the data collected.

This is particularly true when applied to **longitudinal** research, where measurements or observations are made over a long period. By allowing passive measurement of health-related factors, we can avoid the bias that can occur when we rely on people self-reporting. The widespread use and acceptance of fitness trackers and smartwatches means that these can also be used for monitoring physiological data in subjects that are part of a cohort study or clinical trial.

AI/machine learning

Big data is a term that you hear more often these days, usually to describe the massive amounts of data collected by companies such as Amazon, Facebook and Google. However, there are many areas of science and healthcare that now generate very large data sets, for example:

- DNA sequences, such as from the Human Genome Project
- proteomics – the study of the proteins produced by the body and how the presence or absence of some proteins is correlated with various diseases
- high content imaging – the automated collection and analysis of microscope images to provide information about processes within cells and tissues
- results from clinical trials
- epidemiology data looking at the spread of diseases.

Machine learning is a branch of **AI** (artificial intelligence) that uses computers to imitate the ways in which humans learn. The **algorithm** (a list of rules that a computer follows to solve a problem) can be trained to interpret a sample set of data and then automatically improve the algorithm. This can go through many rounds so that, eventually, the computer is able to interpret very large data sets.

This approach has been used to develop algorithms that can help to interpret medical images, such as those from CT scans or radiography for use in diagnostics, particularly of cancer.

Case study

Research carried out by Moorfields Eye Hospital, DeepMind (part of Google) and UCL uses AI to help identify diseases that can lead to blindness. AI and machine learning technology was trained on thousands of historic eye scans, previously used by specialists to diagnose the disease, to identify features of eye disease and recommend how patients should be referred for care. The AI system can recommend the correct referral decision for more than 50 eye diseases with 94 per cent accuracy. This is the level of accuracy achieved by world-leading eye experts.

Some of these applications come under the general heading of **bioinformatics**. This describes the use of tools and software methods to analyse and process large data sets, such as DNA and protein sequences, genetics, cellular organisation and the mutations that lead to cancer. A related area is **systems biology**, which looks at the chemical and enzyme interactions and pathways within the cell.

Mobile technology and applications

Smartphones and high-speed mobile data networks such as 4G and 5G mean that devices can be connected to the internet almost anywhere in countries with more developed economies – and in many other parts of the world as well. Smartphones and tablets can also be used within a hospital, care home or other healthcare facility to connect to Wi-Fi. This opens up many opportunities for using mobile devices to collect and transfer data.

Health informatics is the use of computer science and AI/machine learning to assist in the management of healthcare information.

Case study

Nottingham University Hospitals NHS trust (NUH) has issued 6500 mobile devices to its clinical staff to allow implementation of electronic observations. Alongside this, NUH developed an app called 'Safer Staffing'. This allows the trust to see the nurse staffing position in real time across more than 85 wards and departments. The nurse in charge of each ward is asked to declare whether they judge that the ward is safe to deliver the required level of care to its current group of patients with the staff on duty. This allows the trust to see where support might be required.

Smartphones also have GPS (global positioning system) receivers built in, which means that they can be used for physical tracking of individuals. This, together with Bluetooth® wireless technology, was used in the NHS COVID-19 app that was the basis of 'track and trace' during the pandemic. If you had been in close contact with someone who tested positive for COVID-19, the app would 'ping' you and you would be told to self-isolate for 10 days. Unfortunately, in the summer of 2021, as lockdown restrictions were eased, the number of people who had to self-isolate grew to such high levels that there were severe staff shortages in areas such as hospitality, transport, retail and healthcare. As a result, there were reports of people deleting the app to avoid being contacted. This was a good example of how technology does not always offer an ideal solution, particularly if it is not properly implemented.

Cloud-based systems

Cloud computing means the availability of computers and computer resources, especially data storage, without active management by the user. Cloud-based systems usually have functions distributed over multiple locations or data centres connected via the internet. This distribution can make cloud-based systems more robust, because they will usually recover quickly if a single server fails. However, it can also make them more susceptible to malicious attacks, as described in A5.9.

An important application of cloud-based systems in healthcare is the use of **electronic health records (EHRs)**. Because cloud-based systems are accessible to anyone with the appropriate permission, patient data can be available to anyone who needs it across the whole healthcare system. NHS Digital is implementing systems like this – see A5.1 for an example.

As well as making patient data available to all appropriate healthcare professionals, cloud-based systems mean easier data sharing for further analysis, for example, by AI/machine learning systems that can be used in diagnosis or to gain insight into treatment and prevention of disease.

Digital information management systems

In 2018, the Department of Health and Social Care published a policy paper entitled 'The future of healthcare: our vision for digital, data and technology in health and care'. This addressed many of the areas that we have covered here, such as the use of cloud-based systems, AI/machine learning and mobile technology.

Digital information management systems also offer the advantage of a digital audit trail. This can show, for example, who accessed a patient's records and when, which can help ensure patient confidentiality and protect their personal information (see A5.6).

Data-visualisation tools

One problem associated with the vast amount of healthcare data that is now available is the difficulty in understanding the complexity of numerical and text-based data. The simplest data visualisation tool is the graph (see A5.2). Modern technology makes it possible to take multiple data sources, often from cloud-based systems such as patient databases, and present them in a way that makes it easier to understand and interpret them.

Every modern hospital will use data-visualisation tools to manage its in-house processes, such as:
- maintaining electronic medical records to track and monitor patient health
- avoiding diagnosing errors by eliminating human error
- accessing information about patients' demographics (factors such as age or ethnic background) and lifestyles (diet, exercise, smoking, alcohol consumption, etc.) in order to improve care and treatment of the patient.

Test yourself

1 Explain what is meant by the following terms:
   a Artificial intelligence (machine learning).
   b Cloud-based systems.
2 Give one advantage of the use of mobile technology in healthcare data capture.
3 Give one advantage of using digital information management systems.
4 Give one advantage of using data visualisation tools in healthcare.

A5.6 How personal information is protected by data protection legislation, regulations and local ways of working/organisational policies

We have covered the various applications of modern technology in the health sector. We all know that electronic systems are not always secure from all the hacking incidents that have happened over the years. To maintain faith in these different uses of modern technology, patients and staff must feel confident that their personal information will be protected.

This applies equally to personal information stored on paper or in electronic format. In this section we will look at the ways in which personal information is protected. These issues are covered by legislation as well as local policies that most organisations have in place.

Data Protection Act 2018 (DPA 2018)

The DPA 2018 revised earlier data protection law passed in 1998, but its main purpose was to implement the GDPR of the European Union (EU). Although the UK has now left the EU, the legislation remains in force at the time of writing.

The DPA 2018 controls the use of personal information by organisations, businesses or the government. The Act is enforced by the Information Commissioner's Office (ICO), which is funded by a charge on **data controllers**, who must register with the ICO. A data controller is a person or organisation that stores or processes personal information, either on paper or electronically. Personal information is defined quite widely, so it includes aspects such as:

- your name and telephone number
- your National Insurance or passport number
- your location data, such as home address or smartphone GPS data – i.e. your smartphone tracking you
- online identifiers such as email or IP address.

However, there are other types of personal data that are also covered by the Act and GDPR, including:

- biometric data such as facial images or fingerprints that can be used to identify you
- health data that can reveal information about your physical or mental health status
- genetic data that can also give information about your health or physiology
- ethnic origin
- political opinions and religious or philosophical beliefs
- sexual orientation.

UK GDPR

The GDPR came into force within the EU in 2018 and has been retained in domestic law as the UK GDPR. It provides a set of principles with which any individual or organisation processing sensitive personal data must comply. There are six legitimate reasons why an organisation may process your personal data:

- You have given your consent for them to process your data for a specific purpose. This must be explicit (for example, you tick a box on an online form saying 'I wish to receive emails from you') rather than default – where you must untick a box to opt out of receiving emails.
- The processing is necessary to fulfil a contract you have entered into. For example, if you place an online order with a company, it has the right to process your personal data so that it can deliver the goods or inform you if there are any delays.
- There is a legal obligation for them to process your data. For example, an employer is obliged to pass employee salary details to HMRC for tax purposes.
- The data processing is necessary to protect you or someone else. For example, if you are admitted to A&E with life-threatening injuries following a road traffic accident, disclosure to the hospital of your medical history is necessary to help save your life.
- Processing is necessary to perform a task in the public interest or for an official function. This applies to any organisation that exercises an official authority that has a clear basis in law. Examples include local authorities, courts and the criminal justice system, and other agencies carrying out duties that are laid down in law. For example, a government agency has statutory powers to research the online shopping habits of consumers. The agency can ask retailers to share the personal data of a random sample of their customers so that it can carry out this function.
- Processing is necessary so that the organisation can pursue its legitimate interests. For example, companies and other organisations need to maintain personnel records.

GDPR includes a number of rights of individuals, for example:

- the right to be informed about the collection and use of their personal data
- the right to have access to their personal data
- the right to have incorrect personal data corrected or completed if it is incomplete
- the right to have personal data erased – but only in certain circumstances
- the right to restrict the processing of their personal data – again, only in certain circumstances
- the right to data portability, so that they can copy or transfer their personal data between different systems, for example, if you have accounts with different banks you may be able to view the details of all of them in a single app
- the right to object to processing of personal data, for example, you can refuse to have your personal data used for direct marketing (sometimes called 'spam' or 'junk mail').

GDPR also regulates the processing by organisations outside the EU of personal data of people within the EU. It prevents transfer of data from within the EU to countries outside the EU (such as the UK) unless those countries have appropriate safeguards in place. If you think about the widespread use of cloud computing described in A5.5, you can see how easily personal data might be transferred across national borders. Partly as a result of this, the GDPR has become the model for many national data protection laws around the world.

Local ways of working/organisational policies

The DPA 2018 and GDPR are not the only things that regulate and protect personal information. Many organisations have their own ways of working, rules and policies that ensure compliance with legislation and regulations, as well as to protect the organisation (for example, from reputational damage if personal information is misused). The ICO expects organisations to have their own policies in place to ensure that they and their staff will comply with GDPR.

Examples of this include:

- ensuring that data is stored securely (electronically or paper-based) so that there will not be any loss of data or loss of confidentiality
- restricting the use of mobile devices as ways in which personal information may be misused or divulged. This helps to ensure confidentiality
- preventing potential conflicts of interest. For example, organisations must appoint a Data Protection Officer (DPO) who should be relatively senior in the organisation. However, there are some senior jobs (such as Chief Executive, Chief Medical Officer or Head of Human Resources) that would be in conflict with the job of DPO, so the roles should not be combined.

Reflect

Clearly it is essential to protect personal information and ensure patient confidentiality is maintained, but can that be taken too far? A patient's medical history must be kept secure when sharing it with another provider involved in their care, but is there a risk that vital information could be missed if the medical history is kept confidential? Could keeping data confidential from those who need to know put effective care at risk?

Test yourself

1 Name the legislation and regulation that protect personal information.
2 Which of the following are examples of the acceptable processing of personal information?
   a Collecting contact details from social media so that you can advertise a new patient support group.
   b Providing a patient's medical history to a life assurance company without the patient's permission.
   c Providing a patient's medical history to an intensive care doctor when the patient is in a coma.
   d Transferring a database of patient data to a computer located in an unknown country for analysis.

A5.7 How to ensure confidentiality when using screens to input or retrieve information or data

Having considered the legal framework around the protection of personal data, we should also consider ways in which confidentiality could be lost or compromised through carelessness or poor practice.

Computer systems should always have password-protected access. This helps to protect sensitive personal information from being accessed by anyone without authorisation. It also means that it is possible to see who accessed what information – it provides an audit trail if there is a data breach.

That means that we should always protect login details and passwords. Unfortunately, many organisations insist on the use of passwords that are difficult to remember (although not always difficult for computer hackers to guess). It should be obvious that 'Password1234' is not a secure choice. Nor is it good practice to write your login details on a sticky note on your computer screen. Sadly, both of these – and other – poor practices are widespread.

We should always log out of a system when leaving the screen (PC, laptop, tablet, etc.) so that someone else cannot come along and access sensitive personal information. Alternatively, if you have a password-protected lock screen (which is a basic computer security feature), you should lock the screen when you leave it, even if it is just for a minute.

You should also be aware of your surroundings when dealing with personal information. When you get money out of a cash machine, you should be cautious of 'shoulder surfers' trying to see your PIN as you enter it. The same can be true in the workplace where it might be possible for unauthorised people (colleagues, patients, visitors) to overlook you and your screen. For this reason, privacy screen filters can be useful as they make it difficult to read information on screen unless you are directly in front of it.

Another area that is important is the use of secure internet connections. We should all be aware that regular email is not secure – it passes through so many different systems, any one of which could be compromised. Another potential insecurity is the use of public Wi-Fi, such as in coffee shops. Any information shared over a public Wi-Fi network can be intercepted easily. Your home Wi-Fi may not be completely secure either. For this reason, many organisations now use VPN (virtual private network) to access the internet or exchange information safely and privately.

A5.8 The positive use of, and restrictions on the use of, social media in health and science sectors

It is becoming more common for employers to look at the social media of prospective employees. So think carefully before you post pictures of wild nights out on your social media! Of course, social media can offer many advantages as well.

Positive uses

Social media can play an important part, particularly in:
- awareness campaigns and disseminating information – this could be as simple as reminding social media followers to follow common-sense health practices, but it is also possible to target relevant population groups with specific health messages

- correcting misinformation – there is a lot of health misinformation on social media. This might be simply untrue statements. There can also be misinformation in the form of facts presented without context, or in the wrong context. Unfortunately, people are often more inclined to believe information that supports their existing prejudices
- crisis communication and monitoring, for example, during an epidemic or pandemic (such as COVID-19). More people now get their news from social media than from newspapers.

PHE uses a wide range of social media to share news and information about its work as well as public health incidents – the COVID-19 pandemic, for example.

Other uses of social media in the health and science sectors include:
- monitoring public health. People post about everything online, including their health. Hashtags such as #flu can show when diseases are spreading in new locations. Health data from social media has improved prediction of infectious diseases such as flu
- data gathering
- establishing patient support networks. This can be particularly effective among younger people
- recruitment – both of new employees or of subjects for clinical trials and other health research
- marketing by commercial and healthcare organisations.

Restrictions

We have already seen how social media can be a useful tool but can also have disadvantages. To try to minimise the disadvantages, we need to follow some basic rules when using social media in the health and science sectors.
- Do not post sensitive or personal information about yourself or others on social media. This should be common sense, but it is also likely to be covered by an organisation's code of conduct.
- Maintain professional boundaries when interacting with individuals outside the organisation. Again, it should be common sense, but it can sometimes be easy, in the context of social media, to be less professional than you would be face to face.

Do not share inaccurate or non-evidence-based information. In fact, it is a criminal offence under the Care Act 2014 for care providers who 'supply, publish or otherwise make available certain types of information that is false or misleading'.

More employers in the health and science sectors are implementing social media policies to avoid risks associated with misuse of social media by their employees. These risks can include:

- reputational damage
- infringement of the intellectual property rights of others, e.g. by posting copyright material such as images, videos or music
- liability for any discriminatory or **defamatory** (reputation-damaging) comments posted by employees
- possible unauthorised sharing of confidential information.

A5.9 The advantages and risks of using IT systems to record, retrieve and store information and data

Compared with paper-based systems, IT systems have huge advantages – some of which were discussed in A5.5. However, while being aware of the advantages, we must also guard against the risk associated with IT systems.

Advantages

The advantages of IT systems in the health and science sectors include:

- ease of access, sharing and transferring data – particularly with greater use of cloud-based systems
- the speed of data analysis as well as the ability to use AI/machine learning (see A5.5)
- greater data security, for example, when it is password-protected
- standardisation of data. This can be a barrier to be overcome when transferring information and data from old, particularly paper-based, systems to modern IT systems because the older data is likely to be in a variety of non-standard formats. However, moving forward, standardisation can be a great advantage
- the ability to have continuous and/or real-time monitoring of data. This can be particularly

important during disease outbreaks such as epidemics and pandemics
- cost and space saving. This is another reason why cloud-based solutions are becoming more widespread
- integrated working, making for greater collaboration between colleagues. It also supports safeguarding practices.

Risks

We have covered some of the risk factors, particularly security breaches – accidental or malicious – that can compromise patient confidentiality. However, there are other risk factors to consider.

There is the potential for corruption of data, making databases unusable. This is one reason why all IT systems need robust procedures for backup of data – although making a backup of corrupted data can just mean that you have a corrupted (and so useless) backup. Organisations need to be aware of the risks of data corruption as a result of ransomware attacks. This is where malicious software is installed on a system (usually by someone in the organisation clicking on a link in an email or on a website) which then corrupts or encrypts all the data held on the system. Some organisations have paid millions of dollars in Bitcoin to ransomware criminals so that they can retrieve their data.

Have you ever called a company, only to be told 'I'm sorry, but our system is down'? That might be annoying when you are trying to find out when you will get the item of clothing you ordered. But when you are trying to access critical patient information, it can be much more serious. This is why many healthcare systems have built-in redundancy – there may be multiple computers handling access to data so that a single failure does not bring the whole system down.

It is hard to think about how we managed without modern IT systems. But, do we always appreciate the risks as well as the benefits? We have seen the importance of following policies and procedures, but do we always think about whether that is enough? Are the systems that we use sufficiently secure? Are there procedures to monitor and enforce the policies that are in place? These are questions for management, rather than individual health professionals. However, it is good to ask ourselves these questions, if only to reinforce the need to comply fully with required procedures and policies.

A5.10 How security measures protect data stored by organisations

We have already looked at how security measures can help to preserve patient confidentiality and keep personal information secure. However, that is not the only reason that data must be protected by appropriate security measures. Loss of data can be catastrophic in a modern healthcare setting.

Malicious operators (hackers) may try to access sensitive patient information, or they might have some other criminal intent, such as installation of ransomware. State-sponsored groups have been implicated in various attacks on computer systems in recent years – this has been given the name 'cyber warfare' – and healthcare systems have been targeted. In May 2020, the International Committee of the Red Cross called for governments to take immediate and decisive action to prevent and stop cyber attacks on hospitals, healthcare facilities and organisations, research organisations and international authorities that provide critical care and support for healthcare. This followed cyber attacks against medical facilities in the Czech Republic, France, Spain, Thailand and the United States as well as international organisations such as the WHO.

There are various ways in which an organisation's data can be protected:
  ▷ Controlling access to information, for example, through levels of authorised logins and passwords. This can also mean that some staff can access information but not change it ('read only' access). This helps protect against inadvertent change to or deletion of data by inexperienced staff.
  ▷ Allowing only authorised staff into specific work areas so that they cannot physically access sensitive computer equipment. In some organisations, the USB ports on computers are disabled so that data cannot be transferred onto USB memory devices.
  ▷ Requiring regular and up-to-date staff training in complying with data security. Technology and the methods used by cyber criminals change rapidly, so staff need to keep up to date.
  ▷ Making regular backups of files. Often there will be multiple backups so that there is no risk of replacing 'good' backup data with corrupted or encrypted data.
  ▷ Using up-to-date cyber security strategies to protect against unintended or unauthorised access. Organisations need to employ good cyber security staff or consultants and make sure that their recommendations are taken seriously.
  ▷ Ensuring that backup data is stored externally, for example, cloud-based or on (ideally multiple) servers elsewhere. If there is a fire in the IT suite, you need to be able to get up and running again quickly and this will not be possible if all the backups are stored in a cupboard in the IT suite.

A5.11 What to do if information is not stored securely

Sometimes things go wrong, often due to human error or carelessness. If you discover sensitive information that is not stored securely, you need to take action. This could apply equally to paper-based and electronic patient data.

The first step must be to secure the information where possible. This is mostly likely to be the case with paper records left lying around. However, a colleague might have gone home and left their screen running or still be logged into a sensitive system. In such cases you should log them out and shut the screen down.

Having taken immediate action to secure the information, you must then record and report the incident to the designated person. This will depend on your organisation's policies and procedures, but it might be your line manager or a specialist computer security person or department.

Research

In 2016 the Care Quality Commission (CQC) published a report into whether personal health and care information is being used safely and is appropriately protected in the NHS. The review is available from the CQC website at https://www.cqc.org.uk/publications/themed-work/safe-data-safe-care (or search online for 'safe data safe care').

Think about what the CQC found. Can you see any lessons for your own workplace or professional practice?

There were six recommendations made. Can you see evidence of them being applied in your own organisation?

Project practice

You have been asked to prepare a report on data collection, recording and handling in your organisation.

You should research:
- ▶ The methods used to collect data.
- ▶ The types of data that are collected.
- ▶ The way(s) in which data is stored, analysed and presented.

You should also consider how your organisation addresses:
- ▶ the use of technology to assist in recording and reporting of data
- ▶ the ways in which data is shared amongst those who need to have access to it
- ▶ how personal data is protected
- ▶ procedures and policies that apply to ensuring confidentiality and security of personal data.

Prepare an evaluation of the methods used, covering:
- ▶ ways in which procedures ensure accurate collection and recording of data
- ▶ how information sharing is helped or hindered by the procedures in place
- ▶ areas that might be improved, such as:
  - – reducing the time needed to document activities
  - – improving sharing of information
  - – increasing security of data and personal information (concentrate on how the systems are used, rather than on the detail of the systems themselves).

Present your report as a written document or slide presentation, infographic or scientific poster.

Discuss your findings with the group.

Write a reflective evaluation of your work.

Assessment practice

1 For each of the following, state whether it is qualitative or quantitative data:
   a Blood pressure readings.
   b Responses to a questionnaire about patient perception of the quality of their care.
   c Height.
   d Hair colour.
2 For each of the following, state whether it is continuous or discrete data:
   a Number of weeks that a pregnancy lasts.
   b Number of children in a family.
   c Average number of children in families.
   d Size of the UK population.
   e Blood glucose concentration.

3 Suggest **two** security measures that can be used to protect personal data stored on a computer network.
4 Suggest a disadvantage of automated data collection.
5 Discuss the advantages and disadvantages of using surveys to collect healthcare data.
6 Discuss the impact of machine learning on the role of healthcare practitioners.
7 Discuss the importance of having policies to prevent unauthorised access to patient data.
8 Discuss the ways in which security measures can protect stored data.

A6: Data handling and processing

In the previous chapter we looked at the difference between data and information, as well as the different types of data and how they are managed.

We will now look in more detail about the ways in which data is handled and processed.

Learning outcomes

The core knowledge outcomes that you must understand and learn:

A6.1   the stages of data handling and processing

A6.2   the difference between qualitative and quantitative data

A6.3   the advantages and limitations of different methods of data storage and recording

A6.4   the purposes of software systems used for data capture in scientific settings

A6.5   the difference between systematic and random data errors

A6.6   how to minimise errors occurring in a scientific setting

A6.7   the different methods of data processing and analysis in science environments

A6.8   ways to present data in the appropriate format

A6.9   the purpose of statistical techniques when analysing data

A6.10   how to review data and make decisions based on that review

A6.11   the consequences of bias in data analysis

A6.12   how to prevent or reduce bias in data evaluation

A6.13   links between sample size and effective statistical analysis

A6.14   how to order numbers by relative size in a data set

A6.15   how to ensure proportionality while scaling up or down quantities in a formulation.

A6.1 The stages of data handling and processing

You will remember from Chapter A5 that we made a distinction between data and information – that data is a collection of values (words, numbers and so on) whereas information is data that has been processed in a way that we can read, understand and use.

This chapter will cover the process of obtaining data and converting it into useful information.

The stages of this are:

▶ **Collect** the data: this was covered in sections A5.1 and A5.2.
▶ **Record** the data: this was touched on in section A5.2, while use of new technology in data recording was covered in section A5.5.
▶ **Analyse** the data: this will be covered in more detail below.
▶ **Interpret** the data: analysis and interpretation are the stages where data is converted into useful information.

A6.2 The difference between qualitative and quantitative data

This was initially covered in section A5.2.

▶ **Qualitative** data is often subjective (meaning that it cannot be measured consistently – see section A5.4 for an example). It is usually text-based (rather than numeric and categorical), meaning that it is divided into groups or categories.
▶ **Quantitative** data is objective (meaning that it can be measured consistently) and numerical, meaning that it can be defined as a value.

If you consider the data collection methods we looked at in section A5.1, then you can see that a focus group will always produce qualitative data whereas a survey might produce qualitative data (people's opinions about the service their GP offers, for example) or quantitative data (the number of people who have visited their GP in the last six months).

Other types of data, such as official statistics or the results of laboratory experiments, will usually be quantitative.

A6.3 The advantages and limitations of different methods of data storage and recording

In section A5.2 we looked at things to consider when selecting a method of collecting and storing data. We will now look more closely at two of the main methods of data storage and recording data that are used in the science sector.

Physical lab notebooks

As the name suggests, these are paper-based systems.

Advantages

▶ They are safe from computer failure – nobody ever has to wait for their lab notebook to reboot or charge.
▶ Unlike computerised systems, a paper notebook cannot be accessed by external hackers; it can be physically locked away to keep it inaccessible.
▶ They can be used in conditions that would be unsuitable for computers or tablets, such as in extreme environments or areas where there is limited infrastructure.

Limitations

▶ They can be accessed by anyone in the workplace; this might be an advantage in some situations but can also compromise security or confidentiality.
▶ They can be altered without changes being tracked, although some notebooks have features that can reduce this risk to an extent – for example, by having numbered pages and by users dating each entry.
▶ They cannot be shared or searched easily.
▶ You cannot make a backup of a written notebook easily. They can be lost, damaged and degraded over time; fire or an acid spill could destroy them in minutes.

Laboratory information management systems

You can think of **LIMS** as an electronic filing cabinet, in the sense that it contains lots of different types of data and information in dedicated files. It is a software-based system that supports almost the entire operations of a modern laboratory.

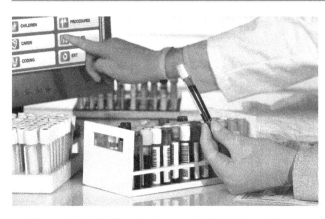

▲ Figure 6.1 LIMS using barcoded labels to allow sample tracking through all stages of analysis

Early systems were used to track samples, particularly in analytical laboratories. There are now many different types of LIMS with varying degrees of functionality, including:

▶ management and tracking of the workflow and samples, often using barcoded labels
▶ data storage and analysis
▶ integration with modern analytical instruments, allowing automated collection, storage and analysis of data
▶ features that facilitate audits and compliance with internal or external regulations.

See section A2.11 for more on compliance with internal or external regulations.

Advantages

▶ They enable data visualisation (see sections A5.2 and A5.5) and customisable reports to be generated quickly and easily.
▶ Data is shared easily as LIMS are usually located on a computer network or cloud-based storage.
▶ LIMS is essentially a database and so can be searched.
▶ LIMS can be accessed remotely. This is useful when users in different physical locations need access to the data.
▶ Cloud storage, if used, ensures the safety of data from physical damage and provides an automatic backup.
▶ LIMS can be used to highlight errors in the system or the data.

Limitations

▶ As with any IT solution, LIMS can be accessed by hackers if IT security is not sufficiently robust.
▶ They can be vulnerable to technology failure. This could be a simple hardware failure or software flaws that lead to loss or corruption of data.

▶ They can be very expensive. There will be various costs associated with implementing and running LIMS, such as the purchase of hardware and software licences, staff training, IT support and other maintenance costs. This can typically add up to tens of thousands of pounds – you can buy a lot of lab notebooks for that price!
▶ LIMS requires maintenance to ensure that they keep running efficiently. This might involve maintenance of the hardware and connections with lab equipment, or the need for regular software updates.
▶ LIMS requires an internet connection for synchronising data across different sites.

See Chapter A5 for a discussion of different aspects of IT security.

A6.4 The purposes of software systems used for data capture in scientific settings

Even without a full LIMS, every modern laboratory will use software to capture data, even if that only involves entering data into a spreadsheet.

Scientific settings

There are likely to be different requirements for data capture depending on the scientific setting. Analytical laboratories usually perform routine or repetitive analyses, such as measuring the concentration of different substances in the blood or the presence of bacteria or other contaminants in water samples. This means the data capture requirements are unlikely to change very much. In a research laboratory, on the other hand, the type of data captured and stored might change as research projects develop.

Data generated in laboratories is most likely to be quantitative and so is well suited to storage and analysis in a spreadsheet. In other settings there is likely to be a need to capture qualitative data, such as patient records, focus group or questionnaire responses and so on. This will usually involve human input of data rather than automatic data capture, and the data would normally be stored in a relational database (where tables of data can be cross-referenced) rather than a spreadsheet.

Clinical trial data usually involves huge quantities of quantitative data, such as the results of tests monitoring the response to a drug or treatment.

However, it is also likely to involve a great deal of qualitative data, such as patient records. For this reason, there are various types of clinical trial management software solutions on the market.

Data sharing

Software systems usually allow data sharing with other scientists or stakeholders as appropriate. This might include collaborators in a project or auditors or inspectors in organisations that are subject to various levels of regulatory control.

Commercial sensitivity

Many organisations have data that is **commercially sensitive**, meaning that disclosing it to someone outside of the organisation could harm the interests of the organisation. This might include data from research, product development or production processes. Cost or performance data is another area where competitors might gain an advantage if they were able to access the data. This is another important reason for having good data security.

Ease of analysis

Software systems usually enable easier analysis and interpretation. For example, it is relatively straightforward to produce a wide range of attractive and informative charts from data stored in spreadsheets.

For larger data sets, particularly those involving qualitative data, relational databases can be used to organise the data and make connections between data or identify patterns or trends.

A6.5 The difference between systematic and random data errors

When collecting and analysing data, we must always consider the possibility of errors in the data. When we make a measurement, there is an uncertainty in the measurement. This is inevitable and is why we use the term 'uncertainty' rather than 'error' (which sounds like a mistake was made). This will be discussed in more detail in section A8.7.

For now, we are going to think about types of error, that is, when something went wrong or you did make a mistake. These errors usually affect the precision or the accuracy of the measurements – this is also covered in more detail in section A8.7.

Systematic errors

These are consistent errors; hence, we say they are **systematic**. They can be caused by flawed design or execution of experiments, or problems with equipment. Systematic errors usually affect the **accuracy** of the measurements, that is, how close they are to the true value, but the measurements may still be precise. Here are some examples:

▶ A calibration error can occur when a thermometer reads +0.5 °C above the true temperature. For example, if a solution was 12 °C, the thermometer with the calibration error would record 12.5 °C. This means that all readings will be inaccurate by +0.5 °C.

▶ A similar error can occur with equipment such as a pH meter that must be calibrated before use (see section A6.6).

▶ A zero error can occur when using a balance if you do not **tare** it (set it to zero) correctly between each weighing.

Systematic errors can also occur in qualitative data and are usually caused by poor design, for example of a questionnaire. Response bias can occur if a questionnaire incorporates a **leading question**.

See section A5.1 (page 64) for an example of how a leading question could alter the response to a survey.

Key term

Leading question: a question that pushes the respondent to reply in a particular way; it almost contains its own answer.

Random errors

These are caused by unpredictable or unknown changes during an experiment. Random errors usually affect the **precision** of the measurements. Examples include:

▶ Something causing interference with electronic equipment that is being used for measurement.

▶ Natural variation in real-world experiments. You might be testing the memory capacity of individuals and carry out the tests at different times of the day. However, some people perform better in the mornings while others perform better in the afternoons. The measurements you take will not, in those circumstances, reflect the true memory capacity for each person.

▶ Some random errors can be the result of operator error, such as misreading a scale on a ruler, measuring cylinder or an analogue electronic measuring device (one that has a needle moving across a scale).

A6.6 How to minimise errors occurring in a scientific setting

Which type of error can cause more problems, random or systematic?

Random errors can usually be reduced by taking repeated measurements. This can help identify anomalous results (ones that do not fit in with the pattern of other results) or outliers. Having repeated the measurements, you will normally calculate a mean or average (see section A6.9). You can then exclude any anomalous results when calculating a mean, and the mean will then be closer to the true value than the individual measurements – this was discussed in section A5.2.

Systematic errors can be more difficult to identify, so usually we just have to be aware and remember that they can occur and take steps to eliminate or at least reduce them.

Using controlled variables

Section A5.2 covered **independent variables** (what we change in an experiment) and **dependent variables** (what changes as a result). There was also an example of a **control (or controlled) variable**. Control variables are any factors that could affect the measurements.

Examples of control variables include:
▶ keeping temperature constant when investigating the effect of concentration of reactants on the rate of a reaction, because temperature can also affect the rate of reaction
▶ using test subjects of the same age and state of health when investigating the effect of a new drug on blood pressure, because these factors may also affect blood pressure.

All well-designed experiments will involve a control or control group. In a laboratory experiment, the control will be exactly the same as the test group or groups, with the exception of the thing that is being tested. For example, if you are investigating the effect of an enzyme reaction, you will usually have a control reaction using boiled enzyme because boiling denatures (inactivates) the enzyme. Everything else will be kept the same, so that you know any results will be due to the enzyme. This will be expanded on in section A10.2.

In clinical trials there will usually be a **placebo group**. This is where one group receives a **placebo** – a 'dummy'

drug that appears identical to the one being tested; it could be a sugar pill or saline injection. This will show that any effects seen in the test group are due to the drug and not just to attention being paid to the patient.

Staff training and monitoring

Earlier chapters have explored the importance of staff training in various contexts – following SOPs, meeting regulatory requirements or organisational policies, for example. The same is true for minimising errors. Operator error, such as misreading a scale, was mentioned in section A6.5 as a type of random error. Errors of this type can be removed, or at least reduced, by proper staff training.

Monitoring is also important. We have seen how it is necessary to keep following an SOP properly rather than finding short cuts. The same is true with carrying out experiments or analyses or using experimental equipment. Monitoring helps to ensure that equipment is being used correctly and experiments are being performed correctly, making it less likely that errors will be made.

Chapter A2 described the importance of audits in ensuring continued compliance with SOPs and GMP. This is particularly important in highly regulated science settings, such as the pharmaceutical industry.

Maintenance and calibration of equipment

You will come across this topic in different contexts throughout your course. Maintenance and calibration of equipment are important features of GLP and GMP, which were covered in Chapter A2. The importance of maintenance of equipment will be covered in section A8.6 and calibration in section A8.7 and again in section A10.1.

For now, it is enough to understand the importance of maintenance and calibration of equipment as ways of minimising both systematic and random errors.

Calibration is essentially the process of comparing measurements. One measurement is of a known size or correctness, for example a reference standard. The other is made using the device or instrument being calibrated. The calibration process involves adjusting the reading of the device or instrument to match (usually) the known value of the reference standard. It should be obvious that a badly maintained instrument may not give accurate results. But even a well-maintained piece of equipment is likely to need calibration – sometimes each time it is used. Calibration methods will depend on the equipment being used, but they all usually involve use of a standard.

Correctly storing materials

Some of the consequences of incorrect storage will be covered in more detail in section A8.10.

Many materials used in experiments or analyses are sensitive to light, temperature or moisture. Incorrect storage can lead to **degradation** (breakdown) of the material. How might this affect the results of an experiment or an analysis? What types of error could result? Think about this whenever you are working in the lab.

Using automated processes

Many types of experiment involve repetitive pipetting, measuring, dispensing or other tasks. The problem with any repetitive task is operator fatigue – after a while you might start making mistakes or being careless. Automated equipment, as long as it is well maintained and possibly properly calibrated, does not suffer from fatigue or carelessness, and should be a source of fewer errors.

Automated processes are widespread in the science and health sectors. Figure 6.2 shows an automated analyser being used in a microbiology laboratory, but there are many other types of automated analyser, particularly in laboratories carrying out routine, high-throughput analysis (that is, dealing with a large number of tests or samples) such as in a big hospital or busy QC department.

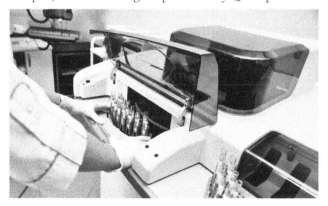

▲ Figure 6.2 An automated analyser being used in a busy department

Automated analysis equipment will often be integrated into LIMS so that results can be uploaded and processed automatically.

Routine analysis is not the only process that can be automated. Some types of chemical synthesis involve highly repetitive tasks that are well suited to automation. One example of this is oligonucleotide synthesis; oligonucleotides are synthetic DNA or RNA molecules that are used in a wide range of applications in molecular biology, including the PCR test for COVID-19.

For more on DNA and RNA see sections B2.2 and B2.3.

Good experimental planning

Experiments usually set out to answer a question, such as: 'What is the effect of temperature on the rate of a reaction?' We get a **valid** result if our experiment gives the real answer to that question and not one that has been skewed by sources of error or uncontrolled variables.

So, it is important to design an experiment that will give valid results. We must think about all aspects of the experiment, including sources of error, and ensure that the conclusion from the experiment must be the real explanation.

An experiment will usually set out to test a theory or an idea – we call this a **hypothesis**. A well-designed experiment may disprove the hypothesis or support the hypothesis. Note that we can never definitely prove a hypothesis – in science we have ideas, develop them into hypotheses we can test, and design experiments to test those hypotheses. If the experiment disproves our hypothesis, we have to think again. If the experiment supports the hypothesis, we can refine our ideas and maybe think of new experiments to test our hypothesis. In this way, our knowledge and understanding grows, but we have not proved it is correct. This fact about science really annoys a lot of people who want simple answers to questions, such as: 'Can you prove this treatment is safe?' The only scientific answer is: 'We cannot prove that it is 100% safe, but all the evidence shows that it causes little, if any, harm.'

In conclusion, good experimental planning will cover several steps. You will have formulated a hypothesis about how the variables that you are investigating are related, for example that there is a positive correlation between them. This hypothesis must be one that can be tested by experiment. You should then consider some, or all, of the following:

▶ Identify the **dependent** and **independent** variables.
▶ Consider any **control** variables that must be accounted for in the design of your experiment.
▶ Ensure that you have addressed all possible sources of error, for example by making sure that the equipment that you are using is well maintained and fully calibrated, where necessary.

▶ Wherever possible, you should follow clear and concise SOPs as these will reduce the chance of random errors.

▶ Gather the data – this must be **relevant**, **accurate** and **precise**. If you are **sampling** (i.e. not measuring the whole of a population) then your sample must be **representative**. There is more about this in section A6.9.

You are then ready to process and analyse your data in order to support or disprove the hypothesis. This is covered in the following sections.

> ### Practice points
>
> It is good practice to think about any experiment or other procedure that you carry out and check that you are taking all steps to ensure that you will get valid results. This is also a key principle of GLP and GMP so that we can obtain consistent and valid results, whether it is an analytical procedure or a manufacturing process.

A6.7 The different methods of data processing and analysis in science environments

There are various ways in which you can process and analyse data, but they all have factors in common. In the end, you are trying to turn raw data into useful information from which you can draw valid conclusions.

Tabulating raw data

Putting raw data into a table is the starting point for data processing. If you look again at section A5.2, you will see that creating a table of your raw data imposes a degree of order and consistency – for example, the layout of independent and dependent variables, repeats and processed data (for example, calculation of a mean). This can help ensure that you have all the necessary data before you go any further.

If you look back to the example, you will also see how **tabulating** the raw data (putting it into a table) can sometimes show up anomalies or inconsistencies in the data.

Using specialist software to analyse large data sets

Spreadsheets, such as those created in Microsoft Excel software, can deal with most types of data analysis and even have built-in statistical and other analytical tools. Specialist analytical software might be needed for really large data sets, however. Some of these are cloud-based solutions.

Very large qualitative data sets can also be analysed by specialist software. This might be necessary if you are analysing the results of a huge survey or a large patient database. Some of these systems incorporate AI or machine learning to provide analysis and insight.

Graphical/statistical analysis

Presentation of data in the form of graphs was covered in section A5.2 and will be addressed further in the next section. Presenting data in the form of a graph can help to identify trends or find correlations. It can also help identify anomalous data or show how much variability there is in the data. This can be enhanced by use of statistical analysis to quantify the variability and to test the significance of any similarities or differences; this is covered in section A6.9.

Identifying trends in the data

At its simplest level, data analysis might show a decreasing or increasing trend in the data. We could see that by simply drawing a graph. Figure 6.3 shows a graph of the first ionisation energies of the elements sodium (Na) to argon (Ar) in the third period (row) of the periodic table. Ionisation energy is the amount of energy needed to remove an electron from an atom to form a 1^+ ion.

If we look at this graph, we can see that there is a general trend – the ionisation energy increases from Na to Ar as the atomic number increases. However, there are 'dips' at Al and S. Analysis of this data and the trends that it reveals has given us important insights into the structure of atoms.

A similar graph can be used identify trends in the population over time (Figure 6.4).

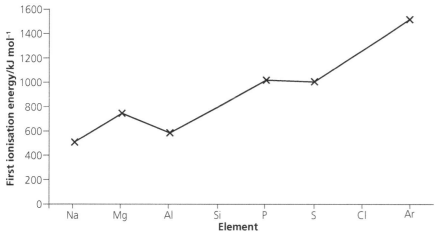

▲ Figure 6.3 First ionisation energies of the elements of Period 3

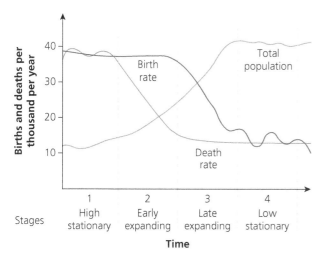

▲ Figure 6.4 Demographic transition in human population

If you look at the graph, you can see a trend in the total population. It is relatively stable in stages 1 and 4 but increases during stages 2 and 3. If you look at the curves showing the birth rate and death rate, you should be able to explain the shape or trend of the total population. This is known as a demographic transition.

Look at Figure 6.4 again.

What factors do you think caused the large changes in birth and death rates?

If you look at the birth rate in stage 4 and the death rate in stage 1, you will see fluctuations. What do you think caused these?

Do you think the total population will remain relatively stable?

Drawing conclusions if appropriate

Having analysed the data, you may then be able to draw conclusions. This might be based on trends or patterns that you observe. You would usually design an experiment to test a hypothesis (see section A6.6) and you might be able to conclude that your experiment supported the hypothesis or disproved it. You might then carry out a statistical analysis (see section A6.9) to determine how much your experiment supported the hypothesis.

Finally, you should do a reality check. For example, if you look back at section A5.2 (Does correlation imply causation?) you will see that, just because there is a strong correlation between two variables, this does not mean that one causes the other. It is unlikely that ice cream consumption causes death by drowning, even if rates for both trend in the same direction.

A6.8 Ways to present data in the appropriate format

The following methods of presenting data were covered in some detail in section A5.2:
▶ table
▶ scatter graph
▶ line graph
▶ bar chart.

There are two more methods you should be aware of in addition to these.

Box and whisker plot

This is a very effective way to display a lot of information. Figure 6.5 shows an example.

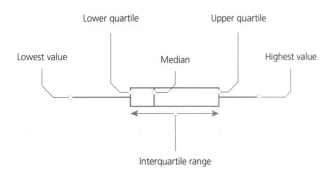

▲ Figure 6.5 A box and whisker plot

From Figure 6.5 you can see that this plot shows the lower and upper quartiles – taken together, these show the interquartile range. It also shows the median and the full range (lowest and highest values). See section A6.9 for an explanation of these terms.

Flow charts

A flow chart is a diagram that represents a workflow or a process. It can also represent an algorithm – a step-by-step approach to solving a task. Figure 6.6 shows an example of a flow chart.

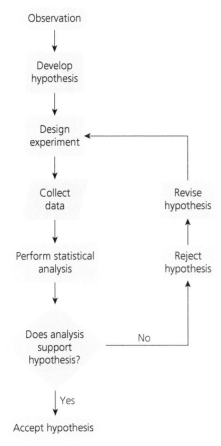

▲ Figure 6.6 A flow chart illustrating the scientific method

Flow charts can be used in many different contexts. The flow chart in Figure 6.6 illustrates the scientific method:

▶ Make an observation.
▶ Formulate a hypothesis.
▶ Design an experiment to test the hypothesis.
▶ Collect the data.
▶ Carry out a statistical analysis.
▶ Make a decision – does the statistical analysis support the hypothesis?
  – If yes, accept the hypothesis – end point.
  – If no, reject and revise the hypothesis, design a new experiment and continue.

The following table shows the symbols/shapes used in flow charts and their meanings.

Shape	Name	Description
	Terminal	Indicates the start or finish point
	Process	Represents a series of operations
	Input/output	Indicates a process that inputs or outputs data or displays results
	Decision	Show a conditional operation, for example a yes/no question or true/false test
	Flowline	Indicates the order of operation of the process; this can be a simple line, but arrows make more complex flow charts easier to follow

▲ Table 6.1 The meanings of symbols/shapes used in flow charts

A6.9 Statistical techniques when analysing data and their purposes

Sometimes you can look at a set of data and draw conclusions right away. However, are those conclusions correct? One common example is the outcome of a clinical trial. Let us suppose we are investigating a new drug to lower blood pressure. We have two groups: one (the test group) receives the new drug and the other (the control group) receives a placebo. We measure the blood pressure of all the participants over a period of three weeks and then analyse the data. Because there are so many in each group, we calculate the mean blood pressure for each group. At first sight, it looks as if the drug has worked because there has been a slight decrease in the mean blood pressure of the test group and a slight increase in the mean of the control group. But is the effect really due to the drug, or is it simply the result of random variation? The only way we can be sure is to perform a statistical analysis.

Let's say you wanted to survey the **population** of the UK. You could be investigating resting heart rate in the population or opinions about the health benefits of exercise. You could use a pulse meter for the resting heart rate and a survey or questionnaire for the opinions. However, it is not practical to measure or survey the entire population. That is why we need to take a **sample**. We want our sample to be representative, meaning that we can apply our conclusions from the sample to the whole population.

How large should the sample be? What other considerations do we need to account for? Would you expect different results if you used the following for your sample:
- a small rural village
- a large city
- a large county?

Key terms

Population: in statistics, a population is the collection of all objects or measurements; all the people in the UK represent a population.

Sample: a subset of a population that is selected for a particular investigation; the size of the sample and how it is chosen can affect the outcome of the investigation.

Mean and median

We saw in section A6.5 that one way to reduce the effect of random errors is to repeat measurements and then calculate a **mean** or average. The mean is simply the sum of the quantities divided by the total number of quantities.

The **median** represents the middle value of a data set. You calculate it by arranging all the values from smallest to largest and then identifying the middle value.

Both mean and median estimate where the centre of a set of data lies. However, sometimes one or the other might give a more useful estimate or way of describing the centre of the data set:
- ▶ The mean gives a better estimate when the distribution of data is symmetrical and there are no outliers.
- ▶ The median gives a better estimate when the distribution of data is uneven (skewed) or where there are clear outliers.

This can be illustrated using the following example of the distribution of salaries in two cities.

In Figure 6.7, the distribution is fairly symmetrical – this is known as a **normal (or Gaussian) distribution**.

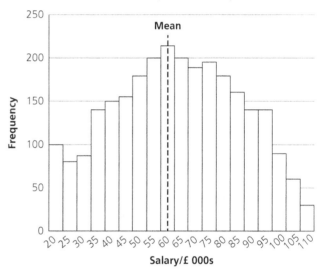

▲ Figure 6.7 Symmetrical distribution of salaries in a single city

Key term

Normal (Gaussian) distribution: a symmetrical distribution of quantities about the mean value; it is also known as a bell curve because of its shape.

First, a couple of points to note. This is an example of a histogram or frequency diagram, where the salaries are grouped into bands of £5000. Also, if you notice the label on the x-axis you will see it says 'Salary/£ 000s' – this means the units are thousands of pounds. You will see this convention used in other, more scientific contexts.

If you look at Figure 6.7, you will probably agree that the mean gives a good indication of the 'typical' salary in this city.

Figure 6.8 shows a different city where the distribution of salaries is 'skewed' – most salaries are concentrated at the lower end, but there are a few high-end values.

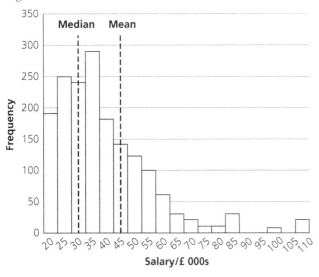

▲ Figure 6.8 Skewed distribution of salaries in a single city

Talking about 'average salaries' can be misleading. If you look at the median and mean, you should be able to see that the median gives a much better estimate of the 'typical' salary. The mean has been influenced by the small number of high-earning individuals, whereas the median gives a more representative estimate of the centre value.

Quartiles and percentiles

Just as the median separates a data set into two parts (the first being less than the median and the second being greater than the median), quartiles separate a data set into four (quarters, hence the term quartiles). The first quartile will contain the smallest quarter of the data, the second quartile will contain the next smallest quarter of the data and so on.

Percentiles take the same approach but separate the data into hundredths. Quartiles and percentiles are often used in population studies or growth charts.

Standard deviation

Experimental or observed data usually has a normal distribution, as shown in Figure 6.7. In such cases, the **standard deviation** is a measure of how variable the data is. This will be clearer if you look at Figure 6.9.

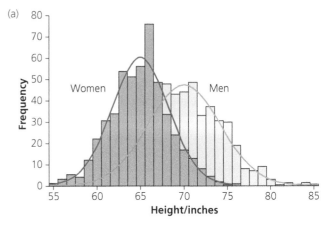

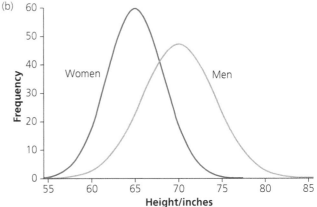

▲ Figure 6.9 (a) Histograms showing the observed distributions of the heights of men and women; (b) The same data with histograms removed, showing the underlying normal distributions fitted to the observed data

You can see immediately that the mean height of men is greater than that of women. Notice that the curve for women's height in Figure 6.9b is less broad than the curve for men's height. This means there is more variety in men's heights – we say that this data is more **variable**.

Standard deviation can also be used to measure the dispersion of a set of values from the mean. This will be clearer if you look at Figure 6.10.

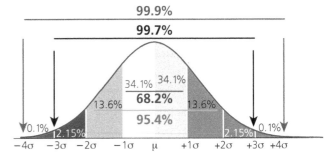

▲ Figure 6.10 A generic normal distribution that shows the relation between the mean, μ, the standard deviation, σ, and the percentages of the areas under the curve between different intervals

This curve shows that 68.2% of the values lie within ±1 standard deviation (σ) of the mean value, μ, and that 95.4% lie within ±2 standard deviations of the mean value. By the time we get to ±4 standard deviations we have included 99.9% of the data.

We can use this as the basis for measuring probability. Look again at Figure 6.10 and you will see that there is a 95.4% probability that any single item of data lies within the range ±2 standard deviations of the mean value.

There are two formulae you will encounter for calculating standard deviation, depending on whether we are dealing with a whole population or just a sample of that population.

For a whole population, we use the term **population standard deviation** and the symbol σ. The formula is:

$$\sigma = \sqrt{\frac{\sum (a - \mu)^2}{N}}$$

Where:

σ = population standard deviation

a = each individual data value in the population

μ = mean of the population

N = population size

You could put this equation into words as: the population standard deviation is the square root of the sum of the squares of the differences between each data value and the population mean divided by the population size.

Fundamentally, the larger the standard deviation calculated, the greater the dispersion of values away from the mean.

In science, most of the time we deal with a sample of the population rather than the whole population. In such cases, the **sample standard deviation** (s) is given by the slightly different formula:

$$s = \sqrt{\frac{\sum (x - \bar{x})^2}{n - 1}}$$

Where:

s = sample standard deviation

x = each individual data value in the sample

$\bar{x}$ = mean of the sample

n = sample size

Again, we could put that into words as: the sample standard deviation is the square root of the sum of the squares of the differences between each data value and the sample mean divided by the sample size minus 1.

If you are starting to feel overwhelmed with maths, you will be relieved to know that a good scientific calculator will allow you to enter your data and then it will calculate the mean and standard deviation. Spreadsheets, such as Excel, include a whole range of statistical functions that you can use on your data. This is another reason why automated data collection, as described in section A5.2, can be so useful, because you can very quickly transfer your data to Excel or other spreadsheet software for analysis. The software will even draw graphs and charts for you.

> ### Practice points
>
> You will often see data presented as mean ± (plus or minus) standard deviation, or ± 2 standard deviations. This allows you to see immediately how variable the data is, which is very useful.
>
> You might also encounter the use of this type of data presentation to decide whether differences between two mean values are **statistically significant**. A common 'rule of thumb' is that, if there is no overlap between the means ± 2 standard deviations, then the difference between the mean values is significant. Strictly speaking, this is incorrect. You should really use something called the **standard error of the mean** to do this. However, using standard deviation in this way can be useful sometimes for estimating.
>
> The best way to determine whether the difference between two means is significant is to use the t-test (see page 97).

> ### Key term
>
> **Statistical significance:** this means that differences between data sets are not due to chance (random variation).

Range

The range, as the name suggests, is simply the difference between the lowest and highest values in a data set. You might see range bars on a graph – typically a histogram. The histogram bar shows the mean while the range line or bar shows the maximum and minimum values in the data set. The range can be highly influenced by outliers in the data, however, whereas standard deviation is not. So, although range can be quick and easy to calculate, standard deviation is a better measure of the variability of the data.

Chi-square test

The Chi-square (pronounced 'kai square') is a way to test the significance of the difference between observed and expected results. You use Chi-square when you have categorical data, that is, it is divided into groups or categories. It is used to compare differences in frequencies between groups and so you can only use Chi-square when you have numbers, not mean values. Chi-square is often used when a theory or hypothesis predicts a particular ratio, that is, gives you the expected results. You can then use it to compare your actual (observed) results and work out whether any difference between observed and expected is simply due to chance (this is known as the **null hypothesis**) or if there is a statistically significant difference.

The formula for Chi-square, χ^2, is:

$$\chi^2 = \sum \frac{(O - E)^2}{E}$$

Where:

χ = the Greek letter 'Chi'

O = observed value (for each category)

E = expected value (for each category)

Once again, putting this into words: Chi squared is the sum of the squares of the difference between observed and expected values, divided by expected value for each of the categories.

It is easier to see how this works if we make a table of the type you might use to calculate Chi-square. In this case, we have done a genetics experiment using fruit flies and we want to find out if the difference between the observed results and the predicted ratios (based on genetic theory) are just due to random variation (chance).

In this experiment, we bred flies with long wings and flies with vestigial (very short) wings. All the offspring had long wings. We then interbred these and found that 401 had long wings and 119 had vestigial wings. Genetic theory says that there should have been a ratio of $3:1$ of long to vestigial. Did our results conform to the theory?

Category	O	E	O – E	$(O – E)^2$	$(O – E)^2/E$
Long wings	401	390	11	121	0.310
Vestigial wings	119	130	–11	121	0.931
Totals	520	520			χ^2 = 1.241

▲ Table 6.2 Observed and expected results from a genetics experiment using fruit flies

We obtain a value for χ^2 by adding up the numbers in the final column.

The next step is to use this value to decide whether to accept or reject the **null hypothesis**. To do this, we need a reference table of χ^2 values (Table 6.3).

df	p values							
	0.99	0.95	0.90	0.50	0.10	0.05	0.01	0.001
1	0.0016	0.0039	0.016	0.46	2.71	3.84	6.64	10.83
2	0.02	0.10	0.21	1.39	4.61	5.99	9.21	13.82
3	0.12	0.35	0.58	2.37	6.25	7.82	11.35	16.27
4	0.30	0.71	1.06	3.36	7.78	9.49	13.28	18.47

▲ Table 6.3 Chi-square values

Null hypothesis: a statistical tool that states that any differences between the observed and expected results or between mean values are due to chance; statistical tests are used to decide whether to accept the null hypothesis (that the differences were due to chance) or reject it (deciding that the results were not due to chance and the differences are statistically significant).

In Table 6.3, 'df' stands for degrees of freedom. This is the number of categories of data, minus 1. So, in this case df = 1, so the χ^2 value calculated will be compared to values within the first row. The p values represent probabilities: $p = 0.50$ means a probability of 50%, $p = 0.10$ means a probability of 10% and so on.

In statistics, a probability of 0.05 (5%) is considered the cut-off point, so $p < 0.05$ means the results are not due to chance (we reject the null hypothesis) while $p > 0.05$ means that they are due to chance (we accept the null hypothesis). For this reason, the values of χ^2 in the $p = 0.05$ column are called the **critical values**.

Having calculated a value for χ^2, it is simply a matter of comparing it to the values in the table. In this case, the value of 1.241 lies between 2.71 and 0.46 and so we can accept the null hypothesis and conclude that $p > 0.10$, or that there is a > 10% probability that the differences between observed and expected were due to chance. This is good enough for us to conclude that any differences between the observed results and the theoretical ratios were simply due to chance. So, although the observed results are not exactly what the theory predicts, any differences are not statistically significant.

t-test

The *t*-test, sometimes called the Student's *t*-test ('Student' was the pseudonym of the statistician who first published it), is used to determine if there is a significant difference between the means of two groups of data, X and Y.

The formula for the *t*-test is quite complex:

$$t = \frac{|\bar{x} - \bar{y}|}{\sqrt{\frac{(s_x)^2}{n_x} + \frac{(s_y)^2}{n_y}}}$$

Where:

$\bar{x}$ = the mean of sample X

$\bar{y}$ = the mean of sample Y

s_x = the standard deviation of sample X

s_y = the standard deviation of sample Y

n_x = the number of individual measurements in sample X

n_y = the number of individual measurements in sample Y

$|\bar{x} - \bar{y}|$ means that we take the absolute value of the difference between the two mean values, that is, we ignore the sign.

As with Chi-square, we need to calculate degrees of freedom, in this case using the following formula:

$$df = (n_x - 1) + (n_y - 1)$$

Having calculated values for *t* and the degrees of freedom, we can consult a table of *t* values.

As with Chi-square, it is easier to follow with a real example, as in the case study that follows.

Case study

An experiment is conducted to compare heart rates after 15 minutes of exercise on either a treadmill or a rowing machine. Two groups of sports science students were asked to either run or row for 15 minutes. Their heart rate (pulse) was then measured. The null hypothesis was that there is no significant difference between the mean heart rate of the two groups.

The results are put into a spreadsheet, and this is then used to calculate the mean ($\bar{x}$), standard deviation (s), s^2 and $s^2 \div n$ for each group. The table shows the spreadsheet.

Treadmill (sample X)				Rowing machine (sample Y)			
Subject	Heart rate (bpm) = x	$x - \bar{x}$	$(x - \bar{x})^2$	Subject	Heart rate (bpm) = y	$y - \bar{y}$	$(y - \bar{y})^2$
1	155	0.80	0.64	1	155	−3.70	13.69
2	156	1.80	3.24	2	166	7.30	53.29
3	158	3.80	14.44	3	162	3.30	10.89
4	152	−2.20	4.84	4	159	0.30	0.09
5	152	−2.20	4.84	5	158	−0.70	0.49
6	153	−1.20	1.44	6	162	3.30	10.89
7	155	0.80	0.64	7	153	−5.70	32.49
8	156	1.80	3.24	8	159	0.30	0.09
9	153	−1.20	1.44	9	160	1.30	1.69
10	152	−2.20	4.84	10	153	−5.70	32.49
$\bar{x} =$	154.2	$\sum (x - \bar{x})^2 =$	39.56	$\bar{y} =$	158.7	$\sum (y - \bar{y})^2 =$	156.10
$s_x =$	2.097			$s_y =$	4.165		
$s_x^2 =$	4.400			$s_y^2 =$	17.344		
$s_x^2 \div n =$	0.440			$s_y^2 \div n =$	1.734		

▲ Table 6.4 Results of experiment to compare heart rate after exercise on either a treadmill or a rowing machine

We can now take the figures from the spreadsheet and use them to calculate a value for t.

$$\left(s_x^2 \div n_x\right) + \left(s_y^2 \div n_y\right) = 2.17$$

$$\sqrt{\left(s_x^2 \div n_x\right) + \left(s_y^2 \div n_y\right)} = 1.47$$

$$\bar{x} - \bar{y} = 4.5$$

Based on the original equation:

$$t = \frac{\left|\bar{x} - \bar{y}\right|}{\sqrt{\dfrac{\left(s_x\right)^2}{n_x} + \dfrac{\left(s_y\right)^2}{n_y}}}$$

$t = 4.5 \div 1.47$

$t = 3.06$

$df = (10 - 1) + (10 - 1) = 18$

The final step is to look up our value of t in a table of values of t and find the corresponding p value at 18 degrees of freedom.

Degrees of freedom	Decreasing value of $p \rightarrow$			
	0.10 (10%)	0.05 (5%)	0.01 (1%)	0.001 (0.1%)
1	6.31	12.71	63.66	636.60
2	2.92	4.30	9.92	31.60
4	2.13	2.78	4.60	8.61
6	1.94	2.45	3.71	5.96
8	1.86	2.31	3.36	5.04
10	1.81	2.23	3.17	4.59
16	1.75	2.12	2.92	4.02
18	1.73	2.10	2.88	3.92
20	1.72	2.09	2.85	3.85
25	1.71	2.06	2.80	3.73
30	1.70	2.04	2.75	3.65
40	1.64	1.96	2.58	3.29
	$p > 0.05$	$p < 0.05$	$p < 0.01$	$p < 0.001$
	Results are not significantly different	Results are significantly different	Results are highly significant	Results are very highly significant

▲ Table 6.5 t values

If you look along the line for 18 degrees of freedom, you see that our value of t (3.05) lies between 2.88 ($p = 0.01$) and 3.92 ($p = 0.001$).

So, first of all we can reject the null hypothesis because our value of t is greater than the critical value (at $p = 0.05$); remember that this is the 'cut-off' point – a probability > 0.05 (or $> 5\%$) is not significant while a probability < 0.05 ($< 5\%$) is considered statistically significant.

Not only do we reject the null hypothesis, but we can conclude that the difference between the two mean values was significant at $p < 0.01$, or that there was $< 1\%$ probability that the differences were due to chance.

Reflect

Think about the experiment described in the case study. Instead of just two groups of students, we could have matched the groups for age and fitness (these are control variables). Would that have given more valid results? Why?

We could also have used one group of students on the treadmill and, after a suitable time for recovery, used the same group of students on the rowing machine. Do you think this would have given more valid results than using matched groups of students? Why?

Research

In experiments where we have two sets of results from the same individuals, we have what is known as paired data. We can still perform the t-test, but we use a different formula.

Search for 'Student's t-test for paired data' to find the formula you need.

Spearman's rank

Scatter graphs were covered in section A5.2. If you look back to that section, you will see that scatter graphs can show a positive correlation, a negative correlation or no correlation between two variables. **Spearman's rank correlation coefficient** is used to put a numerical value on the strength of the correlation.

As with other statistical tests, you start with a null hypothesis, that is, that there is no correlation between the two variables.

The formula for Spearman's rank correlation coefficient (r_s) is:

$$r_s = 1 - \left(\frac{6 \times \sum d^2}{n^3 - n} \right)$$

Where:

d = the differences between ranks (this will be explained below)

$\sum d^2$ = the sum of the squares of the differences between ranks

n = the number of pairs of items in the sample

To calculate Spearman's rank correlation coefficient, you do the following:

1   Plot the results in a scatter graph to see if there is a correlation. The easiest way to do this is to enter the data into a spreadsheet and used the scatter graph function. The correlation may be positive or negative.

2   Write a null hypothesis, usually 'there is no significant correlation between …'

3   Rank each set of data. The largest number is given rank = 1. If you have two values the same, do the following:
    a   If they are, say, the 5th rank, then give them the rank of 5.5 (the mean of 5 and 6).
    b   The next value would then be given rank = 7.
    c   If there are three 5th largest values then give them the rank of 6 (the mean of 5, 6 and 7). The next value would be given rank = 8.
    d   This can be used for any number of equal values.

4   Calculate the differences in the ranks (not the data values). This gives you d for that rank.

5   Square the value of d to give d^2; this removes any negatives signs.

6   Add all the values for d^2 to give $\sum d^2$.

7   Insert the values for $\sum d^2$ and n into the formula above and calculate a value for r_s.

The value of r_s can range from -1 to $+1$. A positive value indicates a positive correlation, so as the values of x increase the values of y also increase. A negative value indicates a negative correlation, so as the values of x increase the values of y decrease. A value of 0 indicates no correlation.

The question that must be asked is 'how significant is the correlation?' Once again, we use a table of critical values of r_s and look up our value for r_s (ignore the plus or minus sign) in the table using the row corresponding to the number of pairs in our experiment.

▶ If our value of r_s is equal to or greater than the critical value in the table, we reject the null hypothesis and conclude that the correlation was significant.

▶ If our value of r_s is less than the critical value in the table, we accept the null hypothesis and conclude that the correlation was not significant.

Case study

A student wants to test a hypothesis that there is a positive correlation between height and mass, i.e. that as height increases, mass will also increase. They measure the height and mass of 12 subjects and get the following results:

Height (cm)	Mass (kg)
175	95
170	84
169	68
182	90

160	55
172	88
163	68
157	50
176	92
173	88
190	85
188	86

They plot these results as a scatter graph (Figure 6.11). This is step 1 of the procedure above.

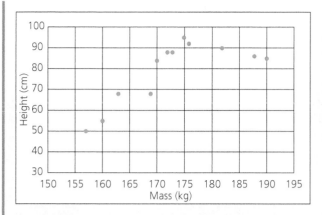

▲ Figure 6.11 Scatter graph of the height and mass data

Looking at the scatter graph, the student believes that they could draw a straight line of best fit through the points up to a mass of about 175kg, which would show a clear positive correlation between height and mass. However, at masses of >175kg the line would curve down again. They decide to use Spearman's rank correlation coefficient to help them decide if the positive correlation is statistically significant.

They write a null hypothesis 'there is no significant correlation between height and mass' (step 2).

They prepare the following table (steps 3–6):

Height (cm)	Rank*	Mass (kg)	Rank**	d***	d²****
175	5	95	1	4	16
170	8	84	8	0	0
169	9	68	10	−1	1
182	3	90	3	0	0
160	11	55	11	0	0
172	7	88	4.5	2.5	6.25
163	10	68	9	1	1
157	12	50	12	0	0
176	4	92	2	2	4
173	6	88	4.5	1.5	2.25
190	1	85	7	−6	36
188	2	86	6	−4	16
				$\Sigma d^2 =$	82.5

▲ Table 6.6 The critical values of the Spearman's rank correlation coefficient (r_s) at $p = 0.05$ and $p = 0.01$

Notes:
* This is step 3 for the heights. The largest height (190cm) is given rank = 1 and the smallest height (157cm) is given rank = 12.

** This is step 3 for the masses, with the largest mass (95kg) given rank = 1 and the smallest mass (50kg) given rank = 12. There were two masses of 88kg and these were the 4th largest masses, so they were given rank = 4.5 (the average of 4 and 5) and the next smallest mass (86kg) was given rank = 6.
*** This is step 4, calculation of the differences in the ranks (not the actual values) to give d.
**** This is step 5, calculation of d^2 followed by step 6, addition of all the values of d^2 to give Σd^2.

Finally, step 7 involves substitution of the values in the equation (there are 12 pairs, so n = 12):

$$r_s = 1 - \left(\frac{6 \times 82.5}{(12^3 - 12)} \right)$$

The calculated value for r_s is 0.7115. This can be compared to the critical values of r_s using the following table.

Number of pairs of measurements	Critical values	
	p = 0.05 (5%)	p = 0.01 (1%)
5	1.000	–
6	0.886	1.000
7	0.786	0.929
8	0.783	0.881
9	0.700	0.833
10	0.648	0.794
11	0.618	0.755
12	0.587	0.727
13	0.560	0.703
14	0.538	0.679
15	0.521	0.654
16	0.503	0.635
17	0.488	0.618
18	0.472	0.600
19	0.460	0.584
20	0.407	0.570
30	0.362	0.467

Looking at the critical values for 12 pairs of measurements, the calculated value for r_s is greater than the critical value at p = 0.05 but less than the critical value at p = 0.01. From this the student rejects the null hypothesis and concludes that it is >95% certain that there is a positive correlation between height and mass.

Look again at the scatter graph. Does the student's conclusion surprise you? Why do you think there might be an apparent difference in the correlation up to 175kg and above 175kg? How could the student improve this investigation?

A6.10 How to review data and make decisions based on that review

In section A6.6 we saw the importance of good experimental planning. To obtain valid results, we need to ensure that each step in the process leads us to have confidence in our data and, therefore, to be able to make valid conclusions. So, we must:

- formulate a hypothesis that can be tested (by experiment or observation)
- design an experiment or observation to test that hypothesis
- ensure variables are properly controlled and that our experiment or observation is a 'fair test'
- carry out the experiment or observation while eliminating or minimising any sources of error
- collect and tabulate our results and then perform the appropriate statistical test.

In this section, we will look at what we do with the results and statistical analysis.

Interpreting the statistical analysis against the original hypothesis/performance criteria

We have seen, in the previous section, how a statistical test will incorporate a null hypothesis ('any differences were due to chance') that we then accept or reject. From this we can decide whether the experiment or observation supports our hypothesis or not. Remember, we cannot 'prove' a hypothesis, all we can do is keep trying to disprove it.

If the hypothesis was incorrect and the null hypothesis is accepted, we need to revise our hypothesis and design new experiments or observations to test the revised hypothesis. If the hypothesis was supported, we can then think of new experiments to further develop our understanding. This is the basis of the scientific method illustrated in the flow chart in Figure 6.6.

Reflect

Look again at the experiment above where we compared heart rate on the treadmill with the rowing machine. You will see that the standard deviation for the rowing machine is larger. That means that the data for the rowing machine is more variable. You can probably tell that by looking at the raw data. But if you did not have access to

the raw data, you would still be able to draw that conclusion from the standard deviation.

So, what is the reason for the greater variability of the data? We do not know the full details of the experimental procedure. However, it is reasonable to assume that the treadmill was set at the same speed for each subject. That means that they would all be working about as hard as each other. What about the rowing machine? Did each subject work at the same rate? Would this affect the variability of the data? How could we modify the procedure to remove this source of variability?

Comparing data with predicted/similar results in published work

In section A6.7, under the heading 'Drawing conclusions if appropriate' (page 91), we mentioned the need for a 'reality check'. We should, of course, try to remove all possible sources of error from the experiment or observation, but we should also ask ourselves: 'Does this result make sense in the light of everything I know?' That was why we probably concluded that the correlation between ice cream sales and deaths by drowning did not imply that eating ice cream makes you more likely to drown.

We can achieve this by comparing our data with the work of other scientists who have published their research. Our results might support their conclusions or be consistent with their predictions. If that is the case, we can have more confidence that our results and conclusions are valid.

Checking tolerance levels

We encountered the concept of uncertainty in section A6.5 and discussed the importance of calibration in section A6.6; this will be expanded on in section A8.7.

For now, we need to think about the acceptable amount of uncertainty in a measurement, even when an instrument is calibrated. This acceptable uncertainty is known as the **tolerance**. Tolerances are often written or stamped on the instrument, or they might be included in an SOP – the acceptable range of tolerance for instruments used in a process, for example.

When we review our data, we must take account of tolerance levels. This will help us decide how close a measurement is to the true value. If tolerance levels are too wide, this might lead us to question the validity of our conclusions.

Deciding on next steps

Having reviewed our data, we might decide that we need to collect more data to give us more confidence in our conclusions.

Alternatively, if we have confidence in our data and conclusions, then we can move to the next stage, which might be:

▶ to publish the data and conclusions in a research paper or internal report

▶ to share the results with the client (in a contract testing laboratory, for example).

A6.11 The consequences of bias in data analysis

So far in this chapter, we have looked at the importance of obtaining valid results and the steps we have to take to help achieve that.

But what if the results are not valid? What happens if **bias** creeps in?

Key term

Bias: anything that prejudices or influences the results of an experiment or observational study in one direction.

In the past, it is estimated that only half of all clinical trials that were completed had the results published. Moreover, trials that produce positive results (i.e. showing the new drug in a favourable light) are twice as likely to be published as those producing negative (unfavourable) results. This is an example of **reporting bias**.

Partly as a result of this, the Medicines and Healthcare products Regulatory Agency (MHRA) now requires all new clinical trials to be registered and the results must be published within one year of the end of the trial.

In the next section we will consider the sources of bias and how to prevent or avoid it. For now, we will just consider the consequences of bias.

Inaccurate findings inferred from the results

Let us suppose you are studying the effect of temperature on the rate of an enzyme reaction – possibly a protein-digesting enzyme that is being investigated for use in a new biological washing

powder. If there is a fault in the temperature control, for example a faulty water bath or thermometer, you might conclude that the enzyme is stable at higher temperatures than is in fact the case.

Case study

An example of experimenter bias was demonstrated by Rosenthal and Fode of the University of North Dakota in 1962. A group of psychology students were asked to judge the ability of rats to learn their way through a maze. One group of students was told that their rats were 'bright' and the other group was told that their rats were 'dull'. The students who analysed the 'bright' group judged the performance of their rats more highly than the students with the 'dull' group.

In fact, the humans were the subjects of this experiment, not the rats. The two groups of rats were chosen randomly with no differences in characteristics. The group that expected their rats to perform well influenced, possibly unconsciously, the rats to perform better.

▶ What can you learn about good experimental design from this work?

▶ Do you think experimenter bias is more or less likely in studies that involve observation rather than measurement?

Wasted time and resources

If we think again about the protein-digesting enzyme for a biological washing powder, a decision might have been made to incorporate the enzyme into the new product. It could then take many months and a lot of expenditure on production and further testing before it was realised that the product was ineffective because the enzyme was actually heat-sensitive and, therefore, inactive at the temperatures used for washing.

Using the same example of a biological washing powder, let's imagine a researcher has been investigating a new enzyme that is less heat sensitive. Initial results showed that this enzyme is less effective in removing stains from clothing. However, the researcher was convinced that the new enzyme would be an improvement and so they continued to repeat the experiments many times looking for sources of error, but without success. Eventually, after wasting considerable time and resources, they were forced to conclude that the results did not support their hypothesis. Eventually they had to revise their hypothesis and design new experiments, for example to investigate the effect of different detergents in combination with the new enzyme. In this case it was bias on the part of the researcher (they were convinced of their hypothesis) rather than any bias in the data analysis.

Damage to reputation

The manufacturer of the biological washing powder would probably have discovered that there was a problem before it was actually sold to the public, so they would not have got a reputation for selling a product that did not work as well as it should. That might not always be the case, and bias in data analysis can easily lead to loss of reputation, whether it involves a company or a single researcher.

Case study

Study 329

This was a clinical trial carried out in North America from 1994 to 1998 to study the antidepressant paroxetine in 12- to 18-year-olds diagnosed with major depressive disorder. The results of the study were published in a medical journal in 2001.

It was later discovered that the publication was not, in fact, written by the authors listed on the paper. The company that produced the drug and carried out the trial had hired a public relations (PR) company to help publicise the new application of the drug (to treat teenagers). The paper was actually written by the PR company and made inappropriate claims about how effective the drug was. It also played down safety concerns, including increased suicidal behaviour in those receiving the drug. These negative results were included in the paper but were not accounted for in the conclusion.

In 2003 the MHRA analysed Study 329 and other studies of paroxetine and found that there was no evidence of effectiveness in children or adolescents, but that there was 'robust evidence' of the drug causing suicidal behaviour. The MHRA launched a criminal investigation into the company's behaviour, but eventually did not bring charges. However, in 2012, the company was fined $3 billion by the US Department of Justice for withholding data on paroxetine, unlawfully promoting it for the under-18s and preparing a misleading article on Study 329.

Risks to health and safety

The paroxetine case study illustrates how bias and selective reporting can lead to increased risk of harm in patients.

Anyone who manufactures or imports more than one tonne of a chemical per year into the UK must register the substance and provide information about risk and safety evaluation. It is important that any tests give accurate and unbiased evidence of safety otherwise there may be a risk to anyone who comes into contact with the chemical, or products incorporating the chemical.

A6.12 How to prevent or reduce bias in data evaluation

Having seen the negative consequences of bias in data analysis and evaluation, we have to consider how to prevent or reduce it. Some of this will be achieved by improving experimental design and eliminating systematic errors. A lot can be achieved by addressing issues with data collection and analysis.

Research

Cochrane (previously known as the Cochrane Collaboration) is a UK-based charitable organisation that collects, organises and evaluates the results of medical research findings so as to improve evidence-based medicine. The Cochrane website, www.cochrane.org, has more information about the organisation and how it works.

Do you think there is a need for Cochrane? What can you learn about COVID-19 from Cochrane?

Ensuring sufficient sample size and appropriate sampling techniques

In sections A6.5 and A6.9, we saw the importance of performing repeat experiments or observations and calculating a mean, perhaps even excluding anomalous results. Clearly, large samples should give more accurate results. But how large is large enough?

One way to determine this is to use what is known as a **rolling mean** (or **moving average**). The procedure is:
▶ Take a small number of measurements and calculate the mean.
▶ Then take another measurement and recalculate the mean.
▶ If the new mean is different, then take another measurement and recalculate the mean.
▶ Keep repeating this process until adding a new measurement does not change the mean.

Once the mean remains constant, there is no point in adding further measurements.

Comparing to known standards and literature values

By comparing the results of our own experiments to those of known standards or literature values, we have a way of validating our methods of data collection and/or analysis. Having done that, we can then apply our methods to novel or unknown situations with more confidence that we will obtain valid results.

Sending out results for peer review

Have you ever produced a piece of written work, checking it many times, only to have someone point out a mistake or typo? Frustrating, is it not? This illustrates an important principle – that we tend not to see the flaws in our own work that a fresh pair of eyes will find.

This is the basis of **peer review**, which has come to underpin most forms of publication of research and is considered the 'gold standard' of academic publication. Peer review means sending the draft or manuscript of a research publication to other experts in the field (reviewers). The process is usually organised by the publishers or editors of the journal that the manuscript has been submitted to. The reviewers will then look for flaws or weaknesses in the design of the experiments, the analysis of the data or the conclusions reached. The manuscript may be accepted, but more likely it will be returned to the author(s) for revision. This might involve performing additional experiments, further analysis of the data or revising the conclusions. Once this has been done, the revised manuscript will be reviewed again before being published. In some cases, the manuscript will be rejected by the publisher.

Peer review might be considered the gold standard, but it is not without its faults. Some journals have a reputation for being less rigorous than others in their review process. Academic research is also highly competitive. It is quite possible that the reviewer is in competition (for research funds, employment or simply prestige) with the author of the paper that they have been asked to review. In such circumstances, would a reviewer always be impartial?

Another problem with peer review is that it can be very slow, sometimes delaying publication by a year or more. This became an issue during the COVID-19 pandemic, when any delay in publishing data on the disease could lead to hundreds or thousands of avoidable deaths.

The scientific community is now exploring alternatives to peer review. Some research is now often published online ahead of formal publication. The research community can get faster access to the work, but they can also examine the work and comment on it.

Do you think this risks sacrificing the quality of published research for speed of publication? Do the benefits of rapid dissemination of information outweigh the disadvantages? How could you judge whether research published online pre-peer review is actually of high quality?

Using critical experts to independently review the data

This is very similar to the process of peer review but is not necessarily part of the publication process. Independent review might be used by a commercial organisation to help them make judgements about the work that is being done. For example, a company might want external experts to review and evaluate the data generated by the company's own scientists before committing significant resources to a project.

Blind analysis

If you think back to the study by Rosenthal and Fode that showed how investigators could influence the outcome of an experiment (in that case, using 'bright' or 'dull' rats), it is clear that experimenter bias can be a significant source of error.

For this reason, double-blind, randomised clinical trials are considered to be the most effective form of testing of new drugs or treatments.

Most clinical trials involve a test group (who receive the new drug) and a control group (who receive a placebo or dummy drug). This is necessary because patients can get better simply by being looked after, and patients on clinical trials are usually better looked after than patients in general because the clinical staff need to make sure that everything is done properly.

However, human nature can mean that the clinical staff take greater care of the test group than they do of the control group; they want the trial to show the new drug is as effective as they hope it will be. This could be completely unconscious, but it can affect the outcome of the study. Therefore, the trial is conducted

double-blind, meaning that the clinical staff do not know who receives the drug and who receives the placebo, and thus cannot treat them differently.

This is even extended to the analysis of the trial data. It is only once the analysis is complete that the trial is **unblinded**, meaning that the researchers find out who received the drug and who received the placebo.

Blind analysis has also become widely used in particle physics. The first example of this was in 1932 where physicist Frank Dunnington measured the ratio of charge to mass (e/m) of the electron. In his experiment, e/m was proportional to the angle between the electron source and the detector. Dunnington asked his machinist to choose an arbitrary angle close to, but not exactly, the 340° that was needed. Only when he had finished the analysis and was ready to publish his results did he accurately measure the 'hidden' angle.

There are many ways that we can allow experimenter bias to influence the results of experiments and the analysis of those results. Blind analysis is a highly effective way to remove this bias.

Using informatics tools to analyse data

Informatics is the science that defines how information is technically captured, transmitted and used. Section A5.5 looked at the use of new technology in the recording and analysis of data. The techniques described there, such as AI/machine learning, cloud-based systems, digital information management systems and data visualisation tools can all contribute to improved data analysis – particularly when dealing with very large data sets.

A6.13 Links between sample size and effective statistical analysis

Sections A6.9 and A6.10 dealt with the performance and interpretation of statistical tests. A statistical analysis allows us to accept or reject the null hypothesis. But what if there is an error in that process?

In general, a larger sample should give a more accurate mean. This should ensure we can draw valid conclusions. But how many is enough? In the previous section we looked at the use of a moving average as a way of deciding when our sample was large enough. That might not always be possible. There are various

constraints that might apply, such as:
▶ cost
▶ time
▶ availability of samples or subjects
▶ ethical considerations, for example should we withhold treatment from a control group of patients?

A small sample might include too many anomalous results or outliers. On the other hand, a large sample might introduce greater complexity and be more time-consuming. Although the results are more accurate, the benefits might not outweigh the costs.

Statistical power

The **statistical power** of a study or experiments is how likely the study is to distinguish between an actual effect and one of chance. In a statistical test we either accept or reject the null hypothesis. But there are two ways this could go wrong:
▶ We could reject the null hypothesis incorrectly (that is, when the null hypothesis is actually true); this is known as a **type I error**.
▶ We could accept the null hypothesis incorrectly (that is, when the null hypothesis is actually false); this is known as a **type II error**.

The statistical power of a test gives the likelihood of rejecting the null hypothesis when it is false. The means of calculating the statistical power can be quite complex (it is almost always performed using a computer), but it does allow you to judge when the sample size is sufficient to reduce the risks of error when accepting or rejecting an experimental hypothesis.

When we collect a sample of the whole population, the sample mean is likely to be different to the population mean – this is the **margin of error**.

We saw in section A6.9 that a probability, or p, of 0.05 is considered to be the cut-off point when judging statistical significance. This is sometimes referred to as the 95% confidence limit – if you look at Figure 6.10, you will see that, in a normal distribution, about 95% of the data lies within ±2 standard deviations of the mean. When deciding on the size of a sample, we need to decide how close we want the sample mean to be to the population mean. We usually choose a sample size so that around 95% of the data lies within 2 standard deviations of the mean, thereby meeting the 95% confidence limit.

In general, we have seen how sample size is factored into each of the statistical techniques covered which then specify the accuracy of the results returned from the investigation (for example, standard deviation).

A6.14 How to order numbers by relative size in a data set

In section A6.9 we saw how we can calculate the median or mid value of a data set. To find the median, you first need to put the numbers in order, from smallest to largest. You need to do the same thing when calculating Spearman's rank correlation coefficient.

It is straightforward with simple numbers, such as the following:

23, 36, 73, 21, 39, 195, 87

But the process is a little more complicated when you have numbers in standard form, such as these:

1.5×10^5, 2.65×10^4, 2.2×10^{-3}

In this case, the numbers get smaller from left to right. Or when the numbers are decimals less than 1:

0.005, 0.05, 0.00005

In the case of the decimals, it is probably easiest to convert them all to standard form before putting them in order. This would give the following order (again, getting smaller from left to right): 0.05 becomes 5.0×10^{-2}, 0.005 becomes 5.0×10^{-3}, 0.00005 becomes 5.0×10^{-5}.

> You will find more about standard form in Chapter B1 Core science concepts: Chemistry (page 204).

A6.15 How to ensure proportionality while scaling up or down quantities in a formulation

A **formulation** is a combination of chemicals in specific quantities that do not react with each other but produce a mixture that has the desired properties. Examples of formulations include:
▶ pharmaceuticals, such as tablets or liquids (for example, cough medicines) or injectables (for example, insulin)
▶ vaccines
▶ cleaning agents
▶ paints
▶ fertilisers
▶ cosmetics, shampoos, deodorants.

The formulation will follow a particular formula (like a recipe) and it is important that it is produced in the same way every time so that it will always have the correct composition and the required properties. For example, with a pharmaceutical it is important that the formulation always contains exactly the right amount of the active drug substance.

To make more or less than the amount in the formula, you need to make sure that you keep the same proportions. So, if you want to make ten times the amount, multiply all the quantities by ten; if you only want half the amount, divide all the quantities by two (which is the same as multiplying by 0.5).

Table 6.7 shows a formulation for making 100 g of a shower gel, as well as how it can be scaled up or down.

Ingredient	Formulation for 100 g Mass (g)	Scale up to 1 kg Mass (g)	Scale down to 50 g Mass (g)
Distilled water	57.50	575.00	28.75
Sodium gluconate	0.20	2.00	0.10
Cleansing blend	33.50	335.00	16.75
Glucose	1.60	16.00	0.80
Glycerin	4.00	40.00	2.00
Gluconolactone	0.90	9.00	0.45
Squalane	1.80	18.00	0.90
Fragrance	0.50	5.00	0.25
Total	100.00	1000.00	50.00

▲ Table 6.7 Formulations and how they can be scaled up and down

Project practice

You have been asked to research LIMS and make recommendations for use in your organisation. Use your placement as an example, if appropriate.

Research the types of LIMS available by searching online.

Identify the needs of your organisation and how LIMS might meet those needs.

Consider how LIMS might improve data capture, analysis and reporting in your organisation.

Prepare a report in either Word or PowerPoint that summarises your findings and makes recommendations.

If your organisation is already using LIMS, prepare a critical evaluation of how it meets the organisation's needs and whether alternative (perhaps newer) systems might offer advantages.

Write a critical review of your work.

Assessment practice

1 Describe the **most** effective way to process the data from the following:
   a The results of a clinical trial of a new anti-cancer drug showing reduction in size of tumour.
   b Market research asking people about their views on the safety of vaccines.
2 Give **two** advantages and **two** limitations of the following methods of data recording and storage:
   a Physical lab notebooks.
   b Laboratory information management systems (LIMS).
3 One method to measure viscosity (a liquid's resistance to flow) uses a small metal sphere falling through the liquid in a measuring cylinder. The time it takes for the sphere to drop between two marks on the cylinder is measured using a stopwatch. This method was used to compare the viscosity of two different liquids.

   a Suggest **two** control variables that should be used in this experiment.
   b Using the stopwatch can be a source of error. Explain whether this would be a systematic error or a random error.
   c Suggest ways that you could minimise errors in this experiment.

4 Fertilisers that are high in potash (potassium) are known to encourage flowering in plants. A grower wanted to compare the effect of two fertilisers on the number of flowers produced in a two-week period. One fertiliser was high in potash and the other was low in potash. Ten plants were given the high-potash fertiliser and another ten plants were given the low-potash fertiliser. The results are shown in the following table (SD = standard deviation).

Potash content	Number of flowers per plant										Mean	Median	SD
High	10	10	8	19	4	14	16	25	28	12			7.56
Low	9	12	12	10	15	15	16	12	9	10			2.58

   a Complete the table by calculating the mean and median for each group.
   b Explain why the median gives a better estimate of the central value for the high-potash group.
   c The grower concluded that the use of high-potash fertiliser did not increase the number of flowers per plant. Evaluate this conclusion.
5 For each of the following, explain which statistical test would be most appropriate:

   a An investigation into the relationship between number of cigarettes smoked per day in the UK and deaths from lung cancer.
   b A filling machine should produce packs of beads containing five different colours in equal quantities. You have been given a pack of beads where the number of beads of each colour is not equal. You need to identify whether there is a fault in the packing machine.

c   Two filling machines are producing the same packs of beads. You measure the mean mass of the packs produced by each machine. You need to work out if one machine is producing significantly heavier packs.

6   You find that the machines in 5c are producing packs of significantly different mass.

 a   Explain how you could use tolerance levels to investigate the cause of the problem.

 b   Suggest **one** other variable that could affect the mass of the packs of beads.

7   A company is planning a clinical trial of a new anti-asthma drug. There will be 100 participants in the trial and each participant will be given either the drug or a placebo. The researchers will know who receives the drug or placebo although the participants will not. The participants will not be screened for other drugs in their system. Evaluate this planned trial.

8   Researchers are increasingly publishing their results online or in the form of press releases ahead of publication in traditional peer-reviewed journals. Discuss the risks and benefits of this approach.

9   You are working for a small company developing a new product for use in the automotive industry. The work is commercially sensitive and so you must keep it confidential. However, you are concerned that sampling bias during development and testing might affect the results.

 a   Describe the possible consequences of this sampling bias.

 b   Discuss appropriate ways to determine whether the results during development and testing are valid.

A7: Ethics

Ethics is a system of moral principles. These principles govern how we make decisions and how we lead our lives. Ethics is sometimes described as moral philosophy. Our concept of ethics has been derived from religion, philosophy and culture, but our understanding of ethics and ethical principles does not depend on any one belief system, religion or even lack of religion. Ethics incorporates ideas such as human rights but, in this chapter, we will focus on how ethics guides our professional conduct while working in the science sector.

Science has a huge impact on people's lives – think about the contribution made by science during the COVID-19 pandemic, to start with. Because of that, we need to be aware of the effect that we might have on others. We must consider the impact our work might have and respect those who are affected by what we do: our colleagues, our clients, members of the public – potentially the whole of humankind.

As you work through this chapter, you will find that we are bringing together some of the themes addressed in earlier chapters.

Learning outcomes

The core knowledge outcomes that you must understand and learn:

A7.1 the key aims of ethical scientific practices as outlined in 'Rigour, Respect, Responsibility: A Universal Ethical Code for Scientists 2007'

A7.2 how to demonstrate integrity in a scientific setting

A7.3 the purpose of codes of practice within organisations

A7.4 the importance of respect in the workplace

A7.5 how intellectual property rights apply to scientific settings

A7.6 what may be considered as IP within the science sector.

A7.1 The key aims of ethical scientific practices as outlined in 'Rigour, Respect, Responsibility: A Universal Ethical Code for Scientists 2007'

The Universal Ethical Code for Scientists was published in 2007 by Professor Sir David King, who was Chief Scientific Adviser to the UK government at the time.

The code has three main themes:
- rigour, honesty and integrity
- respect for life, the law and the public good
- responsibility in communication – listening and informing.

It is a public statement of the values and responsibilities of scientists, including those working in the social sciences, natural sciences (biology, chemistry, physics and so on), the medical and veterinary sciences, as well as engineering and mathematics.

The code is not mandatory or written in law. Instead, scientists, institutions and organisations are encouraged by the authors of the code and organisations that have adopted it to observe the code and reflect on how the guidelines may relate to their own work.

The code does not replace the professional codes of conduct that relate to specific professions, professional societies or organisations, such as those discussed in section A1.4.

Case study

The Royal Academy of Engineering developed its own Statement of Ethical Principles, a code of ethics that applies to all types of engineers. The statement is consistent with the Universal Ethical Code but focuses on issues that are specific to the engineering profession. The Academy has also worked with the Engineering Professors' Council to advise university engineering departments on how to incorporate ethics teaching into their degree courses.

We will now look at the three main aims of the code.

To foster ethical research

The public is increasingly demanding and expects developments in science to be ethical and to serve the public good. While it is vital that scientists carry out valuable research, they must do so in a way that brings the public with them. The following case study on genetically modified (GM) crops illustrates what can happen when the public is not supportive of a new development.

▲ Figure 7.1 GM crops are an area of ethical discussion in agriculture and science

Case study

GM crops were designed to improve yields and efficiency in agriculture and, thereby, feed more of the world's population. However, GM crops have been widely rejected, in Europe at least, for various reasons, including:
- People were concerned that very large agricultural companies only wanted GM so that they could sell more of their expensive seeds or weedkillers.
- They were concerned that some of the techniques to create GM crops might introduce potentially harmful chemicals into the crops.
- They were worried that genes introduced into crop plants might 'escape' into other plants.
- Some people argued that the world produces enough food for everyone already, and that the problem lies with the fact that it is not distributed equally.

Do you think the research into GM crops is ethical? Do you think that people's concerns are valid, or did the companies just do a poor job of explaining the advantages of GM?

The alleged link between the MMR vaccine and autism described in the case study on page 75 saw a drop in vaccination rates from 91% to 80%, and a huge rise in the number of cases of mumps, even though the research this was based on was discredited (see section A5.4 for details).

Many areas of science have ethical issues associated with them, for example:

- the use of animals in medical research or product testing
- the use of human embryos in stem cell or fertility research
- animal and human cloning
- development of GM 'super crops'.

Think about the different ethical issues involved – you could do this in a class discussion. Here are a few questions to consider:

- Is it acceptable to use animals in medical research to save human lives, but not in testing cosmetics?
- Is it acceptable to inflict suffering on animals if the experiments lead to great benefit for humans?
- Is it acceptable to use human embryos in research? What about cell lines derived from a legally aborted fetus, such as HEK cells that are used widely in scientific research?

There may not be easy answers to these questions and opinions may vary widely. One purpose of the code is to encourage scientists to ask themselves these questions.

Any clinical research must first be approved by a hospital ethics committee. One question they will ask is: 'Are the potential benefits of the research worth the risks or the potential for harm to the patients involved in the research?'

Case study

In 2011, two research teams carried out experiments to genetically modify the H5N1 avian flu virus so that it could be transmitted directly between mammals, including humans. Currently, people can only catch this lethal virus by direct contact with infected birds. The researchers felt that their work could benefit society by providing public health officials with information that would help them monitor dangerous mutations and thereby prevent disease outbreaks. They also felt that their research could be used to develop vaccines or treatments for avian flu. They submitted their work for peer-review and publication.

The research was funded by the US National Institutes of Health (NIH) but they were worried that publication of the results could lead to a global pandemic through the accidental contamination of laboratory workers or even deliberate misuse of the work (terrorism).

Publication was delayed while the research was reviewed. The work was eventually published when the researchers included more information about the public health benefits and biosafety measures.

- Do you feel that the NIH had valid concerns, or should the researchers have been allowed to publish their work at the time?
- Can you think of other ethical reasons for delaying, or even preventing, the publication of a study?

This case study applies to scientific research, but the work of any scientist could have an impact on other people.

Outcomes of GM research and the discredited research into the MMR vaccines, along with the ethical code, prove the importance of ethical considerations when undertaking new research all the way through to its publication.

To encourage active reflection among scientists on the implications and impact of their work

Sometimes it is possible to get lost in your work. You might become so engrossed that time passes almost unnoticed, particularly when you are absorbed in something you love doing.

The same can be true of scientists. When you become totally absorbed in finding answers to difficult problems or working towards something that you see as being of great advantage to humanity, it can be possible to lose sight of the bigger picture.

It is important, therefore, that scientists remember to think about the wider implications and impact of their work.

To support communication between scientists and the public on complex and challenging issues

There are several ways in which scientists can communicate with and engage the public:

- Discussing the policy implications of research in articles (magazines and newspapers, as well as research journals), press releases and university courses – for example, many universities offer public engagement courses or lectures to the general public.
- Providing expert testimony on the ethical, legal, social or policy implications of research. This might be in the course of public enquiries, court cases or parliamentary hearings.
- Working with non-governmental organisations that deal with the policy and value implications of science and technology.

The term non-governmental organisation is very broad – here are some examples:

- The Royal Society is an independent scientific academy dedicated to promoting excellence in science for the benefit of humanity. Being elected as a Fellow of the Royal Society (FRS) is considered a great honour that is only bestowed on those researchers who have reached the top of their fields.
- Academic societies, such as the Royal Society of Chemistry, Royal Society of Biology or the Institute of Physics, can engage with government or the public to develop policy and further understanding in their particular fields.
- The Campaign for Science and Engineering is a non-profit organisation that argues for more research funding, promoting a high-tech and knowledge-based economy, highlighting the need for high-quality science and maths education (at all levels) as well as scrutinising the mechanisms by which the government uses science and scientific evidence.
- Science Café (also known as Café Scientifique) organises informal monthly meetings where a scientist is invited to talk about their work in a topical or sometimes controversial area. The intention is that, for the price of a cup of coffee or a glass of wine (meetings take place in cafés, bars, restaurants or even theatres), anyone can explore the latest ideas in science and technology.

A7.2 How to demonstrate integrity in a scientific setting

Sometimes it is not enough to be good, you must be seen to be good.

The same is true in science. As well as behaving with integrity, we must be observed and known to behave with integrity. In this section we will consider ways in which we can all do that.

Maintaining high quality ethical and professional standards

This was discussed, in part, in sections A1.3 and A1.4. We can also maintain high ethical and professional standards through:

- Objectivity – just as we should avoid bias in design and analysis of experiments (see sections A6.11 and A6.12), we should avoid bias and prejudice in all aspects of our work.
- Clarity – this covers several aspects, such as clarity in communication so that our work can be understood by other scientists or the general public. It also covers openness and transparency in our work (subject to the need to maintain

confidentiality or protect sensitive information – see page 78 for GDPR, for example).
- Reproducibility – this is central to the whole of science. It is not enough that we can obtain results and repeat those findings; others must be able to reproduce our work too. This was covered in section A5.3 and will be revisited in section A9.11.
- Transparency – we should not knowingly mislead (or allow others to mislead) about scientific matters. This means presenting and reviewing scientific evidence, theory or interpretation honestly and accurately.

Research

Companies and other organisations often have information on their websites that relate to their own organisational codes of practice, although you may not find anything described specifically as a 'code of practice'. Here are two examples:

Sygnature Discovery (www.sygnaturediscovery.com) is a Contract Research Organisation (CRO) providing a range of drug discovery services to clients that include major pharmaceutical companies, small biotechnology companies and even universities. You will find information about the company's values, including how they treat their staff, the standards they apply to working with clients and collaborators as well as their policy towards use of animals in research.

GSK (www.gsk.com) is a major UK-based pharmaceutical company that has world-wide operations. They have published a Code of Conduct that applies to all their areas of operations. You can download a copy from www.gsk.com/media/4800/english-code-of-conduct.pdf.

You can learn from these examples, but it would be useful to find other companies or organisations in your field of work and research their own codes of practice or conduct. To what extent do they all follow the principles of the Universal Code for Scientists?

Following organisational codes of practice

We looked at the purpose of organisational policies and procedures in section A1.1 and covered specific aspects in sections A2.6 (individual responsibilities in relation to the wider team), A3.3 and A3.4 (health and safety), A5.6 (data protection) and A5.11 (information). We also saw the importance of following professional codes of conduct in section A1.4. This subject is also addressed in Chapter A8 Good scientific and clinical practice. You can see that this subject runs through everything that we do in the science sector.

Different companies in different areas of science will have their own codes of practice specific to their industry or area of science. However, they will all have the same general themes that we have covered here.

Following regulatory guidance

Regulatory guidance applies to areas such as:
- health and safety legislation (see Chapter A3)
- GLP (section A2.7) and GMP (section A2.8)
- compliance with internal and external regulations (section A2.11) and how these apply in different working environments (section A2.12).

There may also be specific legislation applying to particular roles or areas of research or scientific work. Your employer or contractor will be responsible for ensuring everyone involved is aware of the requirements of these, but you must make sure you are confident that you understand it yourself.

The Data Ethics Framework (**www.gov.uk/government/publications/data-ethics-framework**) is a set of principles to guide the design of appropriate data use in the public sector. It is aimed at anyone working with data in the public sector:
- data practitioners such as statisticians, analysts and data scientists
- operational staff
- people working to use data to provide insight.

Some of this is covered by the Data Protection Act (2018) and GDPR (2018) (see section A5.6), but the Data Ethics Framework goes beyond this to provide guidance for a wide range of public bodies, including those dealing with sensitive patient information.

Aspiring to excel, not just meet the minimum standards

We should always aim high – even if we don't make it, we will still have achieved more than if we just settle for mediocre. What is true in life is true in science. We should always aim to exceed expectations, to be the best we can and to achieve the highest standards in our scientific practice.

Several large companies have 'exceeding expectations' as part of their mission statement (a summary of the company's aims, purpose and values).

Do an internet search for 'company mission statements with exceed expectations' – you could even add a search term based on your own industry, such as 'healthcare', 'engineering' or 'pharma'.

Think about the companies you find. If you know any of them, do you think that they have a reputation for excellence? Do you think that using a term like 'exceeding expectations' could become meaningless if everyone uses it? Do you think that we should all have our own mission statement, such as 'I will always strive to exceed expectations'?

A7.3 The purpose of codes of practice within organisations

In the previous section, we saw how following organisational codes of practice helps to demonstrate integrity in a scientific setting. We also saw how codes of practice could be found in companies as well as other organisations, such as professional bodies and academic societies.

The main purpose of codes of practice is to define how employees can remain compliant with policies or legislation – you will find numerous examples of the importance of this here and in other chapters.

Research

Does your own organisation have codes of practice? If not, you might be able to find codes of practice that have been produced by a professional body, academic society, trade union or government department that is relevant to your work.
Find a code of practice and think about the reasons for having that code of practice. Then consider the importance of following that code of practice and codes of practice in general.

A7.4 The importance of respect in the workplace

What do you think respect means, and why is it important in the workplace?

Respect can mean a feeling of admiration for something or someone, as well as having due regard for the rights, feelings or wishes of other people. It is this second aspect that is relevant to the workplace and which we will consider here.

Promoting equality and supporting diversity

Equality, diversity and inclusion policies were covered in section A1.1.

Equality means ensuring that everyone has an equal opportunity and is not discriminated against because of their characteristics. This is covered by the Equality Act 2010. The **protected characteristics** that it includes were described in section A1.1.

Diversity is about the differences between individuals and groups of people, while inclusion is about having space for everyone to be themselves and valuing those differences.

There are several ways in which companies and organisations can promote equality and support diversity in the workplace:

▷ Develop an equality and diversity policy, and make sure that all employees have read it.
▷ Do not tolerate offensive language or discrimination, for example sexism or racism, in the workplace. It may be tempting to excuse this as 'banter', but it can create a hostile and bullying environment and will usually be a serious disciplinary matter.
▷ Lead by example – senior management must ensure that their attitudes and behaviour always foster an equal environment.
▷ Provide equality and diversity training for all employees.

Sometimes companies can have good intentions but still fall foul of equality legislation.

Minimising conflict and stress

Bullying behaviour can be a significant cause of conflict and stress in the workplace.

Stress can also be caused when employees cannot cope with the pressure of work. This could be because they do not feel in control of their own work, when their workload is too great or when they do not get the support they need, for example through training.

Ways in which conflict and stress can be reduced include:

▷ Communication: if employees are confused about their job role or what is happening in the organisation it can lead to conflict and stress. Good communication can help reduce this.
▷ Do not ignore tense situations: this can lead to a build-up of tension, making things much worse. Conflict needs to be dealt with quickly and fairly.

▷ Have a formal complaint process or grievance procedure (this was covered in section A1.1).
▷ Create an environment that promotes collaboration and team work.
▷ Ensure that everyone is treated fairly. Sometimes managers can be accused of favouritism – whether it is real or imagined, it can cause a great deal of conflict.
▷ Conflict resolution in the workplace, for example through a conflict resolution meeting or an investigation where serious complaints or allegations are made.

It is important to minimise conflict and stress in the workplace because this can lead to reduced productivity and poor job satisfaction.

Increasing productivity and job satisfaction

When an employer creates a work environment that respects the employee, they are more likely to be more productive. Respect leads to increased trust, and we tend to work better when we have trust in our employer and feel that we are valued. Just as stress causes the body to increase the secretion of stress hormones such as cortisol, respect and a supportive, nurturing environment can have the opposite effect, leading to increases in serotonin and dopamine in the brain. These 'feel good' compounds can lead employees to have greater emotional commitment to the success of the organisation. Put simply, if we feel respected, we tend to work harder – and are happier.

We have just seen how conflict and stress can reduce productivity and job satisfaction, so it is clear that respect in the workplace will lead to enhanced productivity and job satisfaction.

Inspiring individuals to be loyal to the organisation and each other

Employees that feel respected or valued will be more committed to the success of the organisation. This means that the success of the organisation becomes closely linked to their own success and that of their colleagues. In organisations where this respect does not exist, the relationship between the organisation and the employee can become transactional – simply an exchange of the minimum required wages and labour. This does not encourage a positive attitude in the workplace.

Another reason for respecting ourselves and our colleagues is how we want to be remembered. If we felt confident, it is likely that we have been helped, guided and supported by our colleagues. Likewise, we hope that the people we work with will remember that we cared about them (either as manager or co-worker), encouraged them to do their best and contributed to a safe and collaborative culture in which they could participate and grow. It is not just a question of wanting people to remember you fondly – the person that you support today may be the person who can employ you or help progress your career in the future.

Equality and diversity training often focuses on the words and terms that we use, partly because words can be so hurtful. The appropriate language to use can also change over time.

Can you think of examples where language has changed over the years? Some words or terms that were once unacceptable are now commonplace, and others that were previously acceptable are now considered discriminatory or insulting.

A7.5 How intellectual property rights apply to scientific settings

Why are we discussing patents, trademarks and copyright in a chapter on ethics? It is all a question of fairness. If an individual, company or other organisation has spent time and money to develop a novel product, it would be unfair if someone else came along and copied their product and sold it without having to invest the same time and money. **Intellectual property (IP)** rights are designed to ensure fairness when dealing with ideas, inventions or other creations of the mind.

Key term

Intellectual property (IP): creations of the mind such as inventions, literary or artistic works, designs, symbols or names used in commerce.

We are going to look at three types of IP here: patents, trademarks and copyrights. There are other types of IP that you might encounter, however, for example:

▷ **Trade secrets**, where an organisation has confidential information that they might sell or license. Unauthorised acquisition – stealing a trade secret – is regarded as an unfair practice.
▷ **Industrial designs**, such as the shape or appearance of an object, can be registered and obtain protection from being copied. This is similar to copyright but applies to industrial rather than artistic objects.
▷ **Geographic indications** might be relevant if you work in food science. They apply to products, usually food or drink, from a particular geographical area that have qualities linked to their place of origin. Examples include Melton Mowbray pork pies, Parma ham, Stilton cheese and many wines from certain regions, such as Champagne.

Patents

A **patent** is an exclusive right granted for an invention. The invention might be a new product or process that provides a new way of doing something or a new solution to a problem. Note the use of the word 'new' – the product or process must have some aspect that is novel for a patent to be granted.

The patent holder obtains an exclusive right to sell the product or operate the process for a fixed period (usually 20 years) in return for disclosing technical information about the invention. This means that the holder of the patent can stop others from exploiting the invention – this is known as patent infringement. Therefore, a company can profit from its invention and recoup the investment made before they must face competition from others.

An added benefit of the patents system is that knowledge is shared. Other people – individual researchers, companies and other organisations – can benefit from the information disclosed in the patent and use it to make further developments. However, they may need permission from the original patent holder to sell a product or use technology that is based on that patent during its lifetime.

Patented inventions may not always be exploited by the original inventor. Very often licences will be granted to others to use the IP in the patent. This normally involves the payment of a licence fee, either as a one-off payment or as a royalty (a percentage of sales) or both. Patents can even be bought and sold.

▲ Figure 7.2 Geographic indications can be given to products that are only produced in particular regions, for example (from the top) Treviso red radicchio and Gorgonzola cheese from Italy, feta cheese from Greece and Roquefort cheese from France

Most universities now have 'technology transfer' offices. Their function is to advise researchers on patenting any research done in the university that might have a commercial application. The technology transfer office then assists in obtaining the patent and finding a commercial partner who would either buy the patent, pay the university a royalty, or even set up a new company together with the university to exploit the IP (this type of company was mentioned in section A2.1).

A patent has to be applied for – this is known as **filing** – and is then examined before it is finally granted. The process of filing a patent application and obtaining protection for an invention can be slow and expensive. Moreover, patents are territorial rights – they only apply in the country or region where the patent has been filed and granted. For example, there is a UK Patent Office (now called the Intellectual Property Office) and a European Patent Office. Different countries may have different rules about what can be patented.

As trade and commerce becomes increasingly global, it is often not realistic to protect IP in just a single country. Separate patent applications can be filed with the patent office in each country, but this can quickly become inefficient and very expensive – unless you only need to protect trade in a specific region. An alternative is to file a single application under the Patent Cooperation Treaty (PCT) which offers protection in over 140 countries.

One aspect of the patent system is that the invention must be novel. So, if a university research group publishes its work, in a scientific journal for example, it will no longer be possible to obtain a patent for anything that has been mentioned in the publication. We saw the importance of published literature in section A5.4, but the need to get patent protection can delay publication and so delay the spread of information in the science community.

Do you think that universities should benefit financially from the research that they do?

Research is often funded or even carried out by charities, such as Cancer Research UK. Do you think that they should patent the work that they do?

Here are some examples of patents that have been issued in the last 100 years:

▷ The iPhone (2008), although the original patent gives almost no information about the device!

▷ The quadcopter drone, such as those used as toys or for aerial filming, was first patented in 1962. This is one of many examples of a patent that was ahead of its time. It was only when flight controls and other electronic systems, such as cameras and GPS navigation, caught up with the original idea that drones became such a big commercial success – long after the original protection period.

▶ The 3D printer was another invention ahead of its time (1986). Advances in computing technology have made possible the increasingly widespread application of the technique.

▶ The CRISPR-Cas9 gene editing system developed at the University of California at Berkeley was patented in 2014. The system is now used to modify the genomes of crops and livestock, as well as in the treatment of diseases such as leukaemia.

Besides scientific theories and ideas (see next section) there are several things that cannot be patented in most countries. These include:

▶ mathematical methods
▶ plant or animal varieties (although there are other forms of protection, such as plant breeders' rights, that cover novel plant varieties)
▶ discoveries of natural substances
▶ methods of medical treatment (as opposed to medical products).

Trademarks

A trademark is a sign or symbol associated with the goods or services of a company. It enables those goods or services to be distinguished from those of another company. Trademarks are usually registered with a national or regional trademark office. Once registered, the trademark gives the owner the exclusive right to use the trademark or to license it to another company in return for a fee.

Registered trademarks generally last for ten years but can be renewed indefinitely if additional fees are paid.

Copyrights

Copyright describes the right that a creator has over their literary or artistic work. Examples of works that can be covered by copyright include:

▶ literary works, such as novels, plays, poems, reference works (such as this textbook), newspaper and magazine articles
▶ computer programs
▶ musical compositions, films or choreography
▶ artistic work, such as sculptures, paintings, drawings and photographs.

The copyright owner can authorise or prevent use of their work, for example in publications, performances, recordings or broadcasting. These are known as economic rights, because the copyright holder can charge a fee in exchange for granting permission. Moral rights allow the copyright holder to prevent changes to a copyrighted work, for example if it could harm the creator's reputation.

You will often see the copyright symbol © used, together with a date. Look at page ii of this book for an example. Although it is good practice to use this symbol on any work that you want to copyright, it is not essential. Also, unlike a patent or trademark, copyright does not need to be registered.

A7.6 What may be considered as intellectual property within the science sector

Clearly a novel drug or a novel production process could be classed as IP and therefore be granted patent protection. Similarly, an original work of fiction or a musical composition can be protected by copyright, and there are many examples of trademarks, some of which are words (Apple, Amazon, Google) and others that are symbols, such as the Apple logo or the Nike 'swoosh'.

Not all types of IP can be protected by patents, trademarks or copyright, however. Also, it is not always so clear what constitutes IP within the science sector, so we will look at some examples.

Theories/ideas

Theories and ideas are both forms of IP in that a theory or idea is a product of your intellect (mind or imagination). Theories and ideas are the basis for all scientific research as well as for the development of new products. However, it is not possible to patent a theory (particularly a scientific theory) or an idea in most countries.

It is possible to apply for a patent for an idea if you can show how the idea can be, or has been, converted into a product or process. Also, the idea must fulfil the normal requirements of a patent: it must be novel and not previously published. It must also be inventive, meaning that it is not something that anyone familiar with the field could have thought of – the technical term is that it must be 'non-obvious'.

Papers/research

When we carry out research we accumulate knowledge, which is a form of IP. Publication of research in a paper (for example, in a scientific journal) is a way of sharing your work. Any research that is published will be covered by copyright. However, as mentioned above, once any IP is published it can no longer be patented. Copyright prevents a competitor copying a company's publications but it does not stop them from using the IP in those publications.

Experimental results and design

Just like research, the results of any experiment or design of any experiment are forms of IP and add to the sum of knowledge. The records accumulated are also a form of IP. These sorts of results and records cannot be protected by a patent. When research is published, the results of experiments included in that research are then part of the public domain and can no longer be protected by a patent. For this reason, companies may apply for a patent before they publish research in the scientific literature. In other cases they will keep these forms of IP confidential – they are trade secrets. An example of this would be an improved production process. A company may not see any advantage in applying for a patent as they would not want to license the process to a competitor. In such circumstances, the company may choose to operate a 'secret' process.

Bespoke equipment

Bespoke means specially made for a particular customer, user or purpose, rather than being mass manufactured.

Any equipment that is made for a specific purpose is likely to be novel and will be a form of IP. It is possible that it could be protected by a patent or a registered design. In cases where a company or organisation commissions an individual or another company or organisation to develop or produce a bespoke piece of equipment, they will usually have an agreement to ensure that they own any IP in the equipment.

Anything with a potentially commercial application

From what we have covered so far, it should be clear that IP can have a value that is worth protecting, because it has an actual or potential commercial application. This includes things such as:

- products
- formulations (see section A6.15)
- recipes
- computer software or apps.

Whether these forms of IP can be protected and, if so, how they are protected, depends on the type of IP and, in some cases, the country.

Secret recipes

We have mentioned trade secrets several times. Chemical and pharmaceutical companies sometimes choose to keep a production process as a trade secret rather than obtain patent protection, because that might give competitors an opportunity to adapt the process enough to avoid patent infringement but still benefit from the IP published in the patent.

Do you think it is possible to have 'secret recipes' for foods in an age of sophisticated analytical methods? Is this an example of IP or just good marketing?

Project practice

You have been asked to undertake an 'ethical audit' of an organisation. This could be based on your workplace or it could be a fictitious organisation.

Consider the topics covered in this chapter and collect evidence from your chosen organisation about the ways in which they:
- operate ethical scientific practices
- demonstrate integrity in a scientific setting
- use codes of practice in scientific organisations
- show and encourage respect in the workplace.

If you are using a fictitious organisation, use an internet search to find suitable examples – you could use the bullet points above as search terms.

Review the evidence you have collected against the information in this chapter. Make a reasoned assessment of:
- how closely the evidence follows best practice
- any areas for improvement or gaps to fill.

Prepare a written report that could be submitted to management; if you are basing your work on a real organisation, be sure to maintain a positive and constructive tone.

Discuss your work with the group.

Write a reflective evaluation of your work.

Assessment practice

1. What are the key **themes** of the Universal Ethical Code for Scientists?
2. What are the key **aims** of the Universal Ethical Code for Scientists?
3. State **three** ethical qualities that you should maintain in a scientific setting.
4. Katja works for a contract research organisation that is preparing batches of an experimental drug for use in a clinical trial. The left-hand column gives examples of Katja's good working practice. Match each box in the left hand column with the box in the right-hand column that best describes how Katja's work helps to ensure the integrity of the process.

Katja is filling vials with an experimental drug. The clinical trial protocol says that the dose should be ±5% of the stated value, but Katja has set herself the goal of achieving ±2%.	Following organisational codes of practice.
Katja consults the SOP for the preparation of the vials of drug before she starts working on a new batch in case there have been any changes to the procedure.	Following regulatory guidance.
Katja prepares each batch of experimental drug and her work is then checked by a Qualified Person before the batch is released for use in the clinical trial.	Aspiring to excel.

5. Several companies developed new vaccines against the SARS-CoV-2 virus during the COVID-19 pandemic. Describe **two** aspects of this process that could be considered to be intellectual property.
6. By September 2021, the world population was about 7.9 billion people and about 5.57 billion COVID-19 vaccines had been administered. However, according to the World Health Organization, about 80% of populations in high-income and upper-middle-income countries have received a first dose, compared with only 20% of populations in low- and lower-middle-income countries. In just low-income countries, only 1.9% of the population had received even one dose. Vaccination experts warned that new variants of the virus were more likely to arise in unvaccinated populations and that these might not be as well controlled by existing vaccines. At the same time, there was public demand in high-income countries for third (booster) doses to be given to all adults aged over 50.

   Describe **two** ethical issues that scientists face in this situation.
7. The following statements are taken from a staff handbook. For each one, explain how it shows an example of respect in the workplace.
   a. The company has a formal grievance procedure.
   b. The use of discriminatory language by any employee will be a disciplinary offence.
   c. Shared parental leave is available to both men and women working for the company.
   d. The company operates a system of flexitime. Employees may arrive between 8 a.m. and 10 a.m. and leave between 4 p.m. and 6 p.m., and are able to take a lunch break anytime between 11 a.m. and 2 p.m. Employees are expected to work 36 hours a week, averaged over a 4-week period.
8. What would be the most appropriate way of protecting IP rights in the following situations?
   a. A company has a product that has several competitors, but they feel that it has some advantages. The company has come up with a name for the product that they feel illustrates its advantages over the competition.
   b. A company has developed a new app for making it easier to collect and transfer data from a smartphone to a laboratory information management system (LIMS).
   c. A university researcher has developed a novel technology for testing the strength of metal components used in the aerospace industry.

A8: Good scientific and clinical practice

Introduction

If you bake a cake, you follow a recipe. If you strip down and rebuild an engine, you follow a workshop manual. If you have done either of these many times before, then you may not have to keep looking at the recipe or the manual. But if you are new to baking, then failure to follow the recipe carefully is likely to give disappointing results – something that looks and tastes more like a pancake than a light, fluffy sponge. Not paying attention to the workshop manual might result in a few parts left over and a reassembled engine that may never work again.

The same principles apply to our work in science and healthcare. Therefore, we have to follow proper procedures and do our best at all times – particularly when the consequences of getting things wrong can be disastrous. This is why we must always adhere to the standards and procedures that we call good scientific and clinical practice.

Learning outcomes

The core knowledge outcomes that you must understand and learn:

A8.1 the principles of good practice in scientific and clinical settings

A8.2 what an SOP is

A8.3 why it is important for everyone to follow SOPs

A8.4 how to access SOPs for a given activity

A8.5 the potential impacts of not regularly cleaning and preparing work areas for use

A8.6 the potential impacts of not maintaining, cleaning and servicing equipment

A8.7 why it is important to calibrate and test equipment to ensure it is fit for use

A8.8 how to escalate concerns if equipment is not correctly calibrated/unsuitable for intended use

A8.9 why it is important to order and manage stock

A8.10 the potential consequences of incorrectly storing products, materials and equipment.

A8.1 The principles of good practice in scientific and clinical settings

Whatever our work environment, whatever branch of science or healthcare, we must aim to achieve:

▶ consistency
▶ predictability
▶ reproducibility
▶ reliability.

On top of all this, we must ensure the health, safety and wellbeing of ourselves and others.

We can achieve these objectives in several ways, as described in this chapter, but they all share these common principles:

▶ using standard operating procedures (SOPs)
▶ effectively managing calibration and maintenance of equipment and work areas
▶ effectively managing stock
▶ appropriately storing products, materials and equipment.

The following sections will look at how all of these contribute to good scientific and clinical practice.

A8.2 What an SOP is

A **standard operating procedure** or **SOP** is simply a set of steps or instructions in a sequence that are designed to standardise the approach to a process or action, so that everyone learns to do it the same way. SOPs can be used for everything we do in the workplace, including:

▶ receiving goods, booking them into stock and informing the accounts department that the invoice can be paid
▶ cleaning a room in a healthcare facility or a microbiology lab
▶ producing a batch of a pharmaceutical ingredient
▶ analysing that batch to make sure it meets standards of purity, activity and safety.

Some SOPs may be just a few lines of instruction, others will be long and complex. They will always aim to ensure that the process or activity is done correctly and consistently.

Research

Find one or more examples of an SOP. Ideally these should be from your workplace, but if none is available, an internet search for 'SOP in science' should provide numerous examples.

Review the SOPs. Are there any common features? Would they allow you to carry out the task without further information? Do some of them require prior knowledge or training, or refer to other SOPs?

A8.3 Why it is important for everyone to follow SOPs

SOPs are created for a reason – to ensure that processes can be followed consistently, which can be crucial in health and science facilities. Ultimately, if you do not follow an SOP, you might get fired from your job, or at least disciplined. Of course, this is not the main reason we must all follow SOPs – failure to follow SOPs might have legal consequences in some organisations, or someone may be injured as a result – but it does illustrate why doing so is so important.

Maintaining health and safety

In Chapter A3 we looked at various health, safety and environmental regulations. You will have learned the importance of following proper procedures, especially in areas that are tightly regulated. Part of this is to perform a proper risk assessment. Every SOP should include a risk assessment of the process being carried out and the steps taken to minimise harm. Following the SOP is essential to reduce harm to employees, the public or the environment.

Enabling consistency of approach

The key word is 'standard'. An SOP should ensure that a process or procedure is always carried out in the same way. The outcome should be predictable. An SOP helps ensure that everything is done in the same way each time, producing consistent outcomes.

Of course, it is important that quality is also built into SOPs. Consistency is no good if all it means is that the product is always consistently poor. Consistency without quality is not enough.

Meeting any legal or organisational requirements

The science and healthcare sectors are highly regulated – for good reason. Companies and other organisations can see the importance of good-quality SOPs that are strictly followed, but some SOPs will ensure that any applicable laws are followed.

Here is a range of examples of how SOPs might be required for organisational or legal reasons:
- cleaning staff writing their initials on a chart to show that regular scheduled cleaning has been carried out
- following government (specifically, Home Office) requirements for storing and issuing controlled drugs in hospitals, care homes or research laboratories
- storing and disposing of hazardous waste (see COSHH Regulations in Chapter A3 page 35)
- carrying out clinical trials of new medicines or therapies
- obtaining regulatory approval for new medicines, medical devices or treatments.

Upholding professional standards

Membership of all sorts of professional bodies from the Royal Society of Chemistry to the various medical colleges (Royal Colleges of Nursing, Midwives, Physicians, Surgeons, etc.) requires adherence to certain professional standards. By following appropriate SOPs, you can show that you are upholding those standards.

Some regulated activities, such as production of pharmaceuticals, have to be supervised by a **qualified person** (**QP**). European Union regulations state that a medicinal product cannot be supplied for sale without a QP certifying that it meets the relevant requirements or standards. A QP is typically a pharmacist (MRPharmS), chartered chemist (CChem) or chartered biologist (CBiol). These roles are administered by the Royal Pharmaceutical Society, Royal Society of Chemistry and the Royal Society of Biology.

Demonstrating compliance for audit purposes

It is not enough to do something properly – you need to be able to show that you did it properly. Having an experienced person watch over you might be useful, but it will not be enough to show that the job was done correctly when your organisation is audited or inspected months or even years later. Having robust SOPs, following them carefully and recording how they were followed provides documentary evidence that

will satisfy inspection bodies such as the Medicines Agency, Home Office or Environment Agency when they review this at a later date.

Test yourself

1 What is a standard operating procedure (SOP)?
2 Explain two reasons why it is important for everyone to follow SOPs.
3 Give two consequences of not following an SOP.
4 Explain why SOPs often require you to complete a log or record your actions.

A8.4 How to access SOPs for a given activity

Most organisations should have SOPs for all the procedures and processes that are carried out in the workplace. These should be kept in either hard copy or electronic format and be readily available for use. However, if SOPs are not available, you might need to look elsewhere.

Carrying out detailed index searches (for example, via intranet/manual)

If the SOP is available in your organisation, you might need to search for it. This might involve searching an electronic or paper-based index. SOPs may be held centrally or located in the departments to which they relate. SOPs that are used infrequently might require some tracking down.

Finding SOPs in electronic format via your organisation's intranet or computer network should be quite straightforward – particularly if it has good search facilities, ideally based on one of the online search engines such as Google or Bing.

Completing detailed staff induction and ongoing training

If you wrote an SOP for how to run an efficient organisation, staff induction (processes for getting new employees set up and informed about their new role) and regular training would certainly be very prominent. New staff must be aware of SOPs that affect their area of work – it is not good enough to expect them to learn, sometimes the hard way, 'the way we do things round here'. As well as being part of the induction process for new employees, staff need to be kept up to date with the new SOPs or modifications to SOPs. This should be part of a regular training programme for all staff.

Ensuring the SOP is the most up-to-date version

Using a version of an SOP that was written 20 years ago and has been updated once since then can be worse than useless. Much could have happened in the meantime:

▶ Methods may have changed.
▶ Equipment or processes may have been updated.
▶ Regulations may have changed.
▶ Roles and responsibilities may have evolved.

It should be obvious that some form of **version control** is needed (for instance, giving each new updated version a new number – e.g. v2, v3, etc.). But it is not enough to know which version you are using – you need to know that you have the latest version. This can be achieved in a number of different ways:

▶ Controlling the production and distribution of SOPs so that old versions are returned or destroyed when new versions are introduced.
▶ Maintaining a central deposit (electronic or paper-based) that can be accessed but not removed or downloaded.
▶ Maintaining a central index or database so that version numbers of SOPs in use can be checked to ensure they are the most up to date.

Research

It is unlikely that you will have responsibility for production of SOPs, at least in the early years of your career. However, you will almost certainly have to use them. You should always know where and how to access the SOPs that cover your work. Find out how your organisation stores, makes available and maintains its SOPs. Think about the tasks you undertake (or are likely to undertake) in your role and obtain copies of the relevant SOPs. Do they provide sufficient information and guidance for you to carry out the task adequately? If not, do you know where to go to get additional help or support?

Ensuring all relevant documentation has been completed and signed

Although people sometimes complain about 'box-ticking exercises', 'mountains of red tape' or 'bureaucracy', good

documentation is essential to keep track of:

▶ what is done
▶ when it is done
▶ it being done correctly
▶ who does it.

These are all essential to make sure not just that things are done correctly but that we can show they have been done correctly. 'Signing off' the records helps to emphasise the importance of following procedures and taking responsibility. It encourages taking ownership of a process.

If something goes wrong and a product does not meet specification, or someone is injured, there is likely to be an investigation. It is important to have good records and an **audit trail** to ensure that lessons can be learned for the future. Learning lessons, in the spirit of continual improvement, is more important than assigning blame. There are examples of this in sections A3.3 and A3.4.

Test yourself

1 Describe the ways in which an organisation may:
   a keep or store its SOPs
   b make them available to staff.
2 Give two reasons why it is important to use up-to-date SOPs.
3 Describe one way in which organisations ensure that the most up-to-date version of an SOP is the one that is used.
4 Explain why it is necessary to ensure all relevant documentation is completed.

A8.5 The potential impacts of not regularly cleaning and preparing work areas for use

Risks to health and safety

Some of the more general risks in this area are covered in section A3.3 ('Ensuring working environments are clean, tidy and hazard free', page 41). Other risks can be specific to certain environments or workplaces.

Spread of infection

The COVID-19 pandemic has made us all aware of the importance of hygiene in reducing the spread of infection. Different micro-organisms can spread in different ways, but regular cleaning and preparation can reduce routes of transmission:

▶ Contaminated surfaces, known as **fomites**, can transmit bacteria and viruses when touched. These can include:
  – hard surfaces (door handles, handrails, light switches, mobile phones)
  – fabrics such as clothing, towels and furnishings.
▶ Aerosols, produced by breathing, coughing or sneezing, can transmit infected droplets directly between people. This is why good ventilation is so important. Aerosols can also transfer infectious agents to fomites.
▶ Accumulations of waste, rubbish, dirt, etc. can provide a breeding ground for infectious micro-organisms or for vermin that can be **vectors** for disease (see section B1.26, page 195).

Production of toxic/dangerous by-products

Aseptic technique is used in microbiology to prevent contamination of bacterial cultures as this can have serious consequences. For example, if a pathogen contaminates the culture, this could lead to disease or even death of anyone who comes into contact with the culture.

In some chemical processes, failure to clean equipment properly can lead to contamination of the product and could also lead to production of hazardous by-products that are toxic, explosive or harmful to the environment.

> #### Health and safety
>
> Think about the risks we have covered. Do any of these apply in your workplace? Do you need to get more information about precautions to take and procedures to follow? Are there any ways that you could modify your work practices that will help to reduce risks to health and safety?

Invalid results

Contamination can introduce other micro-organisms (bacteria and fungi) that can compete with the one being cultured. This can obviously ruin an experiment. However, there are many other situations in laboratory or clinical settings where contamination, or cross-contamination, can invalidate results.

Here are some examples:

▶ Environmental samples for water analysis can be contaminated during the sampling process or during transport or storage.

▶ Analytical reagents can be contaminated. This is likely to invalidate any analysis performed using them.
▶ DNA samples are at particular risk from contamination – especially cross-contamination. See the case study for more information.

> #### Case study
>
> The polymerase chain reaction (PCR) is an enormously powerful technique for 'amplifying' DNA or RNA. It uses the enzymes involved in DNA replication to take a few hundred copies of DNA or RNA and create billions of copies. PCR is used in many applications in research (genetic engineering), forensics (DNA samples from crime scenes) and diagnosis (the PCR test for SARS-CoV-2 virus). However, it is also highly susceptible to contamination and so precautions must be taken to prevent this. Use of DNA in evidence in a criminal trial could be invalidated if there were any possibility of 'foreign' DNA being introduced.
>
> ▶ What precautions should be taken to minimise risk of contamination when working with PCR? Think about use of SOPs and staff training.

> #### Notice
>
> Have you ever taken a COVID-19 PCR test? Did you notice any signs of an SOP being followed, or precautions being taken to avoid contamination during collecting the samples?
>
> If you have taken a COVID-19 lateral flow test, think about the instructions provided. Do they have the characteristics of an SOP? Is the information provided sufficient for you to carry out the test correctly?

Inefficient working practices

'Time is money' is a saying you are likely to encounter many times in your working life, largely because it is true. Lack of efficiency means that the job takes longer – and that means it is not just your time that is being wasted but other people may be waiting on your work, so they are wasting time as well. This means higher staff costs as they have to be paid for more hours. And inefficiency is not just about waste of time – it could mean a waste of valuable materials or resources, some of which could be incredibly difficult to obtain.

Damage to equipment

Not following an SOP can have many consequences, including breaking essential and/or expensive equipment. Once again, it is likely to cause your organisation extra costs – such as repair or replacement – as well as increasing the time it takes to get the job done.

Think about a workplace you are familiar with. Make a list of all the things that have gone wrong or could go wrong. Now think about the costs and delays that might be incurred as a result.

Test yourself

1 Describe the ways that regular cleaning can prevent spread of infection.
2 Describe how inadequate cleaning can lead to invalid results.

A8.6 The potential impacts of not maintaining, cleaning and servicing equipment

All of the factors that we considered in section A8.5 about not cleaning and preparing work areas apply just as much to not maintaining, cleaning and servicing equipment properly:

▶ Risks to health and safety:
  – increased risk of injury
  – spread of infection.
▶ Invalid results:
  – contamination or cross-contamination.

Reduced function of equipment

Lack of proper maintenance can lead to decreased lifespan of equipment. That is why second-hand cars sold with a 'full service history' are considered better buys. As well as this, equipment being out of service for repair or because it has not been properly maintained is likely to increase costs and cause delays in getting the job done.

For all these reasons, it is quite common for regular maintenance (sometimes called **preventative maintenance**) to be included in various SOPs.

Similarly, there might be SOPs for regular maintenance of some or all key equipment. These could include:

▶ regular checking that all connections are secure to make sure there are no leaks
▶ regular lubrication of moving parts
▶ replacement of batteries to ensure they do not cause damage if they leak

▶ regular calibration of equipment, as described in the next section.

Research

Endoscopes are used to view the inside of the body, typically the gut. An endoscope can be introduced via the nose or mouth to examine the oesophagus, stomach and upper part of the small intestine. Alternatively, it can be inserted via the anus to examine the bowel. Because of the way it is used, the endoscope can become heavily contaminated with pathogens during use and so must be cleaned, disinfected and sterilised between patients.

More healthcare-associated outbreaks of infectious disease have been linked to contaminated endoscopes than to any other medical device.

Source: Centers for Disease Control and Prevention www.cdc.gov/infectioncontrol/guidelines/disinfection/healthcare-equipment.html

Some endoscopes are heat-sensitive, so they cannot be steam sterilised (which is the preferred method). Chemical sterilisation or high-level disinfection must be used instead.

Endoscopes use delicate fibre-optics to provide a view deep inside the body, so they can be easily damaged. Care must be taken in handling, cleaning and sterilising endoscopes to prevent damage.

The NHS provides extensive advice on endoscope decontamination. The Medicines and Healthcare products Regulatory Agency (MHRA) publishes an information sheet on this: (https://assets.publishing.service.gov.uk/government/uploads/system/uploads/attachment_data/file/372220/Endoscope_decontramination.pdf or search the gov.uk website for 'endoscope decontamination').

▶ Why are endoscopes at higher risk of microbial contamination than other types of medical device?
▶ What are the aspects of good scientific and clinical practice covered in this chapter that are included in the MHRA information sheet?
▶ Can you think of examples from your own practice where these principles would apply?

A8.7 Why it is important to calibrate and test equipment to ensure it is fit for use

The **true value** of a measurement is the ideal or perfect value. Except for some of the physical constants used as the basis of SI units (see section B1.62, page 237), we can never determine the true value because of the inherent **uncertainty** in any measurement.

The uncertainty in any measurement reflects the difference between the actual measurement and the true value as a result of the level of **accuracy** of the measuring equipment or apparatus.

> ### Key terms
>
> **Accuracy:** measurements that are close to the **true value**.
>
> **Precise:** measurements that are close to each other, but they may be inaccurate.
>
> **Calibration:** the process of comparing measurements, usually against a **reference standard**.
>
> **Reference standard:** something of known size, mass, concentration, etc. that we can use to calibrate equipment or methods.

Ensuring accuracy of measurements

Accuracy and **precision** are terms that are not always used correctly. Figure 8.1 illustrates this using the analogy of shooting at a target.

High accuracy   Low accuracy   High accuracy   Low accuracy
High precision High precision Low precision Low precision

▲ Figure 8.1 The target analogy of accuracy and precision

Accuracy in measurement depends on the quality of the apparatus used for measurement, the skill of the person using it and how well the apparatus has been **calibrated**. The degree of precision of a piece of equipment is related to the number and size of **random errors** that it generates. The accuracy will depend on whether and how well it has been calibrated.

Calibration methods will depend on the equipment being used, but they usually involve use of a standard. Calibration is essentially the process of comparing measurements. One measurement is of a known size or correctness, e.g. a **reference standard**. The other is on the device or instrument being calibrated.

Calibration is required for:
- any instrument that has moving parts, such as an analogue ammeter or voltmeter, because movement can upset the balance of the instrument
- anything that can be affected by temperature change

- electronic equipment, as the performance of various electronic components can change over time, for example, the glass membrane in a pH electrode can be affected by deposits (dirt, oils and grease, protein, inorganic materials) and should be calibrated before each use.

However, calibration is only as good as the reference standard used, so it is important to take care of standards, for example, by preventing cross-contamination of standard solutions, corrosion of standard masses, etc.

Prolonging the life of equipment

As well as ensuring the accuracy of measurements, calibration and testing equipment will form an important part of any programme of preventative maintenance. This will help to extend the useful life of the equipment.

Meeting legal requirements

It is not so common to buy loose produce in a grocery store these days. It takes a lot longer to weigh out 1 kg of rice than it does to simply pick a packet off the shelf. However, you also have to consider the time and expense in maintaining and calibrating the scales used. There is a legal requirement for all measuring equipment used in trade to be certified to meet certain standards. This ensures that consumers are not cheated and that they can be confident in their purchases.

Now imagine you are not simply buying 1 kg of rice in the supermarket. If you buy a medicine, you need to have confidence that the dosage shown on the packaging is correct. For this reason, all equipment used in every stage in the manufacture of the raw materials and finished products has to be certified so that it meets the necessary standards of accuracy.

> ### Research
>
> Think about your workplace and the types of measuring equipment that are used. Research the types of measuring equipment used and the legal requirements covering the certification and calibration of the equipment. What records have to be kept? One or more of the following pieces of legislation might be relevant:
> - Ionising Radiation Regulations (2017)
> - Weights and Measures Act 1985
> - National or International Standards such as BSI or ISO
> - Control of Substances Hazardous to Health (COSHH) Regulations 2002
> - Environmental Protection Act 1990.

A8.8 How to escalate concerns if equipment is not correctly calibrated/ unsuitable for intended use

The details of the procedures to follow will vary according to the workplace. However, the principles to follow if any equipment is not fit for purpose remain the same. For some critical pieces of equipment, the course of action might be covered in the SOP.

Taking the equipment out of action

This should be obvious, for the reasons discussed in the previous sections. If a pH meter or thermometer is found to be out of calibration or specification, it should be simple to stop using it and find another one. However, it is not always possible to simply throw a switch. Whenever you encounter new equipment or a new task, give some thought to what you would need to do if that equipment had to be taken out of action.

Labelling the equipment as being out of use, if appropriate

It is not enough to simply stop using the equipment yourself, you need to let other people know that the equipment should not be used. Ideally, there should be signs or stickers available that can be used – if not, perhaps you should suggest some. It is important that any sign or label is prominent so it cannot be overlooked. Also, make sure that it cannot simply fall off or be removed too easily.

Reporting concerns to the relevant person, in line with organisational policies and procedures

You might have to report any concerns to your line manager. Alternatively, there may be someone, such as a safety officer, who should be informed of any issues that could have an impact on safety.

You need to make sure that some action is taken, even if you do not need the equipment right away.

Repairs or re-calibration may take time. It may not be your responsibility to actually do that, or arrange for it to be done. However, you should certainly take responsibility for informing the appropriate person so that they can take the necessary action.

Recording concerns according to organisational procedures

Reporting concerns is important, but so is making a record of those concerns. We saw in section A8.3 that you need to be able to demonstrate that an SOP has been followed for audit purposes. The same is true about documenting concerns about equipment being fit for purpose. This might look like 'covering your back' in case there are any repercussions. That may be a benefit, of course, but it is not the real reason that you need to document everything related to equipment – particularly if it has a safety implication.

Key equipment will often have an associated log, either on paper or electronically. This might cover:
▶ who used the equipment
▶ when it was used
▶ what it was used for
▶ any routine calibration or maintenance carried out.

A log of this sort might be the appropriate place to record concerns about the equipment so that everyone using the equipment is aware of any issues, either current or historical.

A8.9 Why it is important to order and manage stock

Case study

In August 2021, Becton Dickinson, a company that makes consumables for the NHS, announced that there was a global shortage of blood collection tubes. This was due to record demand for these tubes, partly because of the increased need for tests for COVID-19 patients. An additional factor affecting the UK was transport and import issues related to Brexit.

As a result of this, the NHS in England and Wales told GP surgeries and hospitals to temporarily stop some blood testing. Patients would only be able to get tests if they were urgent. The tests that were suspended included those for fertility, allergy, and pre-diabetes.

Doctors and nurses in GP surgeries were asked to:
▶ only request blood tests where they were clinically urgent, or time-critical (such as in pregnancy)
▶ when taking blood samples, not send an extra tube 'just in case'
▶ not duplicate tests that may have already been carried out in hospital.

Source: www.hey.nhs.uk/wp/wp-content/uploads/2021/08/bloodTubeShortageCCGs.pdf

In addition, practice managers in GP surgeries were asked to:
▶ review their local stocks to ensure they had a maximum of 3–7 days' supply
▶ not over-order tubes so that the central Pathology stores teams could manage stocks based on expiry dates and availability
▶ return any short-dated blood tubes that would not be used before the expiry date to the central stores.

Source: NHS blood test tube shortage: Doctors 'facing difficult choices' - BBC News www.bbc.co.uk/news/health-58374553

This was an exceptional situation, based on the combination of worldwide and local factors. Nonetheless, can you see aspects of how normal stock control was managed?

What are the lessons for proper ordering and managing of stock?

Ensuring sufficient supply of required consumables and materials

Clearly, it is important that you have access to sufficient stocks of all the items you need to carry out your work. You may not be responsible for the purchase or ordering of **consumables** or **materials**, but you will almost certainly be responsible for ensuring that you have access to everything you need. That may mean knowing how to get them from stores or ensuring you have your own stock in your work area.

▲ Figure 8.2 Make sure you always have the right amount of stock

Think about the things you use every day in the workplace, such as gloves, paper towels, cleaning materials, etc. What impact would it have on your work if supplies of any of these ran out? Have you been in a situation where shortage of consumables has prevented you from doing a job, or having to do the job in a less-than-ideal way?

Key terms

Consumables: items that are used and then disposed of. They are mostly single-use but might be re-used in some circumstances.

Materials: include items such as ingredients or components used in the manufacture of a product.

Ensuring that materials are used before their expiry date

We are used to seeing 'use by' or 'best before' dates on most of the items we buy in the supermarket. Do you know what they mean? In general, they mean what they say. Some products might become harmful if kept beyond the 'use by' date. On the other hand, something eaten after the 'best before' date may not be as tasty but it is unlikely to cause you harm.

In a science or healthcare environment you are more likely to come across products with expiry dates. Some of these will be more important than others.

If the expiry date on a pack of self-adhesive envelopes has passed, the adhesive may not be fully effective and you might have to resort to sticky tape to seal the envelope. That is inconvenient but is not dangerous and is therefore minor in the overall scheme of things.

However, most of the items that you encounter in a laboratory/science or clinical setting are likely to have expiry dates that must be respected, including:

▶ medicines that may no longer be sufficiently effective, meaning the patient does not receive the effect of a full dose
▶ packs of fluids, swabs, dressings, etc. that may no longer be sterile and could cause harm to the patient if used.

Reducing the costs of excess stock

Sometimes it is necessary to maintain high levels of stock, such as:

▶ items for which there is a long lead time, meaning we need to order them well in advance
▶ items that are critical to patient care, where running out might harm the patient
▶ pre-mixed solutions, such as buffers
▶ tissue culture media
▶ sterile packs of petri dishes, pipettes, tissue culture bottles, etc. may no longer be sterile and could introduce contamination if used.

However, in your working environment it is possible that excess stock will be wasted for no good reason. That puts a drain on resources and means there will be less budget available for essentials. Implementing robust stock control procedures can save huge amounts of money.

Improving efficiency and productivity

Efficiency and productivity are linked. Whether you measure productivity in the amount of work you get done in a given time or how much product goes out of the door every day, if you can work more efficiently, you will be more productive. However, being productive does not necessarily mean being quick. If you rush things and make a mistake, you may find that everything takes longer and costs more than if you had worked at a more steady, methodical pace. Efficiency and productivity also mean being careful.

Reflect

Think of the ways in which your work could be adversely affected if you do not have the consumables or materials you need. Have you ever had to delay doing something because you did not have all the necessary consumables? Have materials been wasted because they could not be used in time when key items were not available?

Think about all the ways in which poor stock control can reduce efficiency. Think about ways in which your own work could be made more efficient.

Ensure safety of stock

We have considered the importance of good stock control. It should be obvious that stocks need to be looked after, which we will cover in the next section.

Ensuring stock is safely looked after might mean keeping it under the correct conditions, for example, in a fridge or freezer. It might also mean ensuring the security of stock. That might be to prevent unauthorised use or to prevent access to hazardous or controlled substances. For example, the Misuse of Drugs Act 1971 and Misuse of Drugs Regulations 2001 apply strict procedures to storage of **controlled drugs**, including:

▶ ensuring safe and secure storage, including restricting personnel access (for example, by use of locked storage cabinets, keycard access, etc.)
▶ undertaking inventory record-keeping
▶ following sign-in/sign-out protocols.

A8.10 The potential consequences of incorrectly storing products, materials and equipment

Are you the sort of person who reads the small print on jars of jam or bottles of tomato sauce? If so, you may notice that they nearly always have the instruction 'refrigerate after opening'. This is increasingly common, as food manufacturers reduce the amount of artificial preservatives added to foodstuffs. The consequence is that something that may have been kept in a cupboard many years ago should now be stored in the fridge. It probably explains why large fridges are becoming more popular, but it also illustrates the importance of proper storage.

Cross-contamination

Keeping with the domestic theme, we are told that cooked foods should never be stored in the fridge on a shelf below raw meat. Raw meat may be contaminated with harmful bacteria. Although these are destroyed by cooking, fluids from the raw meat may drip down onto the cooked food. The benefits of cooking are lost if cooked foods become contaminated. You will encounter many similar situations in your workplace, so pay attention to procedures for safe storage of any items that might be at risk of cross-contamination.

Breakdown of limited stability products

Many products that you will encounter in a healthcare setting are sensitive to heat or light and need to be stored accordingly.

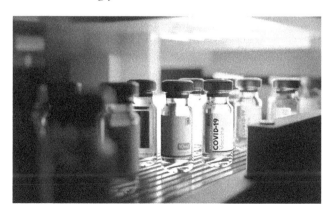

▲ Figure 8.3 COVID-19 vaccine vials in refrigerated storage

Consider the recent example of two COVID-19 vaccines:

▶ The Pfizer vaccine needs to be stored and transported at −70 °C.
▶ The Oxford/AstraZeneca vaccine can be stored at up to +4 °C.

It is clear from this that the Oxford/AstraZeneca vaccine is easier to handle, particularly outside of hospitals or in developing countries that may not have access to ultra-cold storage.

Light-sensitive products are often stored in brown glass bottles (see Figure 8.4), so take care if you need to transfer such liquids to other containers.

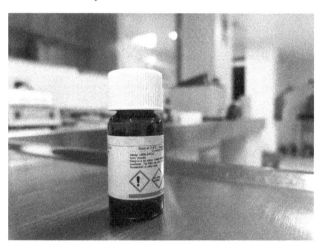

▲ Figure 8.4 A light-sensitive laboratory reagent packed in a brown glass bottle

Products exceeding expiry dates

Stock rotation is a term you are likely to encounter. It means always using the oldest batch first, providing the oldest has not passed its expiry date. Stock should always be kept in a way that makes proper stock rotation possible, rather than in such a way that items are left for months or years at the back of a cupboard. Physically arranging stocks of materials with the oldest batches at the front of a shelf can help, for example.

Loss of samples or degradation of reagents not stored at the correct temperature (−20°C, −4°C, 4°C or room temperature)

NHS guidance to GP surgeries is that blood samples and urine samples should be sent to the lab the same day (see Figure 8.5). If this is not possible, they must be stored refrigerated at 4–8 °C and analysed within 24 hours. Cells in the blood can start to break down and release substances that mean the results may not be valid. It is not just high temperatures that can be damaging. In general, blood samples should not be frozen before analysis as this can also lead to damage to cells.

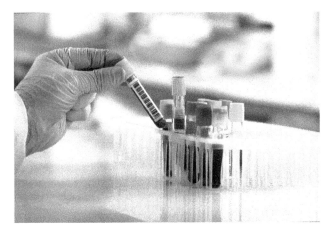

▲ Figure 8.5 Blood samples ready for analysis

The World Health Organization (WHO) has strict guidelines for transport and storage of water samples. Microbiological analysis should ideally be performed within 6 hours or no later than 24 hours after sample collection. Failure to do this means that the results may not be valid.

Correct storage of reagents is equally important. Here are some examples:

▶ Antibodies and proteins usually need to be stored at −20 °C or −80 °C, but conjugated antibodies (antibodies with enzymes attached) must not be frozen and so should be stored at 4 °C.

▶ Antibodies and other reagents that have a fluorescent dye attached must be kept dark to prevent photobleaching of the fluorescent dye, which will make it useless.

▶ Many solids are **hygroscopic**, meaning that they will absorb moisture from the air. When they do so, they are more likely to degrade by **hydrolysis**. This

is when a compound is broken down by water – see section B1.7 for examples of hydrolysis. Also, if the solid has absorbed moisture, you cannot weigh it accurately.

So far, we have described the need for low-temperature storage. What about storage at room temperature? Obviously, it can be expensive to maintain large fridges or freezers, particularly ultra-cold freezers at −80 °C. Therefore, it makes sense to store materials at room temperature where possible. However, there is another reason why it can be better to store at room temperature. Whenever a bottle of solid is removed from a fridge or particularly a freezer, there is a risk of condensation forming on the powder when the bottle is opened. This can also cause damage to the material by hydrolysis. For that reason, it is important that you leave a bottle to reach room temperature before opening. This could take several hours, so if there is no need to store the product at low temperature, it would be better to keep it at room temperature.

Practice points

One useful tip when handling products that require storage at low temperature is to **aliquot**. This means sub-dividing a larger amount into single-use quantities. This will avoid many freeze–thaw cycles that can be damaging to temperature-sensitive materials.

If you do this, you need to make sure that the process of aliquoting does not introduce contamination.

Risks to health and safety

Incorrect storage can have serious, sometimes life-threatening, consequences.

▶ Cultures or samples containing pathogens could be released and that could lead to spread of infection.

▶ Toxic, corrosive or other dangerous chemicals could be released into the lab or into the environment.

▶ Heavy items stored above a safe height could fall and cause injury or could injure someone trying to reach or lift them.

Stock is difficult to locate

What happens when one member of staff goes on holiday and they are the only person who knows that the spare toner cartridges are kept in a particular cupboard in an obscure location? Storage of stocks of consumables and materials should follow a logical pattern so that things can be found easily.

▲ Figure 8.6 Computerised stock control has many advantages

Financial loss

We have focused mainly on the health and safety consequences, but incorrect storage that leads to loss of or writing off stock has a financial implication as well. Think about the high cost of medicines and other clinical supplies and consider how much money could be saved if you reduced losses from incorrect storage.

Research

Think about your workplace and the materials, products or samples that are stored. Do they require any special storage conditions? Is it always clear what these conditions are? For example, are the containers marked or is there information about correct storage conditions in any relevant SOPs?

What would be the consequence of incorrect storage of any of these items?

Project practice

You have been asked to review the way in which SOPs are used in your organisation (you can base this on your own workplace or an organisation you are familiar with). The results of the review will be considered by the management and used as the basis for future action.

You need to complete the following steps:
1 Research a strategy.
  a Carry out a review of the legislation and regulations that affect your organisation.
  b Gather information about clients, customers, stakeholders of different types and how they are affected by your organisation's performance.
2 Plan a project based on the information that you have gathered.
  a Identify who is responsible for production and maintenance of SOPs.
  b How and where are SOPs kept, issued or referred to and by whom?

  c What procedures are in place for documenting and auditing adherence to SOPs?
  d How are SOPs incorporated into induction and training?
3 Present your findings in the form of a written report or PowerPoint presentation. You should include the following:
  a What weaknesses are there in the existing SOPs?
  b Who is responsible for maintaining SOPs and do they have the relevant training and experience?
  c Are all members of staff familiar with the SOPs that affect them?
  d Recommendations for improved systems.
4 Discuss the following questions:
  a How should members of staff be made aware of the importance of following SOPs?
  b Is the culture of the organisation to impose or to encourage the following of SOPs?
  c What are the risk factors if SOPs are not followed correctly?
5 Write a reflective evaluation of your work.

Assessment practice

1 What are the main principles of good practice in scientific and clinical settings?

2 Give **two** advantages and **one** disadvantage of keeping all SOPs on an organisation's intranet/computer network.

3 You notice that an SOP does not include a procedure for calibrating an important piece of measuring equipment. Justify why it should be included in a revised SOP.

4 Which of the following is the correct course of action if you find a piece of equipment is not performing correctly?

   a Put a note on the door of the room containing the equipment.

   b Unplug the equipment, stick a label on the front saying 'Do not use' and inform your line manager.

   c Inform the safety officer when you next meet them.

   d Put a note in the equipment log to say that it is not working correctly.

5 Amir and Sofia are arguing about the terms 'precision' and 'accuracy'. Amir says that calibrating a piece of equipment makes the results more **precise**, but Sofia says this makes the results more **accurate**. Who is correct? Explain why.

6 A package arrives by next-day courier containing several bottles of an enzyme. Explain what should happen to the package.

7 It has been proposed that monitoring equipment should be installed on all temperature-controlled equipment in the laboratory, including fridges, freezers, cell culture and microbiology incubators. The equipment would monitor the temperature-controlled equipment and then send a notification, by email or text message, to a designated person if temperatures fluctuated outside of specified limits. Evaluate this suggestion.

8 What precautions would you take when preparing environmental samples for transfer to an analytical laboratory?

9 You have been asked to review the way in which stocks of consumables are handled in an analytical laboratory. Each floor has a senior technician who is responsible for ordering and maintaining stocks of PPE, analytical reagents, enzymes and sterile packs of plasticware (pipette tips, microtiter plates, etc.). Identify **two** weaknesses in the current approach.

10 Explain the importance of proper stock control.

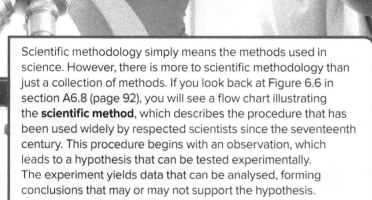

Scientific methodology simply means the methods used in science. However, there is more to scientific methodology than just a collection of methods. If you look back at Figure 6.6 in section A6.8 (page 92), you will see a flow chart illustrating the **scientific method**, which describes the procedure that has been used widely by respected scientists since the seventeenth century. This procedure begins with an observation, which leads to a hypothesis that can be tested experimentally. The experiment yields data that can be analysed, forming conclusions that may or may not support the hypothesis.

In this chapter we will look at how various methods and methodologies contribute to the scientific method, helping us achieve our objectives in science. In the following chapter we will cover a number of specific experimental techniques that you are likely to encounter in your work.

Learning outcomes

The core knowledge outcomes that you must understand and learn:

A9.1 the importance of experimental design and planning when undertaking scientific experiments

A9.2 the importance of a hypothesis/performance criteria in experimental design

A9.3 how planning methodologies contribute to successful experimental design

A9.4 how customer/client requirements may affect the scientific methodology

A9.5 how to provide results and recommendations in appropriate formats to customers/clients

A9.6 how to access and critically evaluate scientific literature and research databases

A9.7 the principles that inform sampling techniques

A9.8 a range of techniques for measuring scientific subject matter at micro and macro scales

A9.9 the need for reliable, verifiable and accurate recording

A9.10 how to isolate and solve problems or inconsistencies in scientific data

A9.11 how to evaluate a scientific methodology and make recommendations for improvement

A9.12 the purpose of International Organization for Standardization (ISO) standards in scientific settings.

A9.1 The importance of experimental design and planning when undertaking scientific experiments

Chapter A6 covered data handling and processing and, in various places in the chapter, we saw examples of the importance of designing experiments correctly to:

- minimise sources of error
- obtain valid results
- be able to draw appropriate conclusions from data
- prevent or reduce bias.

We will now consider some additional aspects of this subject.

Manage time efficiently

You will encounter time management throughout your career (and we will consider it again later in this chapter) because it is an important part of working efficiently. The saying 'time is money' does not just apply to commercial organisations. Costs accumulate whether you are working efficiently or not (see the Reflect box below); any delay in completing an experiment or prolonging it unnecessarily will cost your employer money.

Reflect

Think about the **overheads** (sometimes called 'fixed costs') that an organisation will incur on a daily basis, such as:

- salaries, wages and other staff costs (for example, pensions)
- rent or business rates on premises
- utilities (the cost of heating and lighting, water charges, waste disposal)
- administration costs (human resources, finance and so on).

These costs are incurred whether you are being productive or not.

Can you see how these fixed costs might apply to your own organisation? Does this help illustrate the importance of efficient time management?

One way that good experimental design can contribute to efficient use of time has already been covered. In section A6.12 we saw how a rolling mean (moving average) can be used to ensure that a sample size is large enough without having to keep making measurements unnecessarily.

Good experimental design can also ensure that we have enough repeats or enough measurements to give valid data. It can be very inefficient and wasteful of resources if we must repeat a whole experiment because we find that we did not take enough measurements the first time.

Ensure sufficient resources

Imagine that you are preparing a meal. You consult the recipe and then look in the fridge or cupboard for the ingredients and discover that you only have half of what you need. Unfortunately, it is 5.30 p.m. on a Sunday and the supermarkets are closed. To make things worse, you realise that your only roasting tin was borrowed by your neighbour last week and not returned. Besides which, the recipe is quite complex and requires two pairs of hands at a critical moment and the rest of the household are not home yet. Do you think you are going to manage to make this meal?

Now put that into a scientific context. Before you start you should check:

- that you have adequate supplies of any reagents (chemicals) or other consumables (plasticware, pipette tips, tubes and so on); suppliers can sometimes deliver next-day, but your purchasing department might take longer to process an order

> There is more on the importance of managing stock (including supplies and consumables) in section A8.9.

- that you have the necessary equipment available or have booked time on any shared equipment
- that everyone who is needed for the experiment will be available; if a key person is on holiday or otherwise unavailable, it can stop the entire project.

All these factors need to be taken account of well in advance of starting an experiment.

Ensure safety throughout the experiment

Good experimental design will incorporate a risk assessment, as well as taking all necessary steps to ensure the health and safety of everyone concerned. This was covered in chapters A3 and A4.

Address ethical considerations

When planning an experiment, you need to consider whether it is necessary. If it is not, you will be wasting money and resources that could have been used for something important. You might have to consider other ethical issues, including:

- the use of animals in research
- the use of human subjects in research.

Ethical issues are discussed in more detail in Chapter A7.

Minimise errors

Errors can invalidate an experiment, so we should do all we can to minimise them. One way is to calibrate any equipment that you will use in advance of carrying out the experiment. This, and other ways of minimising errors, was covered in section A6.6.

Test yourself

1 Give **two** reasons why good experimental design is important.
2 Some aspects of experimental design and planning are performed at the start of an experiment or series of experiments. Name **two** things you should do before the start of every experiment, even though you may have performed the experiment many times before.

A9.2 The importance of a hypothesis/performance criteria in experimental design

We saw the importance of a hypothesis in section A6.6, where we looked at the need for good experimental planning as a way of minimising errors. Then, in section A6.9, we used a null hypothesis in a statistical analysis as a way of evaluating the statistical significance of results to see whether they supported our hypothesis.

Remember: a hypothesis is an attempt to explain an observation or a link between two variables.

We might also want to define **performance criteria** when we design an experiment. For example:

- If we are testing a new alloy in a component, do we need it to be twice as strong as the existing material or will a 5% improvement be good enough? This might influence how we test the alloy to look for either a big difference or a small difference in performance.
- If we are screening compounds as a potential new drug using inhibition of an enzyme as a test of activity, do we need it to be active at very low concentrations? If it only has activity at high concentrations, it might not be useful as a drug because the dose would be too high. This might

mean that the drug would be too expensive or might increase the risk of side effects.

Defining outcomes that can be tested

Testing a hypothesis is central to the scientific method. In order to test a hypothesis, we must first define the expected outcomes of an experiment.

Case study

Ignaz Semmelweis was a Hungarian doctor working in Vienna General Hospital in the mid-nineteenth century. The hospital had two maternity wards, one staffed by doctors and the other by midwives. Semmelweis observed that there was a much higher mortality rate from what was known as childbed fever in the doctors' ward than in the midwives' ward. Semmelweis considered several possible causes, including:

- The fever was caused by mothers giving birth lying on their backs rather than on their sides.
- The fever was caused by doctors' unclean hands; it was quite normal for doctors to perform an autopsy on an infected body immediately before examining women who were in labour.

Semmelweis considered how to test each theory:

- If the fever was caused by the mother giving birth lying on her back, then changing the position of delivery should reduce rates of fever. However, when he did this, the incidence of fever did not diminish. The actual observations did not match the expected outcomes.
- If the fever was caused by the doctors' unclean hands, thorough handwashing with a strong disinfectant solution should reduce the rate of fever. When he tried this, the incidence of fever fell dramatically, matching the much lower rate in the midwives' ward. The observations matched the expected outcomes, supporting this explanation.

Sadly for Semmelweis, and for the many pregnant women who continued to give birth in the doctors' ward, his findings were rejected by the medical community, who were outraged at the suggestion that their hands could be unclean, and Semmelweis was dismissed from the hospital.

Semmelweis had no scientific explanation for his findings – it was many years before Pasteur confirmed the germ theory of infection.

- Do you think Semmelweis's experiments were well designed?
- Does the fact that an experiment leads to conclusions that cannot be explained invalidate the results of the experiment?

The scientific method begins with an observation, from which we develop a hypothesis. From there, we design an experiment to test that hypothesis. We might observe that plants in a dry corner of the garden do not grow as well as plants in a part that gets more rainfall. We would then formulate a hypothesis that stated: lack of water reduces the growth of plants. We could design an experiment where we grew identical plants in identical conditions but gave different plants different amounts of water and then measure the rate of growth, or some other measure such as height, after a number of weeks.

If we find that the rate of growth does depend on the amount of water given to the plants, does that explain our observation? Were the plants growing in the two areas the same species, or were they different? Was the light intensity the same in both places? Were the soils similar, with similar levels of nutrients?

These are all outcomes that could be tested, so it is important to define the outcome you will specifically be testing. All these outcomes will be important when considering types of variable covered in the next section.

Deciding on variables

We encountered variables in section A5.2 when considering the best way to present results of experiments. Here we think about the types of variables and how they relate to experimental design.

The value of an **independent variable** does not depend on another variable (hence it is independent). When we design an experiment, the independent variable is something that we change. This means that it should be possible to change the variable using the methods that we have available.

The value of a **dependent variable** will depend on that of another variable, that is, if one changes, so does the other. When we design an experiment, the dependent variable is what we measure. That means it should be possible to measure the value of that variable using the equipment that we have available.

Here are two examples:
▷ In an experiment to study the effect of temperature on the rate of an enzyme reaction, the temperature is the independent variable. We can change this by using a water bath at different temperatures. The dependent variable is the rate of the reaction. We can monitor this by measuring the volume of gas produced or the change in concentration of substrate within a given time period.

▷ In an experiment to study the effect of stretching a spring, we can add increasing mass to the end of the spring (mass is the independent variable). We can then measure the extension of the spring with a ruler – this is the dependent variable.

As well as dependent and independent variables, we also need **control variables** or just **controls**. These are any factors, besides the independent variable being tested, that might have an effect on the dependent variable. We have to design the experiment to exclude any effect of these variables in order to get valid results. For example, the rate of an enzyme-catalysed reaction depends on the temperature, pH, substrate concentration and enzyme concentration. If we want to investigate the effect of temperature, we carry out the reaction at different temperatures. But we must ensure that, at each temperature, all the other variables are kept constant otherwise they will affect the results and we will not know if we are seeing the effect of temperature or pH or concentration.

Clarifying the experiment's objective

A valid result is one that answers the question that the experiment was designed to test. By having a clear hypothesis or clear criteria for success, we make sure that we are designing the experiment to answer the right question. We saw in section A6.6 how good experimental design helps to minimise errors and, therefore, obtain valid results. If we are clear about the objective of our experiment then this will be easier, and we are more likely to get valid results.

Test yourself

1 Describe the difference between an observation and a hypothesis.

A9.3 How planning methodologies contribute to successful experimental design

So far we have seen why we need to design experiments well and how having a clear hypothesis or performance outcome (or outcomes) can help achieve that.

Now, we are going to look at some specifics of how to design a successful experiment.

Objective setting

We saw, in the previous section, the importance of setting objectives to define the purpose and outputs required from any experiment. For example, if we are asked to investigate the stability of a liquid medicine, there are various factors that could affect this, such as temperature of storage, exposure to light, the material of the container or the formulation (other ingredients). Designing an experiment that would investigate all of these factors would be complex and we would risk not getting valid results. Instead, it would be better to design a series of experiments, each of which had a clear objective testing one of the many factors.

Critical path analysis

Critical path analysis is important in many areas of work, not just in the science sector. Any project will involve a series of steps. Some of these steps could be completed at any time but others cannot be started until another step is completed – we say that these steps are **dependent** on others. These **dependencies** can determine the length of critical path, which is the longest path from start to finish. An example of a dependency can be seen in the case study below; Amir's microbiology experiments are dependent on the media being prepared for him by another group.

The process involves the following steps:
- Identify the various tasks. This might involve talking to colleagues, customers, service users and so on.
- Put the tasks in order, from first to last. This will help you identify dependencies. You might find some tasks do not have any dependencies, so they could be done at any stage.
- Allocate a time for each task. This is important, because you need to know how long each task will take before you can identify how long the whole project will take.

Find all the potential paths, or strings of dependencies. The longest path will be the critical path – the one that, if there are any delays in any tasks, will delay the whole project.

You can use a spreadsheet to map out all the tasks and their start and finish times. This method can be used for highly complex projects, such as building a new cruise ship or shopping centre. Fortunately, scientific experiments are usually easier to plan.

Financial forecasting

Financial forecasting does not apply to commercial organisations only – whatever our work environment (commercial, educational, academic research funded by the government or charities, for example) we all have to work within a budget. This means that, before we embark on an experiment or series of experiments, we need to make sure that we are not going to run out of money part way through. The financial forecast will define what is feasible for any given budget and ensures that we can complete the project. After all, if we cannot complete the project, we do not simply fail to reach our objective, we will probably have wasted all the money spent on the project up to the point that the money ran out.

Risk management

Risk management has been covered extensively in Chapters A3 and A4. From what you have learned in those chapters, it should be clear that assessing and managing risks for the workforce has to be an integral part of planning any experiment.

Time management

This is really part of the critical path analysis discussed above. As part of that exercise, you will define timescales and workflows for each part of a project. For a simple experiment, you will probably be able to sketch out a timescale in your lab notebook. For more complex projects, time management will be included in the critical path analysis. We saw in section A9.1 that efficient time management is an important part of effective experimental design.

A9.4 How customer/client requirements may affect the scientific methodology

It is not just high street stores or online retailers who have customers. You might be working in a commercial laboratory that is providing a service under contract, so it is clear who your customer is. Even in a non-commercial organisation (one that is not selling an obvious product), for example a school science lab or companies where you are not interacting with the final customer (pharmaceutical research and development, for example), you will still have internal clients; these are the people whose own work depends on your activities. You might be working in an analytical laboratory providing test results to another part of the organisation – these are your clients.

Defining timescales

Just as you might use critical path analysis to plan your own experiment or project, your work might be part of a larger project with its own plan and critical path. This might mean that you are expected to complete your part of the project in a specified time. If so, you need to account for that in your own planning. If you feel that the required timescale is unrealistic, you need to feed this back to whoever oversees the overall project. If you do not do this and just carry on

with your part, knowing it will not be delivered on time, you risk other people being held up in their own work because you have not completed your part on time.

Representatives from each part of the whole project should be involved in the initial planning of the overall timescale to minimise the risk of this happening. It may be that as the project is started, there are delays which will then delay a project that is dependent on yours. This is why it is important to feedback straight away.

Setting a budget

The amount of money available for your work might be limited by the total budget for a project, or your customer might only be prepared to pay a limited amount for the work. In either case, this will influence the design of the experiments that you can undertake. You might need to find less expensive ways of performing the work needed to achieve the client's objectives.

Specifying scale

In section A6.12 we saw that having a sufficiently large sample size or number of replicates (repeats) was a way to prevent or reduce bias in evaluating an experiment. This is an important part of experimental design. However, your client might have their own requirements and you will have to take account of this.

For example, if you are analysing samples from patients in a clinical trial, the design of the clinical trial will specify the number of replicates or sample size in carrying out the analysis. There may be situations where there are regulatory requirements for a specific sample size or numbers of replicates in an experiment.

Specifying objectives

Your client may well have specific objectives that they need you to achieve. For example, if your laboratory provides a service testing the physical properties of different materials, your customer might require testing to meet the tolerances of their own production process.

A9.5 How to provide results and recommendations in appropriate formats to customers/clients

Section A5.2 covered how the intended audience can influence the way that we select the best method to collect and record information and data. We looked at the different ways of presenting data in section A6.8. In this section we are going to consider that in the context of working with customers or clients.

Answering the brief/research questions

Imagine some of the questions you might be asked in a meeting with your client:

- Did you carry out the experiment as requested?
- Did the experiment work?
- What are your results?

These are all factors that we should consider when we put together our results and recommendations – we need to anticipate any questions that the client might have.

Moreover, a customer or client is likely to have questions about the project brief and research when it is initially proposed. They will want to feel confident that the project will meet the required outcome. If you can anticipate any questions, they are more likely to have confidence in you and the project.

Tailoring language and technical information to the audience

This was covered in section A5.2 – we need to tailor the language we use and the way we present the information to the audience:

- Non-specialist clients will need clear and non-technical language so that they can understand the results and recommendations that we present to them. We must take care to avoid the two major pitfalls: confusing the audience with very detailed technical descriptions on the one hand and patronising them on the other.
- Specialist clients will expect to receive information containing the type of technical language with which they are familiar. This can be important if technical language is clearer and less ambiguous for a technical audience.

Selecting the most appropriate way of presenting data

What is true of language is also true of presenting data. The different methods were covered in sections A5.2 and A6.8, for example visualisations/infographics.

Data visualisation tools are the various ways in which data can be presented in a visual form, such as graphs, charts and infographics.

Infographics can be a powerful tool for presenting data, particularly to non-specialist audiences (Figure 9.1).

WOMEN IN SCIENCE

Proportion of graduates in SCIENCE fields:

1 in 9 men
1 in 14 women

Female researchers by region:

47% Central Asia

44% Latin America & the Caribbean

40% Central & Eastern Europe

37% Arab States

32% North America & Western Europe

30% Sub-Saharan Africa

22.5% East Asia & Pacific

19% South & West Asia

An estimated 58 million primary school aged children are not in school.
31 million are girls.

Women account for 30% of researchers worldwide.

781 million adults across the world are illiterate. Nearly two thirds of them are women.

Adapted from United Nations Department of Economic and Social Affairs (www.un.org/development/desa/publications/graphic/women-in-science)

▲ Figure 9.1 Infographics can be a powerful tool for presenting data

Do you think we are sometimes limited by our own abilities and experiences when we consider the ways in which we can present the results of our work?

For a completely different approach to data presentation, research the technique of sonification, where data is represented as sound.

If that seems slightly crazy, bear in mind one of the earliest and most successful examples of sonification: the Geiger-Müller tube mentioned in section B1.58. This measures radiation and makes a click for every count – the higher the radioactivity, the faster and louder the clicks.

Dr Wanda Díaz-Merced is an astronomer who lost her eyesight in her early 20s. She did a PhD in Computer Science at the University of Glasgow where she developed a technique to convert astrophysical data into sound, allowing her to analyse what would normally be visual data.

Some clients might want to receive our results in electronic format, as a database or spreadsheet for example. This might be because they want to transfer it directly into their own data retrieval and analysis system. It can also allow them to look in more detail at the raw data, if necessary, to gain greater insights.

Highlighting the commercial/business benefits for the customer/client

If we are working in a commercial environment, we might want to add value to the results that we present by highlighting their commercial or business benefits. This might require an element of understanding of the client's business – although we should probably have gained that understanding before starting work, as a way of ensuring that the work we did met the client's needs and objectives.

Examples might include:
▸ You have developed a new, more efficient SOP that saves time and will, therefore, save money.
▸ A different reagent was used that was cheaper and more widely available.

A9.6 How to access and critically evaluate scientific literature and research databases

What was your main source of information about COVID-19 during the pandemic? Television news? Government announcements and press conferences? The NHS website? Social media? Friends and family?

Which source did you think was most reliable? Which source influenced you the most?

In our personal lives, we tend to place most trust in those we know best. But is that the best approach when dealing with sources of scientific information? Do we need to be more critical of information and information sources?

We are often told not to believe everything that we read on the internet. But that does not mean that we should not believe *anything* that we read on the internet! There is now a huge quantity of scientific information available online, much of it accessible through search engines. We just have to be careful and critical when choosing our sources.

Searching for relevant existing scientific research/literature

Searching for relevant information used to require access to databases that might require a subscription (that is, you had to pay to access them), and searching was a specialist skill possessed only by information scientists. It was necessary to identify a relevant database and then select particular key words, terms or phrases and construct a search that would produce the desired results from the enormous amount of data available. That is still often the case when we carry out a search of scientific literature.

However, online search engines have made that process much easier and more accessible, as long as we use them carefully.

A number of scientific databases can be accessed online and many are free of charge. Some of the most prominent are listed here:

- **PubMed** is a database of research publications, mostly on biomedical and life science topics (including related areas of chemistry). It also gives access to PubMed Central, which is an archive of full-text, free-to-access publications and preprints (papers that have been submitted for publication but not yet published).
- **Google Scholar** is a free-to-access search engine that provides links to the full text of a wide range of academic literature, including most peer-reviewed academic books and journals, conference proceedings, university theses and dissertations, and many more.
- **Astrophysics Data System** covers publications in astronomy and astrophysics as well as physics.

> **Practice point**
>
> Many scientific databases are free of charge to access, and many of the published articles are also free of charge. This is known as 'Open Access'; on PubMed, look for 'free article' or 'free PMC article'. However, some publishers will charge for access to the article, unless your employer has a subscription.

The differences between primary and secondary sources

You should be able to recognise the two different types of information sources.
- **Primary sources** are ones that have direct access to the original data or information, for example, journal articles.
- **Secondary sources** are those that give an interpretation of information from a primary source. This includes review articles or commentaries from a researcher on a specific subject.

Do not think that primary sources are more reliable than secondary sources; they are just different. A secondary source might give a more balanced or informed view, particularly if it is based on a range of primary sources, or it might give a misleading view due to misinterpreting the primary source.

Age/relevance of literature

More recent literature tends to be more reliable as knowledge, techniques and methods improve. Also, as knowledge accumulates, our understanding improves and older theories can be disproven.

At the same time, more recent literature may not have had the same level of scrutiny, for example by other researchers trying to reproduce the work.

So, when you are looking at a particular piece of literature, check the publication date and bear that in mind when assessing it.

Reliability of sources

If you look at a publication in any reputable academic journal you may see a statement to this effect: 'The authors declared that they had no conflicts of interest.' You might also see details of the organisations funding the research. This is usually to acknowledge the support provided. However, if there is no explicit statement about conflicts of interest, knowing the source of funding can be important. For example, it might reveal that one or more authors may have had a conflict of interest – such as if research into lung cancer is funded by a tobacco company.

Another way to assess the reliability of a source is to look at what are known as **citations**. When a researcher bases their work on previously published research, they will put a reference to that work in their own publication. This reference is known as a citation. The more times a paper is cited in subsequent research, the greater its influence on the scientific community – we can also have more confidence in the paper because it will have been read and evaluated by many other researchers.

However, it is important to understand that the number of citations does not always indicate the quality or importance of a particular publication or piece of research. Papers that first describe methods that become widely used tend to be cited more often because lots of researchers will use those methods and therefore cite the original paper.

The most cited paper of all time, with over 300 000 citations, is a paper published in 1951 by the US biochemist Oliver H Lowry describing an assay for determining the total amount of protein in a solution. The Lowry Assay is very commonly used and, therefore, widely cited, although Lowry himself said in 1977: 'I really know it is not a great paper.'

Citations are a measure of how influential a particular paper has been, whereas the **impact factor** measures how influential a particular journal is. The impact factor is calculated from the mean number of citations that articles in a journal receive over a two-year period – the higher the impact factor, the more prestigious and influential the journal is thought to be.

Reliability of data

Chapter A6 addressed several ways in which data can be more or less reliable, for example adequate sample sizes, appropriate collection method used, quality of statistical analysis and so on. Just as we can apply these tests to our own data, we can also use them when evaluating any published research.

A9.7 The principles that inform sampling techniques

By now, you should have a good understanding of the need to avoid bias in experiments and how to ensure that sample sizes are large enough to produce valid results. In this section we will look at how we use that knowledge to decide on the most appropriate sampling techniques.

Reflect

Why do we need to sample?

Think of an experiment where you need to estimate the frequency or incidence of something in a population. This might be the incidence of an infectious disease in the UK or the growth of a bacterial culture.

You could count every single person in the UK that is infected with the disease and then divide by the total population of the UK. That would be a huge undertaking. Similarly, you could try to count every single bacterial cell in a culture, but that would be an impossible task – particularly if we have several hundred litres of culture.

Instead, we take a representative sample and count that. We then use this to estimate the total number. We could measure the number of people with the infection in several areas and use that to estimate the total number of infected individuals. With the bacterial culture, we could take a small sample, perhaps just 1cm^3, and count the number of bacterial cells in that sample and then multiply by a factor to get an estimate of the total population. In fact, even when we count the bacteria, we would probably use a haemocytometer (see section A10.6) to count the number of cells in a small fraction of the 1cm^3 and extrapolate from that.

Can you see the need for sampling? What precautions do you think are necessary when sampling?

Whenever we sample a larger population, we must take precautions to ensure that we have a **representative** sample. This means, if we actually counted the whole population, we would get an answer similar to the one based on our sample.

Avoiding bias

This is critical whenever we sample a population. If we wanted to calculate the incidence of an infectious disease across the whole of the UK and sampled only the infectious disease wards of hospitals, we would certainly not have a representative sample – our sample would be highly **biased**.

Similarly, if we want to estimate the total number of bacterial cells in a culture, we need to ensure that the culture is well mixed before we take a sample. If the cells settle to the bottom of the vessel and we take our sample from the top, we are likely to get an inaccurate estimate.

Ensuring a large enough sample size to produce valid results

We have seen the importance of sample size in producing accurate results. We must ensure that the sampling method we use gives us a sample that is large enough to be representative of the whole population.

In section A6.12 we saw how a rolling mean (moving average) can be used to ensure that a sample size is large enough without having to keep making measurements unnecessarily.

Practical constraints

The larger the sample, the more representative it will be and, therefore, we can expect our results to be more valid. It may not always be possible to take as large a sample as we would like, however. It might take too long or cost too much.

In such cases, we have to do the best that we can. When we interpret these results, we need to bear in mind that the smaller sample size might make the results less valid.

A9.8 A range of techniques for measuring scientific subject matter at micro and macro scales

We are going to look at a number of experimental techniques and the equipment you will encounter in the next chapter. For now, we will look at three basic examples that you are highly likely to encounter in your scientific career.

Mass

You might have a set of bathroom scales that you use to weigh yourself (remember, when we talk about weight in everyday life, what we generally mean is mass).

These probably give a reading to one decimal place. A reading of 95.1 kg would be to the nearest 100 g – the true mass could be anywhere between 95.05 kg and 95.14 kg. If you need to determine the mass of a grain of rice, bathroom scales would be useless. You would need to use an analytical balance that gives a reading to two or even four decimal places. See section B1.64 for more information about significant figures.

Length

Length can be measured with a ruler or a tape measure, but what about measuring much larger or smaller distances?

A **laser measure** can be used to measure distances of several hundred metres. One type of laser measure fires a laser pulse at the object whose distance you want to measure and times how long it takes for the pulse to be reflected. A microprocessor multiplies the time taken by the speed of light and divides the answer by two (because the pulse travels there and back) to give a measurement in metres.

An **eyepiece graticule** can be used to measure the size of the very tiny structures you observe under a light microscope. The eyepiece graticule is calibrated with a **micrometer slide** (also known as a **stage micrometer**). The eyepiece graticule is fitted into the eyepiece of the microscope and has a scale marked on it with equal divisions. What each division represents in terms of length will depend on the magnification of the objective lens. The eyepiece graticule is calibrated using the micrometer slide with each different objective lens (see Figure 9.2).

> See section B1.21 for more about light microscopes.

In this case, 1 micrometer slide division = 0.1 mm

$$20 \text{ eyepiece divisions} = 11 \text{ micrometer divisions}$$
$$= 11 \times 0.1 \text{ mm}$$
$$= 1.1 \text{ mm}$$
$$1 \text{ eyepiece division} = 1.1 \div 20 \text{ mm}$$
$$= 0.055 \text{ mm}$$
$$= 55 \, \mu m \text{ (as } 1 \, \mu m = 0.001 \text{ mm)}$$

Having calibrated the eyepiece graticule, you can use it to measure the size of any object observed under the microscope using that objective lens. If you change the objective lens, you will change the magnification, so you need to recalibrate the graticule.

Volume

Volume can be measured in various ways using different types of **volumetric glassware** depending on the scale. For example, measuring cylinders can range in size from 10 cm³ or smaller up to 1 litre or even larger, and have graduations at intervals such as 0.1 cm³ for a small cylinder and up to 10 cm³ for a larger cylinder.

Pipettes can measure a fixed volume, such as 10 cm³, 25 cm³ or 50 cm³. This type of pipette is often used in titration (see section B1.44).

Another type of pipette, the **graduated pipette**, can be used to measure different volumes. This is used, together with a safety filler, to draw up the required amount of liquid. The pipette will have graduation marks at intervals; as with a measuring cylinder, the intervals will depend on the size of the pipette.

With all kinds of volumetric glassware, it is important to read the volume correctly (see Figure 9.3).

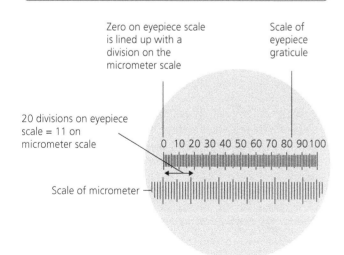

▲ Figure 9.2 Calibrating an eyepiece graticule with a micrometer slide

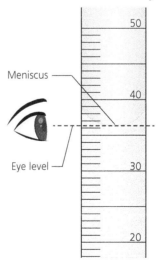

▲ Figure 9.3 Always read volumetric and graduated glassware from the bottom of the meniscus and at your eye level; this reading is 36.0 cm³ to the nearest cm³

For smaller volumes – generally less than $1\,cm^3$ down to as low as $0.2\,\mu l$ $(0.0002\,cm^3)$ – a micropipette is used (Figure 9.4).

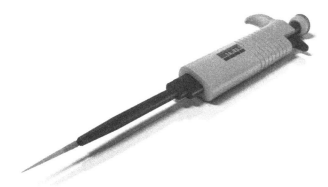

▲ Figure 9.4 A micropipette

These are very easy to use, particularly for repetitive measurements; the required volume is set by turning the knob and it appears on the dial. They also have disposable tips, so that they can be used many times without risk of cross-contamination. It is even possible to have multichannel pipettes allowing simultaneous delivery into 8 or even 12 tubes. These are employed widely with the 96-well micro-titre plates used in a wide range of assays (Figure 9.5).

▲ Figure 9.5 A multichannel micropipette being used to deliver liquid simultaneously into 8 wells in a 96-well plate

A9.9 The need for reliable, verifiable and accurate recording

Chapters A5 and A6 covered various aspects of why it is important to obtain accurate data, how to avoid bias and how to process data to give reliable results. Section A5.2 dealt specifically with considerations when choosing a method of collecting and recording data. In this section, we will look at how this relates to good scientific methodology.

Data or information is repeatable

It is central to the scientific method that experiments should be **repeatable**; if we repeat an experiment in the same way using the same equipment, we should get the same results. This means that we should always repeat experiments enough times to show that we do have repeatable results.

Data or information is relevant to the experimental purpose

A valid experiment is one that answers the question asked (see section A9.2). It follows that valid data recording is where the data or information collected is relevant to the purpose of the experiment – it helps us answer the question that the experiment was designed to ask.

Collecting and recording data can be time consuming (which will have a cost implication most of the time), so we must ensure that the experimental methodology we choose will collect and record only relevant data.

Data or information truly reflects the results obtained

Just as an accurate measurement is one that is close to the true value, accurate recording of results means that the data or information collected truly reflects the results obtained as discussed in Chapters A5 and A6.

A9.10 How to isolate and solve problems or inconsistencies in scientific data

Reflect

Should we eliminate errors from science? The answer seems obvious – of course we should! However, we could argue that making errors is a part of scientific advancement. Staying within the boundaries of established thinking and methods can limit the advancement of our knowledge. Sometimes real advances are made when the new person in the lab does not know how things should be done and does it their own way, only to find that their way works better or gives new insights. Of course, this does not apply where doing things differently could cause serious safety issues!

How important do you think making mistakes is as part of learning and advancement of science? Can you think of times when you have learned from your mistakes?

In this section we are going to look at how we can identify and correct errors in our scientific data. The details of this will depend on the type of experiment, but the principles are always the same, as shown in the table.

Identify and define the problem	This might involve looking at anomalous data or other inconsistencies in the data, including outliers. The results might be inconsistent with previous work or the accepted knowledge in the field. Note that this does not necessarily mean that an experiment or its conclusions are wrong, but it does require investigation.
Investigate and examine possible causes	Are there any weaknesses in the design of the experiment? Has data processing been done correctly? Is the statistical analysis sufficiently robust?
Decide on changes to be made	This might involve redesigning the experiment or repeating certain parts, or simply collecting more data to achieve representative results.
Implement the changes	Make sure that you either keep other variables the same, or consider what new control variables are needed.
Evaluate the impact and continue to monitor any changes	This might be an iterative process – in other words, we keep repeating the experiment and review the data with a view to making further improvements until we are satisfied with the validity and reliability of the conclusions.

▲ Table 9.1 Principles for identifying and correcting errors in scientific data

A9.11 How to evaluate a scientific methodology and make recommendations for improvement

This is better done before undertaking an experiment rather than after, so that you avoid wasting time and resources on imperfect experiments. Of course, you might need to evaluate a published methodology before assessing the reliability of the publication or before using the methodology in your own work.

Laboratories that repeatedly follow the same SOP should periodically evaluate the scientific methodology used and recommend improvements. Methods might have advanced or new equipment might offer increased performance. Also, it is possible that data or results become increasingly inaccurate. The cause of this should be investigated and improvements found.

Reflecting on experimental design

The points that we have covered in this chapter and previous chapters should have given you the tools you need to consider whether the design of an experiment will give valid and reliable results.

Assessing the reliability of methods, and precision, accuracy, repeatability and reproducibility of results

Methods are reliable if they are not affected by errors or bias and give repeatable results. From what we have learned you should now be able to judge whether methods are likely to be reliable.

If they are, we will be able to see that:
- the experiments are well designed with well-defined dependent and independent variables and adequate controls
- samples are representative, meaning sample sizes are sufficiently large and samples are selected randomly to avoid bias
- methods for measurement have a degree of accuracy and precision appropriate to the experiment.

Remember some of the key terms we have covered in previous chapters:
- Results are **accurate** if they are close to the true value.
- Results are **precise** if they are close to each other (that is, within a narrow range).
- Results are **repeatable** if we carry out the same experiment multiple times and get the same results.
- Results are **reproducible** if someone else carries out the experiment, using different techniques, equipment or methods, and gets the same results.

Identifying areas for improvement

We should always critically evaluate our work and identify areas for improvement. You may have noticed that this step is usually the last one in most of the project practice examples.

Think about things such as:

- experimental design – did it achieve what we set out to achieve?
- data collection – did we collect the right data and did we collect enough data?
- data presentation and analysis – did we present the data in a way that made it easier to interpret and did we use the most appropriate statistical methods?
- conclusions – were our conclusions valid?

Making recommendations for future improvement

Having identified areas for improvement, we then need to make recommendations. It is useful to include this in any report or publication so that others can benefit from our experience.

Case study

The Lanarkshire Milk Experiment

In 1931, a statistician with the pseudonym 'Student' (of Student's *t*-test, see page 97) published a critique of a study done in Scotland to compare the effect of either raw or pasteurised milk on the growth of primary school children. The study included 20000 children and involved considerable time, effort and expense (the study cost about £7500 in 1930, which would be about £500000 today). However, there were no solid conclusions because of poor design and execution of the study. Student identified numerous weaknesses, including:

- Instead of being random samples from the same population, the subjects were selected samples from different schools (meaning they may have been different populations).
- The controls were not appropriate to either of the test groups (who received either raw or pasteurised milk).

Student also suggested improvements, such as that the controls and test groups should have been matched for age, sex and physical characteristics. Alternatively, subjects should have been randomly assigned to the control and test groups.

Furthermore, identical twins could have been used as a way of excluding genetic or social factors. Student suggested that 50 pairs of identical twins could have given more reliable results than the 20000 children actually used.

- What do you think the lessons are for the importance of good experimental design?
- This paper was published over 90 years ago – does it still have lessons for us today?

The International Organization for Standardization (known as ISO for equality between the three official languages) was formed in 1947 as an independent, non-governmental organisation to develop and publish worldwide technical, industrial and commercial standards. In many ways an ISO standard is considered to describe the best way of doing something. The organisation has set over 20000 standards covering areas as diverse as manufactured products and technology to food safety, agriculture and healthcare. You will see that we have covered several of these topics already.

- Quality management standards help organisations work more efficiently and reduce product failures. This was covered in section A1.2.
- Environmental management standards help to reduce waste and impacts on the environment, as well as promoting sustainability. This was covered in section A3.1.
- Health and safety standards help to reduce accidents in the workplace. This was also covered in section A3.1.
- Energy management standards help to reduce energy consumption.
- Food safety standards help to prevent food from becoming contaminated.
- IT security standards help to keep sensitive information secure.

Enables accredited laboratories to demonstrate competency and validity through collaborative testing

ISO does not itself carry out certification or accreditation, it produces the standards. Various accredited certification bodies can then certify that a product, service or system meets the requirements of a particular standard. The ISO works closely with two other international standards development organisations: the International Electrotechnical Commission (IEC) and International Telecommunication Union (ITU). This explains why some standards are just ISO while others are ISO/IEC or ISO/ITU depending on the area of application.

For example, ISO/IEC 17025 applies to testing and calibration laboratories and enables them to demonstrate that they operate competently and generate valid results. It also helps to facilitate co-operation between laboratories and other organisations. This generates wider acceptance of test reports, results and certificates between different countries.

Another example is ISO/IEC 15189:2012, a standard that applies to medical laboratories. It focuses on improved patient safety, risk reduction and operational efficiency, which allows for the global acceptance and recognition of accredited test reports. The certificates issued by accredited laboratories are accepted in over 80 countries worldwide.

Facilitates co-operation between organisations by generating wider acceptance of results

The topic of quality control has been covered in section A2.10. The QC department of an organisation will generally test products to ensure that they meet the organisation's own standards or specifications. If you are buying products from that organisation, you will want confidence in the quality of the products. Can you be sure that the testing was done as carefully as your own organisation would expect? By operating to ISO standards, organisations can be confident that another organisation's results can be accepted.

Improves international trade as test reports and certificates can be accepted from one country to another without the need for further testing

Globalisation – the growing interdependence of the world's economies, population and culture – has been in the news a lot in recent years and sometimes for quite negative reasons. Globalisation means that products may travel back and forth across the world several times in the course of their production. This leads to long supply lines that can be disrupted in many places by world events.

In spite of this, it is clear that international trade will become increasingly integrated. ISO standards help to remove the need for repeat testing as goods move across borders, because test reports and certificates produced by accredited laboratories working to ISO standards are accepted internationally.

Specifies the general requirements for the competence to carry out tests and/or calibrations, including sampling

Conformity assessment is a set of procedures that shows that a service meets the requirement of an ISO standard. This involves testing, certification and inspection of the organisation providing the service. Part of this addresses the competence of the organisation in carrying out the tests or providing a calibration service. ISO has a committee that develops standards and addresses issues related to conformity assessment.

Assessment practice

1 Keira is planning and designing an experiment. Explain the reason for each of the following aspects of Keira's plan:

   a Checking stocks of chemicals and solvents needed in the experiment.

   b Starting the experiment by midday at the latest so that a long reaction can be left overnight.

   c Performing a risk assessment as part of the plan.

   d Including adequate controls.

2 Your laboratory has been asked to test several different fuels to determine which produces the most heat energy. You have been given a calorimeter. This burns the fuel to heat a volume of water and measures the temperature increase. This allows you to calculate the energy produced by each fuel.

   a What would be the objective in this experiment?

   b What is the independent variable?

   c What is the dependent variable?

   d Suggest **two** control variables that you should use.

3 When deciding for how long pharmaceuticals can be stored, they are tested for stability at the recommended temperature. For the following testing procedure, identify whether each point is a dependent variable, independent variable or control variable.

   a The pharmaceutical was stored at 25 °C ±2 °C and 60% relative humidity.

   b Samples were taken every three months for the first year and every six months for the second year.

   c The purity of each sample was tested.

4 For the experiment described in question 3:

   a Identify the objective.

   b Assess whether the experiment would produce valid results.

5 You are planning an experiment to prepare a sample of an active ingredient in a medicine. The experiment must be completed within three days. Describe what you would do to ensure that you achieved your objective.

6 You work in a research organisation that tests potential new drugs for safety. At the end of a contract for a large pharmaceutical company, you have been asked to prepare two reports on the work that has been undertaken. One report is for the pharma company's toxicologists while the other is for the senior management who are not specialist toxicologists. Describe how you would present the results differently for the two groups.

7 You are researching an unfamiliar scientific topic. Evaluate the different sources of information that you could use.

8 Explain the difference between:

   a accuracy and precision in measurement

   b repeatable and reproducible experiments.

9 Give **two** benefits of a testing laboratory being accredited to an ISO standard.

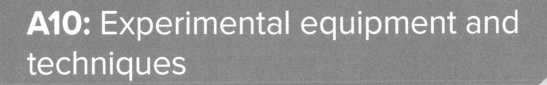

A10: Experimental equipment and techniques

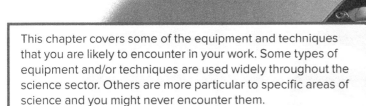

This chapter covers some of the equipment and techniques that you are likely to encounter in your work. Some types of equipment and/or techniques are used widely throughout the science sector. Others are more particular to specific areas of science and you might never encounter them.

This chapter will focus on what different equipment is used for, rather than how it is used – this might vary significantly from one type to another or between different brands.

Learning outcomes

The core knowledge outcomes that you must understand and learn:

A10.1 common causes of equipment and technical faults that may have an impact on scientific results

A10.2 the requirements for positive and negative controls in identifying faults

A10.3 applications of a range of equipment when undertaking scientific techniques

A10.4 the appropriate techniques for handling a range of different substances

A10.5 appropriate equipment to measure accurate results at different scales

A10.6 how to use a light microscope

A10.7 the reasons for using aseptic techniques

A10.8 how to follow aseptic techniques.

A10.1 Common causes of equipment and technical faults that may have an impact on scientific results

We looked at ways to minimise errors in section A6.6, where we saw the importance of staff training as well as maintenance and calibration of equipment. These are important because they can cause faults that have an impact on results. It would also be useful to review section A8.7, where we looked at the importance of calibration, to remind yourself of the difference between **precision** and **accuracy**.

We will consider some specific causes of equipment and technical faults as we look at different types of equipment and techniques in sections A10.3 and A10.4.

User error

There are many different types of user error. The following are some examples.

▶ Not using a piece of equipment correctly (hence the importance of staff training).

▶ Not carrying out a procedure correctly (hence the importance of following SOPs).

▶ Random errors in data, such as making a series of measurements that could be both inaccurate and imprecise.

▶ Systematic errors, such as not properly zeroing a piece of equipment (for example, taring a balance) which might produce precise measurements (i.e. they are close to each other) that are inaccurate (i.e. they are not close to the true value).

▶ Reading the wrong units (mm instead of cm or dm^3 instead of cm^3).

▶ Forgetting the decimal point or recording a result with the decimal point in the wrong place.

If you look at Figure 9.3 in section A9.8 you will see one common type of user error – incorrect reading of volumetric glassware. If your eye is not perfectly level with the meniscus, you will not get an accurate reading.

This is a good point to think about **uncertainty** and **percentage error**. We encountered uncertainty in section A6.5, where we saw that there will always be uncertainty in any measurement. With many types of measurement, this uncertainty is ± (plus or minus) half the smallest division or graduation. For example, a 50 cm^3 measuring cylinder might have 1 cm^3 graduations. This means the uncertainty will be

±0.5 cm^3. As a result, we could estimate a volume to be 10.0 cm^3 or 10.5 cm^3, but not 10.3 cm^3. The uncertainty in the measurement means that the measuring cylinder has a maximum **resolution** of 0.5 cm^3.

> ### Key term
>
> **Resolution:** the smallest change in a quantity being measured that gives a perceptible change in the reading. In microscopy, resolution is the ability to distinguish two points as being separate.

A larger measuring cylinder, say 1000 cm^3, might only have 20 cm^3 graduations. This means that the uncertainty in any measurement will be ±10 cm^3, and this is the maximum **resolution** of the larger measuring cylinder.

Now, let us think about percentage error. We calculate percentage error using the following formula:

$$\text{percentage error} = \frac{\text{uncertainty}}{\text{measurement}} \times 100$$

We can calculate the percentage error if we used these two measuring cylinders – one with a total volume of 50 cm^3 and an uncertainty of ±0.5 cm^3, and the other 1000 cm^3 with an uncertainty of ±10 cm^3 – to measure a volume of 40 cm^3.

For the 50 cm^3 cylinder, the calculation is:

$$\text{percentage error} = \frac{0.5}{40} \times 100 = 1.25\%$$

For the 1000 cm^3 cylinder, the calculation is:

$$\text{percentage error} = \frac{10}{40} \times 100 = 25\%$$

That is a huge percentage error! Obviously, we have chosen an extreme example, but this does illustrate one important type of user error – selecting an inappropriate piece of equipment to make a measurement. Choosing equipment with a low resolution will give greater error.

Think about the example comparing 50cm^3 and 1000cm^3 measuring cylinders. Can you think of any others where the resolution of a piece of equipment is important?

What would be the effect of increasing the volume measured on the percentage error? Can you think of ways that you could use this to reduce the percentage error in an experiment?

Setting-up errors

These can be a source of systematic error, such as if we do not correctly set the zero point, for example if we do not correctly tare a balance. This can cause an **offset error**, where every reading is incorrect by the same amount. Some types of electronic instrument can be affected as they warm up, so this must be allowed for (for example, the operative must wait for the instrument to give steady readings). This temporary variation is known as **drift**.

Poor maintenance (including calibration)

In section A6.6 we saw how maintenance and calibration of equipment was important to minimise errors. Then, in section A8.6, we looked at the potential impact of poor maintenance, particularly the increased likelihood of getting invalid results. The importance of calibration and testing of equipment was covered in section A8.7. It is useful to check any piece of equipment before using it – how to do this will depend on the equipment and may well be covered in an SOP.

Electrical faults

Electrical faults can cause failure of laboratory equipment, meaning loss of function or inaccuracy in measurement, depending on the type of equipment:

▶ If the lamp in a light microscope fails, the microscope will be useless until the lamp is replaced.
▶ If a heating jacket in a distillation fails, the solvent will not boil and the distillation will fail.
▶ Failure of the heating element in an incubator can halt the growth of a microbial culture or even cause it to die.
▶ Failure of a freezer can destroy stocks of temperature-sensitive reagents, such as enzymes. This can be very expensive and years of work can be lost. For this reason, some freezers have built-in warning devices; these could sound an alarm or even send a warning text or email message.

> ### Health and safety
>
> Electrical faults in any equipment, including laboratory equipment, can be hazardous. This is why all electrical equipment should be tested for safety regularly. Overheating, chemical or physical damage can cause insulation materials to break down leading to short circuits. This is another reason why equipment should be properly maintained.

> ### Test yourself
>
> 1 You are using a balance to determine the mass of a sample. The balance shows two decimal places on the display. The mass of the sample is shown as 1.56g. Calculate the percentage error in the mass.
> 2 What type of error or fault would cause the following?
>   a You determine the mass of a series of samples and discover that each measurement is 0.5g larger than it should be.
>   b You are using an electronic thermometer, but there is no reading on the digital display. You have checked that the thermometer is turned on.
>   c You are using a voltmeter and the reading is negative when you expect it to be positive.

A10.2 The requirements for positive and negative controls in identifying faults

We first encountered **control variables** in section A9.2 and controls were covered throughout that chapter. We will now look more closely at the two main types of control variables: **positive** and **negative** controls.

One of the greatest sources of experimental error occurs when we treat an experiment as a 'black box', meaning that we accept without question the results that we obtain. An important part of good experimental design is the need to eliminate **false positives** and **false negatives**.

> ### Key terms
>
> *False positive:* a positive result that is not true.
>
> *False negative:* a negative result that is not true.

Positive control

A **positive control** is anything that we know will produce a known result. It is a way to demonstrate that the experiment, equipment or method is working correctly. As a result, we can be sure that any negative results are true negatives and not the result of an issue with the method, equipment or reagents.

Lateral-flow test kits, such as pregnancy test kits or the COVID-19 lateral-flow test kits, have a positive control incorporated. In these, the appearance of a control line indicates that the sample has actually flowed into the test area so that a negative result is not due to something, such as a blockage, preventing the sample from entering the test area.

Negative control

A **negative control** confirms that no other variable is responsible for positive results in the test.

In an experiment to investigate how the rate of an enzyme reaction changes with temperature or substrate concentration, a negative control would either leave out the enzyme or use boiled enzyme (since boiling destroys enzyme activity).

Clinical trials often include a **placebo** (dummy drug) to act as a negative control. This is because patients can often improve simply because they are being looked after, so the placebo allows us to see the actual effect of the drug.

Case study

A new antibiotic is being tested to determine its ability to kill MRSA (a type of bacteria that is highly resistant to almost all antibiotics). We set up the following tubes, each containing MRSA and a nutrient solution:

▶ Tube 1 contains the new antibiotic. This is the test.

▶ Tube 2 contains vancomycin, an antibiotic that we know will kill MRSA. This is the positive control.

▶ Tube 3 contains methicillin, an antibiotic that does not kill MRSA. This is the negative control.

We incubate the tubes and then count the living bacterial cells in each tube. If the new antibiotic is effective, there will be fewer living cells in Tube 1 than in Tube 3. Ideally, the number of living cells (if any) should be the same in Tube 1 and Tube 2. If there are no living cells remaining in Tube 3 then we know the experiment has failed and must be repeated.

▶ In this experiment, what would be considered a false positive?

▶ What would be a false negative?

▶ What do you think might have caused there to be no living bacteria in Tube 3 after the incubation?

A10.3 Applications of a range of equipment when undertaking scientific techniques

Autoclaves

An **autoclave** uses high pressure steam at 121 °C for 15–20 minutes to decontaminate or sterilise equipment and some consumables, often in a microbiology or clinical lab. Surgical instruments are also sterilised before use by autoclaving.

▲ Figure 10.1 An autoclave being used to sterilise instruments

Many types of clinical or microbiological waste must be sterilised by autoclaving before disposal.

Many procedures now employ single-use items, such as sample tubes, hypodermic needles and other surgical instruments. Some of these are heat sensitive so must be sterilised by gamma irradiation (see section B1.61) or chemicals such as ethylene oxide gas instead.

Prions (infectious proteins that cause Creutzfeldt-Jakob disease) and some bacterial toxins are not destroyed by autoclaving. These must be subjected to chemical treatment (such as with sodium hydroxide) before autoclaving.

Some materials are heat-sensitive and cannot be autoclaved, for example some plastics will melt. In such cases, alternative decontamination and sterilisation methods must be used – see sections A4.17 and A10.8 for more information.

Centrifuges

A centrifuge spins a fluid, usually in a tube or other container, at high speeds to separate the components of the fluid or suspension. A suspension will be separated into a solid **pellet** and liquid **supernatant**. The process of forming a pellet is known as sedimentation, although you might come across the term 'spinning down' or to 'spin-down'. Most centrifuge speeds are measured in **r.p.m.** (revolutions per minute), meaning how many times it rotates in one minute.

A low-speed centrifuge has a maximum speed of 4000–5000 r.p.m. This type of centrifuge is used for blood fractionation, the process by which blood is separated into its components: red blood cells, white blood cells, platelets and serum.

▲ Figure 10.2 A benchtop centrifuge

One application of a low-speed centrifuge in chemistry or biochemistry is when a reaction produces an insoluble precipitate. A centrifuge can be used to obtain the precipitate; this is a useful alternative to filtration when working on a small scale. For example:

▶ The first stage in protein purification is often precipitation using ammonium sulfate (this is known as salting out). The precipitate can then be isolated by centrifugation.
▶ DNA and RNA can be precipitated using 95% ethanol and the precipitate isolated by centrifugation.

High-speed centrifuges operate at speeds up to about 15 000–30 000 r.p.m., while an ultracentrifuge operates up to 65 000 r.p.m. At these speeds, the friction from air resistance can generate a lot of heat, so these types of centrifuge are often operated in a vacuum with refrigeration to prevent damage to the sample.

Their speed is often described in terms of the centrifugal force as a multiple of the force of gravity. An ultracentrifuge can operate at a speed of $150\,000 \times g$ (150 000 times the force of gravity) or more.

In biology, an ultracentrifuge can be used to separate the organelles (see section B1.3) after cells or tissues are homogenised (which breaks the cells open). The different types of organelle sediment at different speeds. An ultracentrifuge can also be used to separate and purify large molecules such as DNA and proteins.

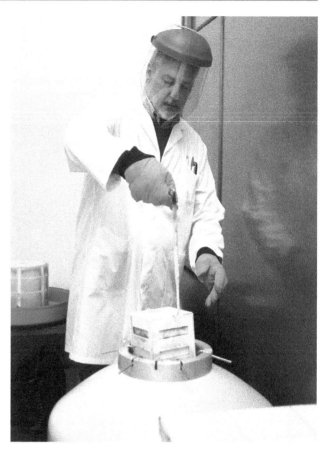

▲ Figure 10.3 Samples being placed in a cryogenic nitrogen container

Cryogenic equipment

Cryogenics is the study of very low temperatures, how to produce those temperatures and how materials behave at these temperatures. By 'very low' we generally mean less than 120 Kelvin (–153 °C).

The type of equipment that you are likely to encounter consists of liquid nitrogen used with double-wall vacuum-insulated containers for storage, transport and experimentation. The boiling point of liquid nitrogen is 77 Kelvin (–196 °C). It can be kept in an insulated vacuum flask, known as a Dewar, for between a few hours and a few weeks, depending on the size and design of the flask.

In biology, cryogenic storage is used for cell cultures, blood, sperm and egg cells, as well as embryos. Liquid nitrogen has many other applications as a coolant, even being used to make ultra-smooth ice cream because the freezing process is so rapid that the ice crystals are very small.

Equipment for bulk storage of liquid nitrogen is expensive, but the equipment for transporting and handling small amounts in the laboratory make it convenient for creating very low temperatures on a small scale. The equipment for cryogenic storage can provide much lower temperatures than ultracold freezers but requires topping up with liquid nitrogen making it less suitable for long-term storage. Ultracold freezers (see page 159) may be more suitable for long term storage at temperatures of −80°C or above.

Even lower temperatures can be achieved with liquid helium. At normal atmospheric pressure, helium has a boiling point of 4.15 Kelvin (−269 °C), that is, just above absolute zero. Liquid helium is used in superconducting magnets, such as those used in MRI scanners (see section B1.52).

Health and safety

Extremely low temperatures can cause cold burns (frostbite), so special insulated gloves must be worn when handling liquid nitrogen. A small splash or even pouring the liquid nitrogen down the skin will not cause immediate burns because the liquid evaporates rapidly and forms an insulating layer – so if you are splashed with small amounts, there is no need to panic – but if the liquid nitrogen pools anywhere it can cause severe burns.

Vessels containing liquid nitrogen can cause oxygen to condense from the air and so the liquid in these vessels can become enriched with oxygen as the nitrogen evaporates. If it comes into contact with organic material, such as proteins or tissue samples, it can cause violent oxidation.

As liquid nitrogen evaporates, it reduces the concentration of oxygen in the air. This can lead to asphyxiation (suffocation), particularly when working in enclosed spaces.

Data loggers

A data logger is an electronic device for collecting, storing and recording data over a period of time. They are usually attached to a computer, allowing data capture and analysis. The advantages of this were discussed in section A5.2; data can be captured continuously, without human intervention, and

analysis of the captured data is easier without the need to transfer between systems.

Examples of the use of data loggers include:
- unattended weather stations, hydrographic recording (for example, water depth and water flow) and soil moisture recording
- road traffic counting
- measuring the temperature of pharmaceutical products (medicines and vaccines) during storage
- process monitoring to ensure compliance with protocols or regulatory requirements
- environmental monitoring.

Digital and non-digital pipettes

Digital pipettes (or micropipettes) and non-digital pipettes (volumetric pipettes) allow accurate and precise measurement and transfer of solutions. These two types, together with their applications, were described in section A9.8.

Fume cupboards

A fume cupboard – you might also encounter the American term 'fume hood' – is a safety measure to capture and remove airborne hazards. They incorporate powerful, high-capacity exhaust fans that draw air through the front opening and discharge via a ventilation system. This maintains a high flow of air through the work area and means that any toxic vapours or other airborne hazards are carried away and vented outside into the atmosphere.

The front of the cabinet incorporates a glass sash that can be raised or lowered for access to the work area. Care should be taken not to raise the sash too high when using the fume cupboard as this will reduce the airflow to below the safe level. Fume cupboards often incorporate alarms to indicate when the sash is opened too far.

Fume cupboards often have an airflow gauge to show the face velocity. This is the rate of flow of air and there will be a recommended minimum face velocity, depending on the type of work being carried out – typically around 0.5 m/s (metres per second) with the sash opened to a working height of about 500 mm.

Some types of fume cupboard, known as **ductless**, are self-contained, meaning that they do not exhaust to the outside. Instead, they contain filters that remove any toxic or harmful substances and return the air to the lab. The filters used must be appropriate to the type of hazardous material being used, so ductless designs are less versatile.

▲ Figure 10.4 A fume cupboard; note that the sash is in the fully-raised position – this would not be recommended for safe working

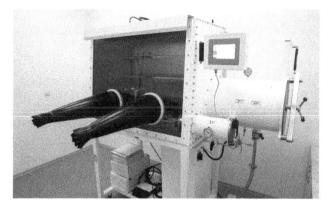

▲ Figure 10.5 A glove box with two gloves

Glassware

You will encounter many different types of glassware used to store, measure, transfer and collect reagents and samples. Some types of volumetric glassware have been discussed in section A9.8. Other types of glassware include:

▶ beakers and flasks
▶ test tubes made from soda-lime glass or heat-resistant boiling tubes made from borosilicate glass (see section B1.33)
▶ funnels, such as filter funnels, for use in filling larger containers
▶ distillation apparatus, such as condensers
▶ glass rods for stirring or spreading
▶ bottles for storing solvents and solutions.

The advantage of glass is that it is **inert** (unreactive) to most solvents and chemicals. Of course, the disadvantage is that it is fragile and easily broken. It can also be heavy. For this reason, many types of glassware have been replaced by plastic alternatives. However, there have been moves to reduce single-use plastics in laboratories more recently, meaning that reusable glass is becoming more common again.

Glove boxes

Do not confuse these with glove compartments in the dashboard of a car! A glove box provides a contained and controlled environment (sealed atmosphere) for manipulating samples, substances and objects. They can be used for handling highly toxic samples that are too dangerous to use in a fume cupboard. They usually incorporate an airlock to create a sealed atmosphere. This is particularly useful when handling highly reactive substances or ones that would be damaged by air, because the box can be filled with an inert gas such as nitrogen or argon. They can also be used to handle substances that are highly sensitive to moisture, as the atmosphere in the box can be made very dry.

Incubators

Incubators are used in hospitals to help keep very premature babies warm and in controlled conditions. In the science sector, incubators are used to provide a controlled and accurately maintained environment (for example, temperature and/or humidity) for certain experiments or lab work. They are widely used in:

▶ microbiology, for growing bacterial cultures in liquid medium or on agar plates
▶ cell culture, for growing human or animal cells or embryos in Petri dishes or culture bottles.

Incubators used for culture of mammalian cells usually maintain a relative humidity >80% to prevent evaporation and a carbon dioxide level of 5% (compared with 0.4% in the atmosphere) to maintain a slightly acidic pH.

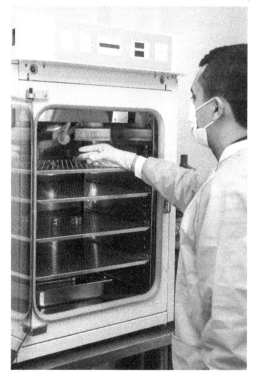

▲ Figure 10.6 Incubator being used for cell culture

157

Microbiological equipment

We will cover the various types of microbiological equipment that you are likely to encounter in section A10.8 when we look at how to follow aseptic techniques.

For now, we will just look at one important piece of equipment used in microbiology – the **biological safety cabinet** (**BSC**), shown in Figure 10.7. These are essential in any laboratory that handles **biohazardous** material. Their main function is to protect the worker, the laboratory environment and materials from exposure to aerosols and splashes that may contain infectious agents. Aerosols may be generated by many processes, such as shaking, pouring, stirring or dropping liquid on to a surface or into another liquid. This means that various laboratory activities may generate potentially infectious aerosols without the worker being aware of it, hence the usefulness of BSCs.

▲ Figure 10.7 A biological safety cabinet (BSC); although it looks similar to a fume cupboard, there are a number of features that make it particularly suitable for use with biohazards

BSCs, when used properly, are highly effective in reducing laboratory-acquired infections and cross-contamination of cultures. They also protect the environment. Different categories of BSC are available depending on the level of protection required.

The location of a BSC in the laboratory is important because air currents generated by people walking past, open windows or opening and shutting of doors can disrupt their operation. Therefore, a BSC should be located as far away from any of these as is practicable.

It is important to use a BSC properly. Movement into and out of cabinets can disrupt airflow, so should be kept to a minimum. Materials placed inside the cabinet should be disinfected with 70% alcohol (ethanol) solution. Work can be carried out on paper towels soaked in disinfectant to absorb splashes. Materials and aerosol-generating equipment (mixers, centrifuges and so on) should be placed as far back in the cabinet as possible. Biohazard bags (for waste) and discarded pipette trays should be placed inside the cabinet to one side – if they are placed outside the BSC, the operator will have to keep moving their arms in and out of the cabinet, which disrupts the airflow.

Another type of cabinet you might come across is the **laminar flow cabinet**, used mostly in cell culture. They look very similar to BSCs but they are designed to protect the cultures being handled from the user or the environment. For this reason, they incorporate a flow of filtered air from the back of the cabinet towards the front, thereby protecting cultures from contamination.

Multimeter

A multimeter is a meter than can measure voltage, current and, therefore, resistance in a circuit. Most multimeters have a digital display rather than the more old-fashioned analogue ones (where a needle sweeps across a scale).

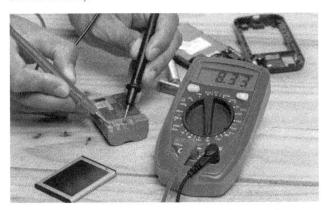

▲ Figure 10.8 A multimeter being used to test the voltage of a rechargeable battery

If you look at a multimeter you will see that it has red and black cables attached to metal probes. These allow measurement of voltage or current across terminals (for example of a battery, as in Figure 10.8) or in circuits. The dial allows you to select a range of different options:

▶ AC or DC (see section B1.48)
▶ current, voltage or resistance.

You will see that the probes on a multimeter are insulated where you hold them, but the pointed metal part can be a hazard. These allow you to get access to places that might be relatively inaccessible, but there is also a risk that you could cause a short circuit by touching both positive and negative (with a DC current) or live and neutral (with AC). It is also critically important that you take care to keep your fingers away from the metal part of the probes, particularly when using the meter with mains voltage, as the shock could kill you.

pH meters

A pH meter, as the name suggests, measures pH (how acidic or alkaline a substance is).

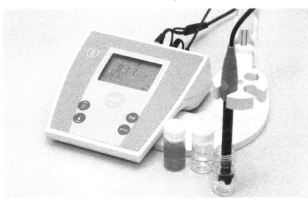

▲ Figure 10.9 A pH meter

A pH meter consists of an electrode attached by cable to an electronic meter. The electrode usually has a fragile glass bulb at the end; this is the part that is inserted into the solution you are measuring, but you must be careful not to break it. The electrode must not be allowed to dry out, so it is kept in a special storage solution (usually 3M potassium chloride solution, but definitely not water).

When using the meter, it is important to calibrate it regularly using buffer solutions of known pH, usually pH 4, 7 and 10. It is also important to rinse the electrode with purified water between measurements to avoid cross-contamination between the substances being tested.

Buffer solutions are used widely in biology and other areas to keep pH relatively constant. As well as using a buffer solution, you may well need to monitor the pH during an experiment using a pH meter.

Portable pH meters are also available. These may not be as accurate as a benchtop model, but they have

the advantage that they can be used in the field (in ecology studies, for example) or in non-laboratory environments, such as a production line.

Refrigerators and freezers

Laboratory refrigerators and freezers provide a controlled and accurately maintained temperature, typically set at:

▶ +4 °C (refrigerator)
▶ −20 °C (freezer)
▶ −70 °C to −80 °C (ultracold freezer).

Ultracold freezers can be an alternative to cryogenic storage if temperatures below −80 °C are not necessary. They are expensive to purchase but can be left unattended and can be fitted with high-temperature alarms in case of malfunction.

Practice points

Remember that when you take anything out of a refrigerator or freezer, condensation will form on it. If you have a container of powder in a fridge or freezer, you must allow it to reach room temperature before opening, otherwise condensation will form on the contents. Some materials will be moisture sensitive, so this could cause them to **degrade** (break down). Also, if you have to weigh out some of the contents, the condensation will make the mass measurement inaccurate.

Another point to bear in mind is that you should open and close the freezer door promptly to avoid moisture getting in and causing ice build up and to maintain the temperature. Remember that lab freezers are likely to be opened and closed many times in a day.

Avoid freeze–thaw cycles

Many biological products (enzymes, antibodies and so on) are kept in solution at −20°C. However, these can be damaged by repeated freeze–thaw cycles, such as if you have a stock solution that you keep using and putting in and taking out of the freezer. To avoid this, it is good practice to **aliquot** the solution. This means to split it into single-use quantities (aliquots) and freeze the aliquots. This way, you only have a single freeze–thaw for each aliquot, ensuring that it will be in good condition.

Scientific balances

We often talk about weighing or weighing out a sample. Strictly, we are determining the mass, rather than weighing, and that is what a scientific balance does: accurately determines the mass of a sample, from very large to very small samples.

A **top pan balance** will typically measure masses up to 500 g and above. Those with a lower capacity, such as the one in Figure 10.10, will measure mass to two decimal places, meaning that they have an uncertainty of ±0.005 g, although this might vary over the range – you need to check the individual balance (it will often be marked on the balance).

▲ Figure 10.10 A top pan two-decimal-place balance

Because of their open construction, top pan balances can be affected by draughts. Analytical balances (Figure 10.11), which are designed for small samples, avoid this by having an enclosure with doors (a draught shield) and are typically four-decimal-place balances – this means that they have an uncertainty of ±0.00005 g.

▲ Figure 10.11 An analytical four-decimal-place balance

Analytical balances can be affected by vibrations, so they are often placed on a concrete slab or specialist anti-vibration tables.

Thermometer

Thermometers monitor temperature or temperature changes. They can be simple, such as the mercury-in-glass thermometer invented by Gabriel Fahrenheit (after whom the Fahrenheit scale is named), or more complex, such as electronic or infrared thermometers. Liquid-in-glass thermometers (such as mercury or alcohol thermometers) rely on the fact that the liquid expands with increasing temperature, pushing it up a narrow capillary tube that has a scale marked on it.

Simple liquid-in-glass thermometers are used for monitoring temperatures in laboratory-scale chemical processes, such as distillation, or for measuring temperature changes in thermodynamics experiments, such as combustion or solution.

Electronic thermometers used in laboratories usually incorporate a **thermocouple** or a **thermistor** attached to an electronic circuit. In these, changes in temperature induce either a voltage (thermocouple) or change in resistance (thermistor) that is measured electronically by the circuit and the temperature calculated.

Electronic thermometers are used in a wide range of applications; this is just a very small selection:
▶ indicating the temperature of a freezer, incubator or water bath
▶ incorporated into environmental monitoring equipment, such as in remote weather stations
▶ wherever continual monitoring of temperature is required – the electronic thermometer can be attached to a data logger.

Infrared thermometers measure the infrared radiation emitted and convert this to a temperature reading. They are mostly used for remote sensing of temperature and so are widely used in restaurants, venues and airports to quickly check people's temperature on entry. In science and industry, infrared thermometers are used to measure the temperature of components that cannot be accessed easily or where they are too hot to touch with any other type of thermometer. They are also used in food processing and testing, where achieving a particular temperature is critical to the safety of the food (to kill pathogens) without risk of contamination by a normal thermometer.

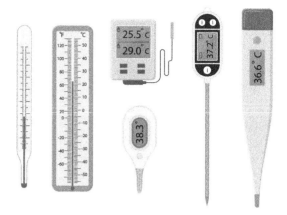

▲ Figure 10.12 A selection of different thermometers

Thermometers can be calibrated using two fixed points, usually the boiling point and freezing point of pure water. A mixture of ice and water left for a few minutes will have a temperature of 0 °C while boiling water will have a temperature of 100 °C at standard pressure.

Test yourself

1 Why you would use the following pieces of equipment when undertaking scientific techniques?
   a An autoclave.
   b Cryogenic equipment.
2 Name the type of equipment you would use for the following:
   a Growing cells in culture.
   b Carrying out a chemical reaction that produces a corrosive gas.
   c Dispensing 0.01cm³ of antibody solution into each well of a 96-well test plate.
   d Packaging a highly toxic liquid into sealed vials.
3 What precautions should you take when using a fume cupboard?
4 What type of glassware would you use for the following?
   a Dispensing 25cm³ of liquid into a flask for use in titration.
   b Storing a buffer solution or solvent.
   c Isolating a precipitate.
5 Chart recorders were used widely to record the output from a piece of electronic equipment. The output would be recorded as a trace on paper. Give **two** advantages of using a data logger instead of a chart recorder for this purpose.

A10.4 The appropriate techniques for handling a range of different substances

Anything that you encounter in the laboratory – or, more widely, in any scientific organisation – must be treated with respect. Even if it is not something that could harm you, there is always the possibility that you could harm it, meaning damage or contaminate it.

For that reason, it is important that you are familiar with the appropriate techniques for handling a range of solids, liquids and gases.

Referring to material safety data sheets

Material safety data sheets (MSDSs) were covered in section A4.18. You will recall from that section that an MSDS should provide all the information that you need to be aware of any hazards associated with handling the substance as well as the precautions that should be taken when doing so.

For example, corrosive substances require particular precautions:

▶ Always use them in a well-ventilated area so that corrosive vapours, fumes, mists or aerosols are removed. For some volatile (low boiling point) corrosives, this means you must only use them in a fume cupboard.
▶ Always ensure that containers are undamaged.
▶ Store them in specialised cabinets that resist attack and incorporate spill trays.
▶ Do not store acids and bases next to each other as they can react with each other, sometimes violently.
▶ Take care when transporting containers of corrosives, following appropriate manual handling recommendations.
▶ Take care when dispensing corrosives to avoid spillages.
▶ Some corrosives generate a lot of heat when mixed with water; if you need to mix a corrosive substance with water, always add the corrosives to cold water in small amounts and with frequent stirring.

Practice points

Remember, one important rule to follow when reducing the risk associated with any procedure is to always try to find a less hazardous alternative.

Using personal protective equipment

The Health & Safety Executive's *5 Steps to Risk Assessment* makes clear that PPE should be used only once all other ways of removing or reducing risk have been implemented (see section A3.2).

Having said that, it is unlikely that the risk in any laboratory will be reduced to zero, so you will always have to wear some form of PPE, even if it is just a lab coat.

Phenol is a good example of a substance where you certainly need to use appropriate PPE when handling it. Phenol is a white crystalline solid with a low melting point (40.5 °C), which means that it is volatile, so care must be taken not to breathe any vapour. Although phenol is classed as a weak acid, it is highly corrosive to eyes, skin and the respiratory tract because of its protein-degrading activity, causing chemical burns.

Phenol also has an anaesthetic action, so if it comes into contact with your skin it can cause severe chemical burns before you notice. For this reason, it is essential to double glove (that is, wear two pairs) with **nitrile** (chemically resistant) gloves even for brief contact. Make sure the gloves are intact and undamaged before starting work. For extended contact you should wear thicker gloves made of **neoprene** or **butyl rubber**.

> ### Practice point
>
> Remember that PPE means **personal** protective equipment. Your PPE requirements may be different to those of a colleague. For example, if you have allergies, you may require a face mask when handling certain substances.
>
> Those with accessibility requirements might also have their own particular PPE needs. For example, if you use a walking stick you may be required to place an additional glove over the handle when you are in the lab.

Using equipment for safe handling

Besides the usual manual handling rules that you should always observe (see section A3.3), there are some specific precautions that you need to take.

Alkali metals (lithium, sodium and potassium are the ones you are most likely to encounter) are highly reactive, particularly with water. That is why they are always stored in mineral oil and handled in a fume cupboard. Small pieces can be removed using plastic tongs or forceps.

Here are some examples of equipment that you might need to use to ensure safe handling in the laboratory:

► Pipette fillers (manual or electrical) to avoid mouth pipetting (see A4.7).
► Specially designed openers for glass ampoules or serum vials. Alternatively, if an opener is not available, you should hold the ampoule in a wad of tissues to prevent injury from broken glass.
► Needles should be avoided, where possible, but if you do use them then they should be removed using forceps.
► Petri dishes should be kept in racks or baskets, rather than stacking them in unsupported piles.
► Use sharps containers for disposal of syringes, needles and disposable lancets.

There are also specific requirements for using equipment to ensure safe handling when dealing with biohazards. There is more information about this in section A10.8.

Applying containment controls

Containment controls, such as fume cupboards, should be used with any substance or reaction that could produce hazardous (toxic, harmful or corrosive) fumes or gases.

Chlorine is a toxic and corrosive gas that can be obtained as compressed gas in cylinders. Smaller quantities can be produced by adding concentrated hydrochloric acid to sodium chlorate(I) (also called sodium hypochlorite, the main ingredient of bleach). This and other similar methods should always be carried out in a fume cupboard.

More hazardous substances may require the greater containment that a glove box provides (see section A10.3).

Biohazards, such as pathogens, human tissues or samples (including ones that might contain prions – see section B1.25), and even animals and plants require specific containment, such as the use of Biological Safety Cabinets (see Microbiological Equipment in section A10.3). Other types of containment measures include the use of pressurised clean rooms (see section A4.10).

Procedures for dealing with compressed gases

As we have just seen, some compressed gases may have specific hazards, such as chlorine which is highly corrosive. Others, such as propane or butane (also known by brand names Calor Gas or Campingaz), are highly flammable.

However, irrespective of their contents, cylinders of compressed gas must be stored at the correct temperature (generally below 52 °C) to prevent excessive pressure build-up. See section B2.34 for more on this subject.

Another potential hazard with compressed gases is the cylinder itself. Large gas cylinders are heavy and must be chained or strapped to a suitable support (such as a wall) to prevent them toppling over. They must be transported using special trollies that also provide support. Another potential hazard is the regulator (valve) attached to the cylinder as this can be easily knocked, creating a risk of the contents escaping, possibly violently.

Test yourself

1 You are using a new chemical for the first time and will carry out a risk assessment first. What document do you consult to get the information you need for the risk assessment?
2 Explain why there can be a greater risk when using phenol compared with other chemicals.
3 What precautions should you observe when storing cylinders of compressed gas?
4 Why should chlorine gas only be used inside a fume cupboard?

A10.5 Appropriate equipment to measure accurate results at different scales

Before reading this section, it would be helpful to read section B1.63 (page 238) on how to convert between units and section B1.64 (page 239) on significant figures.

Kilo

Kilo means one thousand (1000), hence a kilogram is 1000 g or 10^3 g. You could measure on that scale accurately using a top pan balance (see section A10.3)

but bear in mind that some top pan balances may measure a maximum mass of only 500 g or less.

When using a balance of this type, you might find that above a certain mass the display changes from showing two decimal places to one, or even no decimal places.

If you have a mass of 1kg and you place it on a top pan balance and see a reading of 1000.05g, can you be confident in that reading? How many significant figures is this? Is that an appropriate number of significant figures? Does it mean that the mass is not, in fact, 1kg?

Milli

Milli means one thousandth (0.001), so 1 mg is 0.001 g (10^{-3} g). This is the sort of range of mass that an analytical balance is designed for. You will have noticed that analytical balances incorporate a shield with a door so that draughts do not upset the accuracy of the balance. The fact that the display shows four decimal places means that you should be confident that a mass of 5.5 mg is actually 0.0055 ±0.0005 g.

Some analytical balances can be used for masses smaller than 1 mg.

Micro

Micro means one millionth (0.000001), so 1 microgram (µg) is 0.000001 g (10^{-6} g) and 1 micrometre (µm) is 0.000001 m (10^{-6} m). Although some analytical balances can be used in the range of a few tens or hundreds of µg, you are more likely to encounter micro units when measuring very short lengths or distances using a **micrometer**. A **micrometer**, sometimes known as a **micrometer screw gauge**, is a measuring gauge that allows measurements to within 0.01 mm (10 µm) or, in the case of a vernier micrometer, to within 0.001 mm (1 µm). It incorporates a moveable spindle (a very accurately machined calibrated screw) and a fixed anvil mounted on a frame. The object to be measured is placed between the anvil and spindle and the screw is turned by the ratchet knob or thimble until the object is lightly touched by both the anvil and the spindle. The measurement can be read off the graduations on the spindle and the thimble, or directly from the digital display.

Micrometers can be mechanical (Figure 10.13a) or digital (Figure 10.13b).

(a)

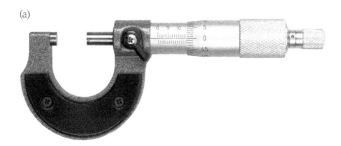

(b)

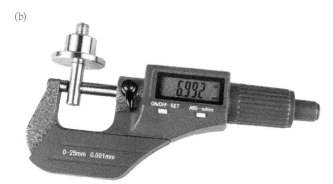

▲ Figure 10.13 Two micrometers: (a) mechanical and (b) digital

Practice point

Be careful about spelling.

A **meter** is a measuring instrument, so we have micrometers, pH meters and multimeters and we will encounter ammeters and voltmeters in section B1.52.

A **metre** is a unit of distance, so we have micrometres (µm), kilometres (km), millimetres (mm), etc.

Nano

Nano means one billionth (0.000000001) so 1 nanogram (ng) is $0.000000001\,g$ ($10^{-9}\,g$). However, it is more likely you will encounter nanometres (nm) in the context of light wavelength (see section B1.54) than units of mass. For example, red light has a wavelength of 625–700 nm, while blue light has a wavelength of 450–85 nm.

One piece of equipment that works at this level of accuracy is an atomic clock. These measure the electromagnetic signals that electrons in atoms emit when they change energy levels. Atomic clocks are the most accurate time (and frequency) standards we currently have. The national standards agencies in many countries maintain a network of atomic clocks that are kept synchronised to an accuracy of $10^{-9}\,s$ (1 nanosecond) per day.

A10.6 How to use a light microscope

One thing to bear in mind about light microscopes is that they can only be used with very thin samples, because light does not penetrate very far in biological samples. We can prepare a thin specimen in different ways so that we get just a single layer of cells on the glass microscope slide:

- by squashing the tissue – this is often done with root tips so that we can observe cell division; a lot of cell division occurs at the tip of roots and shoots, so this is a convenient tissue to use
- by spreading a sample – for example, you spread a blood sample across a slide to create a **blood smear**, where all the cells are visible. This method can be used with bacteria, food samples or other materials.
- by cutting a very thin **section** – this can be done with many plant and animal tissues.
- growing cells on the slide – this includes examining cell cultures to monitor cell growth or cell proliferation. There are specialist microscopes adapted for this purpose.

See section B1.21 for more on light microscopes.

Practice point

There is quite a wide range of light microscopes that you might encounter in your career. These include basic types that you will find in a school or college laboratory, more specialised types used to examine cell cultures, or phase contrast microscopes that can be used to examine living cells without the need for staining.

The steps involved in using light microscopes are, in general:
- Prepare a slide with a thin specimen and stain, if necessary.
- Place the prepared slide on the microscope stage.
- Select the lowest power objective lens and focus on the slide.
- Change to a higher power objective lens, if required, and re-focus.
- Measure or count objects in the field of view.

Preparing slides using different staining techniques

Staining is used in microscopy to increase contrast and make it easier to see cells (see section B1.23, page 193). Unstained cells are almost transparent, making it very difficult to see them.

Staining can also identify particular structures, such as the nuclei (**haematoxylin**), or chemicals, such as starch grains (**iodine**). Some stains can be used to distinguish between live and dead cells, or to identify cells undergoing apoptosis (programmed cell death). These are examples of **simple** stains.

Haematoxylin and eosin (H&E) stain is one of the most widely used stains in histology (the study of cells and tissues under the microscope). For example, a pathology laboratory examining a biopsy sample from a suspected cancer is likely to make a tissue section and stain it with H&E. Haematoxylin incorporates a **mordant** – a metal ion that aids binding of the stain – and stains cell nuclei a purple–blue colour. Eosin stains cytoplasm and extracellular material pink.

H&E is an example of a **differential stain**, where two different stains are used to distinguish different parts of cells or different types of cells. Two other types of differential staining methods are described in section B1.23: **Gram stain**, which can distinguish between different types of bacteria; and **Giemsa stain**, which can distinguish between different types of cells, particularly in blood samples.

Individual cells, such as bacteria, cells in culture or blood cells, can be stained without further treatment, although it is usual to **fix** the cells to preserve them (for example, with formaldehyde).

Tissue samples, however, are too thick to be viewed under the microscope and must be **sectioned** (cut into very thin slices). The procedure involves fixing followed by processing to remove water and replace it with a medium, such as paraffin wax, that will support the tissue during sectioning. The whole tissue sample is then embedded in a support (such as agar or paraffin wax) so that it can be sectioned and mounted on a microscope slide. Once mounted, the section can then be stained.

Altering magnification and focus

Figure 10.14 shows a modern light microscope. The objective lenses have different magnifications and the total magnification of the image is calculated using the following formula:

total magnification = magnification of eyepiece lens × magnification of objective lens

So, if the eyepiece has a magnification of ×10 and the objective lens is ×5 (low power), the total magnification will be ×50 (50 times). If we change the objective lens to one that has a magnification of ×40 (high power), then the total magnification will be ×400.

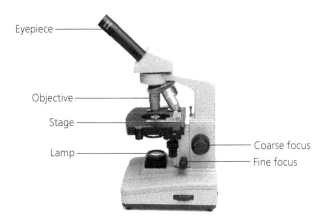

▲ Figure 10.14 The main features of a light microscope

The specimen on a microscope slide is placed on the **stage**, which can be raised or lowered to bring it into focus. This is done using the **coarse focus** knob, which moves the stage a relatively large amount, followed by the **fine focus** knob, which moves the stage a relatively small amount to allow for finer adjustment of the focus. It is important to take care when focusing to avoid crashing the objective lens into the slide. For this reason, you should only use the fine focus knob when using a high-power objective.

The slide is fixed on the stage, but you can move it around using the adjustment knobs to view different areas of the sample.

You might encounter microscopes with a slightly different arrangement to the one illustrated, but the principles should be the same.

Setting scale

You might need to measure the size of objects that you examine under the microscope. This is done using an eyepiece graticule, as described in section A9.8.

Cell counting

One important application of light microscopes is cell counting using a **haemocytometer**. The term means 'blood cell counter', because it was originally used in clinical laboratories to determine the blood cell numbers in a blood sample, but haemocytometers can be used for other types of cells, for example:

▶ the number of viable sperm cells in order to assess fertility
▶ the number of yeast cells, for use in brewing
▶ the number of cells in a culture, when monitoring cell growth or sub-culturing (starting new cultures from an existing culture).

The haemocytometer consists of a thick glass microscope slide with a rectangular indentation that creates a chamber. This chamber is engraved with very fine lines (modern haemocytometers use laser-etched lines) creating squares and rectangles of different areas, the largest being 1.0 × 1.0 mm and the smallest being 0.05 × 0.05 mm. The cover slip is 0.1 mm above the grid, so the volume of each square or rectangle can be calculated (Figure 10.15).

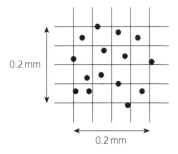

▲ Figure 10.15 Part of a haemocytometer showing cells being counted

When counting the cells, some will be only partially within the area being sampled. It is not practical to work with fractions of cells, so we must use a rule to decide which cells are counted. In Figure 10.15 there are 10 cells wholly within the 16 squares being sampled. There are two possible rules:

1 All cells touching or straddling lines on the left or top of the area are included. This gives a total of 13 cells.
2 All cells touching or straddling lines on the right or bottom of the area are included. This gives a total of 12 cells.

Notice that rules 1 and 2 give slightly different results. Which rule we use does not matter, as long as we use one consistently.

We then repeat the count in several different areas of the slide and obtain a mean. This will average out any differences between the two rules, as long as we

use the same rule each time. The mean is then used to calculate the number of cells per unit volume and then multiplied by any relevant dilution factor to obtain the number of cells in the original sample.

A10.7 The reasons for using aseptic techniques

In section A10.8 we are going to look at the details of **aseptic** technique in the context of a microbiology laboratory. Aseptic conditions are required in many other situations, however, and we will consider the reasons why.

Key term

Aseptic: free from contamination with micro-organisms that might be pathogenic or cause food spoilage.

To avoid contamination of products

Packaging of pharmaceuticals is normally done under aseptic conditions, to prevent contamination with pathogens, following GMP procedures (see section A2.8). **Aseptic processing** involves sterilisation of the product followed by packaging into previously sterilised containers. This is particularly useful for heat-sensitive products such as vaccines, hormones or other proteins because the liquid can be sterilised by microfiltration using previously sterilised membrane filters. These have pores small enough to remove micro-organisms; 0.22 μm is generally considered sufficient to remove bacteria, although nanofilters (pore size 20–50 nm) are required to remove viruses.

The sterile filtered liquid can then be packed, under sterile conditions, into pre-sterilised vials or other type of container. Glass vials can be sterilised by autoclaving (see section A10.3) while plastic or other heat-sensitive containers can be sterilised by irradiation or with chemicals such as ethylene oxide gas.

A similar approach is taken in food production. Aseptic processing allows the food to be properly sterilised in bulk outside of its final container. The sterilised food is then packaged into previously sterilised containers, which are then sealed in a sterile environment. This process is more efficient than sterilising the food after it is packaged. In this way, food spoilage micro-organisms (that make the food go off) as well as pathogens can be destroyed.

To avoid transmission of disease

Contaminated food can cause transmission of disease (food poisoning) when we eat it. Here we are concerned with the way in which infected samples can transfer disease (to humans or animals) while handling those samples in the laboratory.

Following proper aseptic techniques reduces or eliminates the risk of infection from samples that could contain pathogens. When handling any culture of micro-organisms or biological samples, it is always good practice to assume that they could be contaminated with pathogens and handle them as described in the next section.

A10.8 How to follow aseptic techniques

This section deals with aseptic techniques that you would follow in a laboratory dealing with bacterial cultures and other micro-organisms, such as yeasts. Techniques used in cell culture are similar. Aseptic techniques used in a production setting will be different and follow specific SOPs according to the requirements of GMP.

Aseptic techniques are designed largely to prevent contamination of samples and cultures. Contamination can invalidate any results and also risks the introduction of potentially hazardous pathogens. Contamination can arise from the person conducting the procedure, from airborne micro-organisms, unsterilised equipment and laboratory surfaces.

Following proper aseptic technique will also protect you and the environment from contamination or infection with potentially pathogenic micro-organisms.

Flaming equipment

Flaming uses heat to sterilise equipment, meaning the use of a Bunsen burner with the valve open to produce a blue flame. Wire loops are used to transfer micro-organisms from culture to culture or to spread them on agar plates. The loop should be held in the hottest part of a Bunsen burner flame (the top of the blue part of the flame) until it glows red hot. Allow the loop to cool and then use immediately, so that it does not have a chance to become contaminated. Single-use pre-sterilised loops and spreaders (usually made of plastic) are also available.

The necks of bottles and test tubes should be sterilised by passing the neck of the bottle or tube forwards and then backwards through the Bunsen burner flame.

Transfer cultures/samples as quickly as possible with minimal exposure to the air

When transferring cultures, work quickly to reduce the risk of contamination. Bottles or Petri dishes containing bacterial cultures should be opened for as short a time as possible to reduce the risk of bacteria getting in or out.

It is also important not to put any tops or caps down on the work surface as this could cause contamination. Instead, you should follow these steps:

1 Loosen the cap of the bottle or tube before you start to make sure it is easy to remove.
2 Hold the bottle or tube towards the bottom in your non-dominant hand.
3 Wrap the little finger of your dominant hand around the cap and remove it by turning the bottle or tube, not the cap. The other fingers on your dominant hand will be holding the loop that you will have just sterilised by flaming.
4 Do not put the cap or lid down.
5 Flame the neck, as described above, and transfer the sample using the loop.
6 Flame the neck again.
7 Replace the cap using the little finger – again, turn the bottle or tube, not the cap.

Holding bottles and tubes at an angle to prevent contamination

When flaming the neck of a bottle or tube, hold the container as close to horizontal as possible (without spilling any contents) to reduce the likelihood of airborne micro-organisms entering.

When you open a Petri dish you should hold the lid at an angle above the dish and make any transfers as quickly as possible to reduce the risk of contamination.

Sterilising tools

Any tools that you might need, such as spreaders, forceps, pipettes and so on, should be sterilised before use. The method of sterilisation will depend on whether the things being sterilised are heat stable. If they are, then autoclaving (see section A10.3) is the preferred method. Otherwise, items can be sterilised by:

▶ radiation (see section B1.61)
▶ chemical sterilisation using agents such as ethylene oxide gas, perchlorate, peroxide, 70% ethanol or another substance that is compatible with the tools being sterilised.

It is also important to clean and sterilise any non-disposable equipment after use. Any disposable apparatus, samples and cultures should be sealed and sterilised by autoclaving before disposal.

Working in a sterile air environment

Maintaining the sterility of the air is important in many situations. In some cases, a whole laboratory or workroom might have to be kept sterile. In other cases, we need a smaller work area with sterile air. This can be achieved by working in a downflow cupboard or BSC (Figure 10.7), close to a blue flame Bunsen burner. Movement into and out of cabinets can disrupt airflow, so should be kept to a minimum. There is more detail about the use of BSCs in section A10.3 (Microbiological equipment).

Refraining from contaminating any sterile objects by placing them on non-sterile surfaces

Do not place any sterile object on a non-sterile surface. This may mean you need to be very dextrous, for example when flaming the neck of bottles and tubes, as described above. Work can be carried out on paper towels soaked in disinfectant to absorb splashes.

Not consuming food or drink

You should not be doing this in any laboratory. It is highly likely to be dangerous:

▶ Your food or drink could become contaminated with harmful substances or pathogens.

▶ The process of chewing or swallowing will increase the risk of you absorbing airborne contaminants or pathogens.

▶ Your food or drink could contaminate whatever you are working with.

For these reasons, eating or drinking in a laboratory or similar work area will always be prohibited.

Employment law (see Chapter A3) states that you should always be provided with a separate break room or other area where you can eat or drink.

Following correct handwashing techniques

Washing your hands before starting to work is an important part of avoiding contamination of your work area, even if you are going to wear gloves. It is also important that you wash your hands on leaving the laboratory or other work area, even if you have been wearing gloves – so you need to wash your hands before, as well as after, using the toilet!

Since the emergence of COVID-19, we are probably all much more familiar with correct handwashing. Nonetheless, it is good to have a reminder: see Figure 10.16.

Note that, wherever possible, we should use taps that can be turned on and off with the elbows, so that our newly washed hands are not contaminated by touching the tap.

HOW TO WASH YOUR HANDS?

1. Wet hands with clean, running water

2. Apply enough hand wash/soap to cover your hands.

3. Rub hands palm to palm

4. Lather between fingers

5. Scrub between your fingers

6. Rub the backs of fingers on the opposing palms

7. Clean thumbs

8. Wash fingernails and fingertips

9. Rub each wrist with opposite hand

10. Rinse hands and wrists under clean, running water

11. Dry with a single use towel

12. Your hands are clean

HOW TO USE HAND SANITIZER?

1. Apply a palmful of the hand sanitizer in a cupped hand and cover all surfaces.

2. Rub hands palm to palm

3. Palm to palm with fingers interlaced

4. Right palm over left dorsum with interlaced fingers and vice versa

5. Rub the backs of fingers on the opposing palms

6. Rotational rubbing of left thumb clasped in right palm and vice versa

7. Rub fingernails and fingertips

8. Rub each wrist with opposite hand

9. Your hands are clean

▲ Figure 10.16 Handwashing poster showing technique to use

Donning and doffing suitable clothing and PPE

Donning (putting on) and doffing (taking off) suitable clothing and PPE should be done away from the work area. When dealing with biohazards at Category 2–4 (see section A4.6) it is likely that special containment measures (section A4.8) or even pressurised clean rooms (section A4.10) will be required. In such cases, there will be an area for putting on and taking off protective clothing and PPE before entering and after leaving the clean room.

The order in which clothing and/or PPE is donned and doffed is important. This can vary depending on the nature of the hazard and the type of PPE, however, the following is a typical procedure.

Donning

▶ Perform hand hygiene (handwashing).
▶ Put on shoe covers (if applicable).
▶ Put on gown or labcoat.
▶ Put on mask/respirator (if applicable).
▶ Put on eye protection (if applicable).
▶ Put on gloves.

Doffing

This is not simply the reverse of donning – there are some important differences.
▶ Remove shoe covers (if applicable).
▶ Remove gown or labcoat and gloves together. If gloves are removed first, take care not to touch potentially contaminated surface of the gown or labcoat.
▶ Perform hand hygiene.
▶ Remove eye protection (if applicable).
▶ Remove mask/respirator (if applicable).
▶ Perform hand hygiene again.

Preparing surfaces and equipment

All surfaces should be wiped down with disinfectant before starting work. You should also prepare everything beforehand and make sure everything you need is within reach. At the same time, it is good practice to only have the equipment you need in the work area – anything else is likely to cause clutter and increase the risk of contamination.

Minimising human traffic in the area

There should never be unauthorised access to microbiology or biohazard laboratories. In addition, movement of people can create draughts and disturbances as well as increasing the risk of coming into contact with harmful pathogens. For this reason, there should be as little movement of people through the area as possible.

Practice point

Working in a BSC is essential when handling potential pathogens or cultures that could become contaminated, but you have to be careful to use them properly. The location of a BSC in the laboratory is important because air currents generated by people walking past, open windows or opening and shutting of doors can disrupt their operation. This is why a BSC should be located as far away from any of these as is practicable.

Reducing draughts by closing windows/doors

Windows and doors should be closed as much as possible to prevent draughts as these can introduce airborne contamination from outside the laboratory. Air currents from open windows or doors can also affect the operation of a BSC.

Test yourself

1 Why is it important to follow aseptic techniques when handling micro-organisms?
2 What is the purpose of flaming equipment such as tubes and bottles?
3 Why should caps and tops **not** be placed on the work area?
4 Why is it important to exclude draughts when working in a BSC?

Project practice

You have been asked to contribute to an information resource for new starters in your workplace. Your contribution will cover scientific equipment and techniques.

Select five pieces of equipment that are used in your workplace. For each one:
- ▶ Take a photograph (if your workplace permits this) or find an image on the internet.
- ▶ Describe what the equipment is used for and how it is used – you could do this by annotating the image.
- ▶ Describe the precautions that should be taken to reduce risk to the user and/or the environment when using the equipment.
- ▶ Describe the possible sources of error when using the equipment and how these can be reduced or eliminated.
- ▶ Describe the routine checks, calibration or maintenance that are required.

You may need to carry out your own research, either by asking colleagues or by an internet search.

Present your work as a poster or as an illustrated document produced using a word processor or slide presentation software.

If appropriate, prepare a short video that could be used to demonstrate the use of the equipment to a colleague.

Discuss your work with the class.

Write a reflective evaluation of your work.

Assessment practice

1 You have been asked to prepare a buffer solution of pH 4.2. To do this, you have calculated the amounts of ethanoic acid (liquid) and sodium ethanoate (solid) that you need. You measure the amounts of each, mix them and make the volume up to 500 cm^3 in a volumetric flask with distilled water. You then check the pH of the buffer with a pH meter and find that it is actually pH 4.0 and not pH 4.2. Describe the potential sources of error **and** what you could do to identify the source.

2 You are investigating the antibacterial effect of different household cleaners against a specific type of bacterium. You have an agar plate that has been inoculated with a culture of the bacteria. You place disks of filter paper soaked in solutions of the different cleaners on the surface of the agar. After incubating the plate you measure the area around the disks where bacteria have been killed by the solutions. In this experiment, what is the function of the following:
   a a disk of filter paper soaked in a solution known to kill the bacteria
   b a disk of filter paper soaked in distilled water?

3 Discuss the advantages and disadvantages of storing samples in either an ultra-cold freezer or in cryogenic storage with liquid nitrogen.

4 What precautions should you take when using a multimeter?

5 Explain the advantages of using an analytical balance rather than a top pan balance for weighing quantities less than 1 g.

6 Suggest the type of staining method used in the following:
   a A tissue section showing cells with the nuclei stained purple and the cytoplasm stained pink.
   b A slide of a bacterial culture showing two shapes of bacterial cells that have been stained different colours.
   c A section of plant tissue stained red to make the cells more visible.

7 You have been asked to produce samples for a small-scale trial of a new vaccine. Explain what information you would need before deciding on the best aseptic production method.

8 What different precautions would you need to take if you have to package samples of a toxic anti-cancer drug?

B1.1–B1.32: Core science concepts: Biology

Introduction

Biology is the study of living organisms, which makes it an enormous subject! In this chapter we will cover some important basics, such as the structure of cells and the way in which they are organised. We cannot really understand how organisms work without understanding cells, and we cannot understand how cells work without learning about the main types of biological molecules: proteins, carbohydrates and lipids. Exchange and transport mechanisms – the ways in which substances enter or leave – are essential for the working of individual cells and multicellular organisms. Genetics helps us understand how characteristics are inherited and introduces the fourth main type of biological molecules – the nucleic acids – as well as providing a basis for our understanding of evolution. Microbiology is not just the study of very small organisms; it helps us to understand infectious diseases. Finally, immunology helps to explain how our bodies protect themselves against infection.

Learning outcomes

The core knowledge outcomes that you must understand and learn:

Cells and tissues

B1.1 the 3 principles of cell theory

B1.2 the different types of cells that make up living organisms

B1.3 the structure and function of the organelles found within eukaryotic cells

B1.4 the similarities and differences between plant and animal cells in relation to the presence of specific organelles and their function:

B1.5 how eukaryotic cells become specialised in complex multi-cellular organisms

B1.6 how prokaryotic cells differ from eukaryotic cells

Proteins, carbohydrates and lipids

B1.7 the relationship between the structure, properties and functions of proteins

B1.8 the relationship between the structure, properties and functions of carbohydrates

B1.9 the relationship between the structure, properties and functions of lipids

Exchange and transport mechanisms

B1.10 how the surface area to volume ratio affects the process of exchange and gives rise to specialised systems

B1.11 the principles of cellular exchange and the transport mechanisms which exist to facilitate this exchange

B1.12 the advantages of having specialised cells in relation to the rate of transport across internal and external membranes

Genetics

B1.13 the purpose of deoxyribonucleic acid (DNA) and ribonucleic acid (RNA) as the carrying molecules of genetic information and the role they play in the mechanism of inheritance

B1.14 the relationship between the structure of DNA and RNA and their role in the mechanism of inheritance

B1.15 the function of complementary base pairing in forming the helical structure of DNA

B1.16 the process and stages of semi-conservative replication of DNA

B1.17 how this semi-conservative replication process ensures genetic continuity between generations of cells

B1.18 the link between the semi-conservative replication process and variation

B1.19 the difference between genetics and genomics

Microbiology

B1.20 the classification and characteristics (size of cell, type of cell, presence of organelles) of the following micro-organisms

B1.21 the benefits of using light and electron microscopes when investigating micro-organisms.

B1.22 how to calculate magnification from the size of the image and the size of the object

B1.23 the uses of differential staining techniques

Immunology

B1.24 the nature of infection

B1.25 pathogens (causative agents) of infection and examples of resulting diseases

B1.26 the different ways in which pathogens (causative agents) may enter the body

B1.27 how infectious diseases can spread among populations and communities

B1.28 the definition of an antigen and an antibody

Cells and tissues

We can study and understand biology at different levels of organisation. Starting with the whole **organism**, we can move upwards to study the ways in which organisms interact in populations and ecosystems. Alternatively, we can look at the way in which organisms work in increasing levels of detail. The cell is the basic unit of all organisms. We need to learn the structure and organisation of the cell to get a proper understanding of how cells work together and also understand the environment in which the chemical reactions of the cell take place.

Key terms

Organism: an individual plant, animal or single-celled lifeform.

Membrane: all membranes consist of a phospholipid bilayer together with proteins and other components. They are selectively permeable (meaning they let some things through and not others) and can control movement of substances across the membrane as well as being the sites of many important processes in the cell.

Phospholipid: a large molecule formed from a glycerol molecule covalently bound to two fatty acid molecules and a phosphate group. It has a **hydrophilic** (can interact with water) head group (because of the phosphate) and a **hydrophobic** (repels water) tail (because of the fatty acids).

Cytoplasm: is the fluid component of the cell, enclosed by the cell membrane and surrounding the organelles.

Organelles: specialised structures within plant and animal cells that have specific functions. Some types of organelle are also found within bacterial cells.

B1.1 The 3 principles of cell theory

Robert Hooke (1635–1703) was the first person to recognise cells, although the 'cells' in cork that he saw using his microscope were the empty spaces between the cell walls of the cork. Hooke laid the foundations for what we now know as the three principles of cell theory. This states that:

▶ All living things are made up of one or more cells. This means that living things can be **unicellular** (single cells) or **multicellular** (made up of more than one cell).

▶ Cells are the most basic unit of structure and function in all living things. Cells contain many components (nuclei, mitochondria, etc.) but these cannot exist or reproduce on their own.

▶ All cells are created by pre-existing cells, i.e. cells cannot just appear from nowhere. New cells are created from pre-existing cells in the process of **mitosis** (cell division).

B1.2 The different types of cells that make up living organisms

There are two types of cell: **prokaryotic** cells and **eukaryotic** cells. Eukaryotic cells are complex and include all animal and plant cells as well as yeasts, other fungi and algae. Prokaryotic cells are simpler and smaller and include the bacteria. Both types of cell have **membranes**, **cytoplasm** and DNA. However, eukaryotic cells have membrane-bound **organelles**, such as mitochondria or chloroplasts. Also, the DNA is contained within the nucleus. The DNA is bound to proteins known as histones and together they form a complex known as chromatin (see below). In prokaryotic cells, the DNA just floats freely in the cytoplasm, or is found as small circular molecules known as plasmids, and is not associated with proteins.

B1.3 The structure and function of the organelles found within eukaryotic cells

Plasma membrane

Also called the **cell surface membrane**, this is found around the outside of the cell and consists of a **phospholipid bilayer** together with proteins and other components. The **plasma membrane** controls entry and exit of substances into and out of the cell.

Key terms

Phospholipid bilayer: a double layer of phospholipids with the hydrophobic tails arranged towards the middle and the hydrophilic head groups on the outside. It forms the basis of all biological membranes.

Plasma membrane: sometimes called the cell-surface membrane, it is the membrane that surrounds all types of cell; animal, plant and bacterial. Like all membranes, the plasma membrane consists of a phospholipid bilayer together with proteins and other components.

Nucleus (containing chromosomes)

The nucleus is the largest organelle and is surrounded by the **nuclear envelope.** This is a double membrane that has many gaps or **pores**.

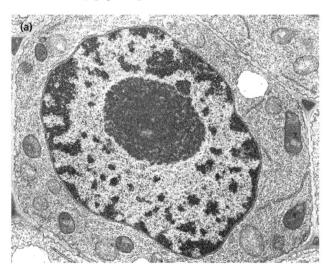

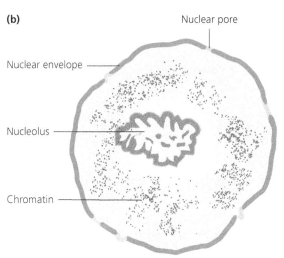

▲ Figure 11.1 A nucleus: (a) an electron micrograph (x 25 000) and (b) a diagram, showing the structure of the nucleus

The nucleus contains the genetic information, in the form of DNA. The DNA is combined with proteins known as histones; this forms the complex known as **chromatin**. The chromatin is coiled and super-coiled to form the chromosomes.

Mitochondria

Mitochondria (the singular is mitochondrion) are the site of **aerobic respiration** and therefore the site of adenosine triphosphate (ATP) production. Aerobic respiration is the process where glucose is reacted with oxygen to produce carbon dioxide and water. As this reaction is exothermic, the energy transferred from this reaction is used to produce ATP, the 'energy currency' of the cell. Almost all processes in the cell that require energy obtain it from ATP.

Like nuclei and chloroplasts, mitochondria are enclosed by a double membrane (envelope). The inner membrane is folded into structures called **cristae**.

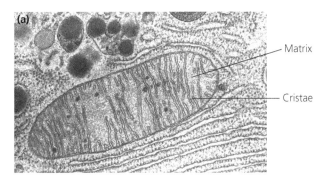

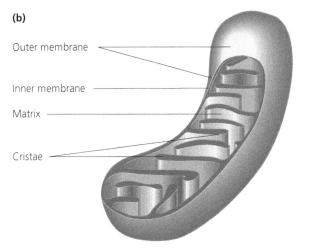

▲ Figure 11.2 A mitochondrion: (a) an electron micrograph (x 1100) and (b) a diagram

Ribosomes

Ribosomes are the smallest of the organelles and are the site of protein synthesis. Some float free in the cytoplasm and make the proteins needed within the cell, whereas others are attached to the rough endoplasmic reticulum. Ribosomes use the information coded in an mRNA molecule to assemble the correct order of amino acids in the protein.

Rough and smooth endoplasmic reticulum

The **endoplasmic reticulum (ER)** is a system of membrane-bound flattened sacs that fills a large part of the cytoplasm. The **rough ER (RER)** has ribosomes attached to its outer surface. Proteins that will be released from the cell or incorporated into the plasma membrane are made on these attached ribosomes and then folded and transported in the RER to the Golgi apparatus.

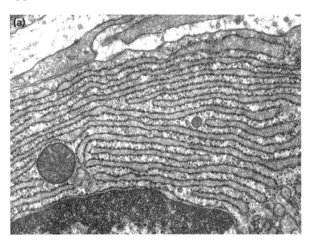

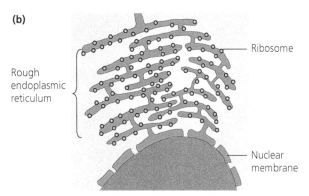

▲ Figure 11.3 Rough endoplasmic reticulum: (a) an electron micrograph (x18 000) and (b) a diagram

The **smooth ER** does not have attached ribosomes and is responsible for synthesising, storing and transporting lipids and some carbohydrates.

Golgi apparatus and Golgi vesicles

The Golgi apparatus is a stack of flattened sacs, known as cisternae (singular is cisterna). Each cisterna is surrounded by a single membrane and filled with fluid. The Golgi modifies proteins that have been transported from the RER, for example by adding carbohydrates to them. These modified proteins are then transported by **Golgi vesicles** that form when the ends of the cisternae are pinched off. These vesicles can form **lysosomes** (see below). Others, called **secretory vesicles**, carry their contents to the plasma membrane where they can be released to the outside of the cell.

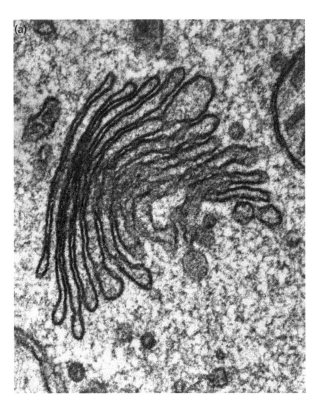

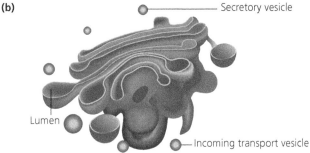

▲ Figure 11.4 The Golgi apparatus: (a) an electron micrograph (x50 000) and (b) a diagram showing the Golgi vesicles and secretory vesicles

Lysosomes

These are the cell's recycling facility. When proteins and other cell components get worn out, they are moved into lysosomes. Digestive enzymes break these down into their constituents, e.g. amino acids that can be re-used to make new proteins. It is important that these enzymes are kept separate from the rest of the cytoplasm because of the damage they could do. Lysosomes are also involved in digestion of invading **pathogens** (bacteria and viruses) that are taken into the cell by the process of **phagocytosis** (see page 198 later in this chapter).

Centrioles

Centrioles are structures made of a tubular protein called tubulin. They are involved in the formation of the spindle in **mitosis** as well as formation of **cilia** and **flagella**. They are not present in many types of plant cells.

Key terms

Pathogen: a micro-organism that causes illness or disease by damaging host tissues and/or by producing toxins.

Cilia: (singular **cilium**) are hair-like structures found on the plasma membrane of some types of cell, particularly in the lungs.

Flagella: (singular **flagellum**) are similar in structure to cilia but are much longer and are involved in propulsion of the cell.

Chloroplasts (in plants)

Like mitochondria, chloroplasts are enclosed by an **envelope** (double membrane) and contain membranes called **thylakoids** arranged in stacks called **grana** (singular is **granum**). The chloroplast is the site of photosynthesis, the process whereby plants and algae use light energy to make complex organic molecules from carbon dioxide and water. There are two stages to photosynthesis. The first stage occurs in the thylakoid membranes which contain chlorophyll and other pigments that absorb light energy as well as proteins involved in the production of ATP. The rest of the chloroplast consists of a fluid called the **stroma**, which is where the second stage of photosynthesis take place.

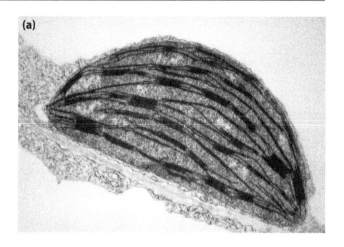
(a)

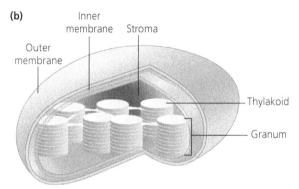
(b)

▲ Figure 11.5 A chloroplast: (a) an electron micrograph (x 13 750) and (b) a diagram

Cell wall (in plants)

Plant cell walls consist mainly of the carbohydrate **cellulose.** Cell walls provide strength and rigidity for protection and support. If animal cells take up too much water they burst, whereas in plants the cell wall prevents this.

Cell vacuole (in plants)

Cell vacuoles are fluid-filled sacs surrounded by a single membrane. Their size varies; in some cells the vacuole almost fills the cell. The fluid is a dilute solution of molecules and ions. The vacuole can be used to store mineral salts, amino acids, sugars and waste products. Vacuoles are also involved in maintaining the water balance of the cell.

B1.4 The similarities and differences between plant and animal cells

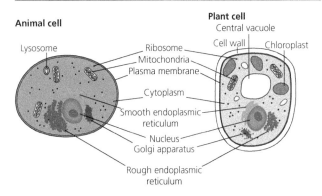

▲ Figure 11.6 Components of plant and animal cells

If you compare the animal and plant cells shown in Figure 11.6 you will see that there are many similarities. In fact, the main difference between the two types is that plant cells have structures such as the cell wall, chloroplasts and vacuoles that are not found in animal cells.

The cell wall provides support to plant cells and helps to keep them rigid. This also means that they have similar shapes. There are some types of highly specialised plant cells, but there is a much greater variety of types of animal cells. Animal cells also have a larger variety of shapes as they do not have a cell wall to keep them rigid and hence the need for skeletons in larger organisms.

> **Key term**
>
> **Stem cells:** undifferentiated (non-specialised) cells that can give rise to one or more types of differentiated (specialised) cell.

B1.5 How eukaryotic cells become specialised in complex multi-cellular organisms

In multicellular organisms, different cells are specialised to fulfil different functions. This is controlled by which genes are expressed (i.e. genes that are switched on and therefore have an effect). Every **somatic** cell (i.e. all cells excluding gametes) contains in its nucleus the whole **genome** (all the genes) of the organism. How the cell functions is determined by which of the many genes are expressed and which are not.

The process by which a cell changes from one cell type to another is known as **cell differentiation.** To understand this, consider how a **zygote** (fertilised egg cell) develops into an embryo and how one single cell gives rise to not just many cells, but many different types of cell.

Stem cells can differentiate to form specialised cells. The human embryo contains stem cells that can give rise to the more than 200 cell types of the adult human body. By the time a baby is born, most cells have already differentiated. This is why the term 'adult cells' applies to cells in babies and children as well as in adults. However, adult stem cells persist and are responsible for cell turnover, e.g. production of red blood cells from bone marrow cells (see Figure 11.7). These bone marrow cells also differentiate to form the various types of lymphocytes involved in the immune response (see Section B1.30). Most epithelial cells (see B1.12) need to be replaced throughout the life of an organism. Examples include cells in the skin, lungs, cornea and intestine. In each case, stem cells are responsible for replacement of worn-out cells.

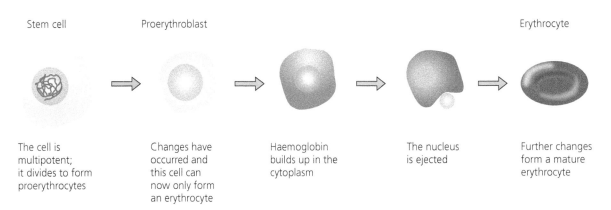

Stem cell	Proerythroblast			Erythrocyte
The cell is multipotent; it divides to form proerythrocytes	Changes have occurred and this cell can now only form an erythrocyte	Haemoglobin builds up in the cytoplasm	The nucleus is ejected	Further changes form a mature erythrocyte

▲ Figure 11.7 Differentiation of red blood cells (erythrocytes)

B1.6 How prokaryotic cells differ from eukaryotic cells

You need to be able to distinguish between prokaryotic and eukaryotic cells based on drawings.

Differences can also be identified on electron micrographs.

Figure 11.8 shows a diagram of a typical prokaryotic cell. The table summarises the differences between prokaryotic and eukaryotic cells.

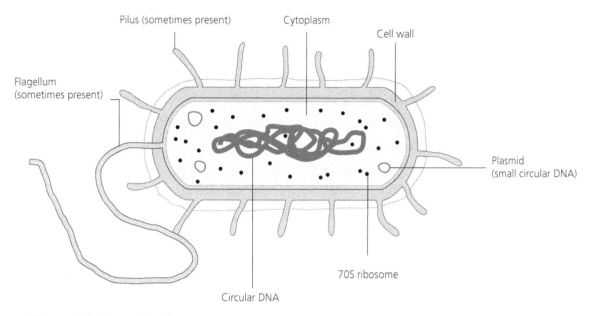

▲ Figure 11.8 A typical prokaryotic cell: some features are not present in all prokaryotic cells.

Prokaryotic cells	Eukaryotic cells
They have cytoplasm that lacks membrane-bound organelles.	They have cytoplasm containing membrane-bound organelles.
They have smaller ribosomes.	They have larger ribosomes.
They have no nucleus; instead, they have a single circular DNA molecule that is free in the cytoplasm and is not associated with proteins.	Chromosomes are linear and contained within the nucleus. The DNA is associated with proteins called histones.
They have a cell wall that contains murein/peptidoglycan, a glycoprotein.	Plant cells have a cellulose cell wall while fungi have a cell wall made of chitin.
They may have one or more plasmids.	There are no plasmids.
They may have a capsule surrounding the cell.	There is no capsule, even in plant cells. Some fungal cells can form a carbohydrate capsule.
They may have one or more simple flagella.	Flagella, where present, are more complex.

Test yourself

1 What are the three principles of cell theory?
2 State two differences between prokaryotic and eukaryotic cells.
3 State two differences between plant and animal cells.

Proteins, carbohydrates and lipids

Proteins, carbohydrates and lipids are three of the main classes of large biological molecules. We will encounter the fourth class in section B1.14 (page 186).

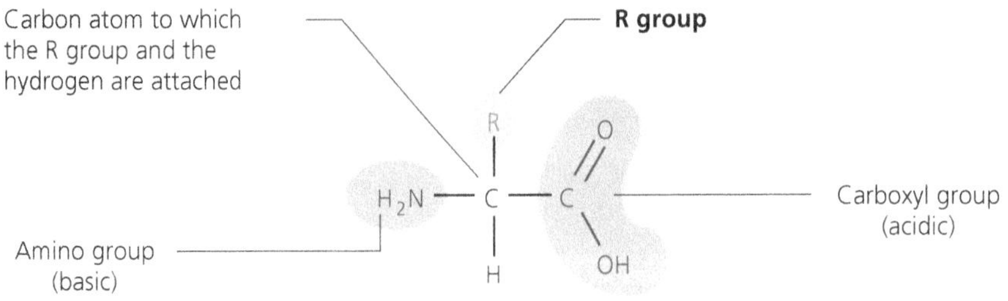

The 20 different amino acids that make up proteins in cells and organisms differ in their side chains (R groups). Below are three examples.

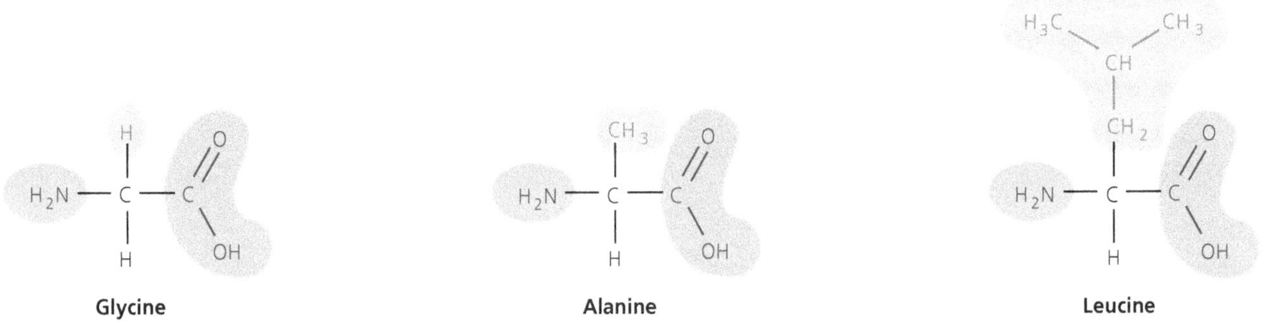

Glycine

Alanine

Leucine

▲ Figure 11.9 The basic structure of an amino acid and three examples of different amino acids.

B1.7 The relationship between the structure, properties and functions of proteins

Dipeptides are **peptides** formed by joining two **amino acids** in a **condensation reaction** (Figure 11.10). **Polypeptides** are **polymers** formed by the condensation of many amino acids, joined by **peptide bonds**.

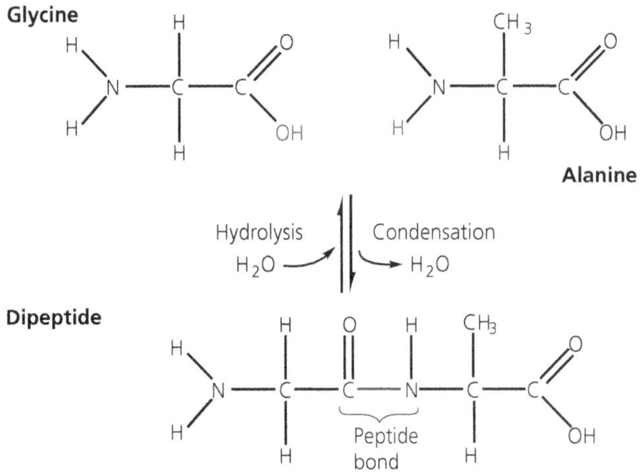

Glycine

Alanine

Hydrolysis H_2O Condensation H_2O

Dipeptide

Peptide bond

▲ Figure 11.10 The formation of a peptide bond between two amino acids in a condensation reaction produces a dipeptide. The reverse reaction is a hydrolysis.

Key terms

Peptide: a compound containing two or more amino acids joined together by **peptide bonds**. A **dipeptide** contains two amino acids bonded together.

Amino acid: a molecule with both an amino group and a carboxyl group. Amino acids are the small molecules (monomers) from which all **proteins** are made. There are 20 naturally occurring amino acids found in proteins and all have the amino and carboxyl groups attached to the same carbon. This carbon also has a hydrogen and another substituent – the side chain, or R-group – which is different in each different amino acid; see Figure 11.9. Amino acids are the monomers from which all proteins are made.

Protein: a **polypeptide** with a recognisable three-dimensional structure. It may contain more than one polypeptide chain.

Condensation reaction: a reaction between two small molecules to produce a larger molecule and water; most large biological molecules are formed by condensation reactions.

Polypeptide: a **polymer** of amino acids joined together by peptide bonds.

Polymer: a long molecule made from many small molecules called **monomers**.

Functional proteins, such as fibrous proteins or globular proteins, contain one or more polypeptide chains. The sequence of amino acids is different in each different protein. It is the amino acid sequence which will determine how the polypeptide chain folds up. This, in turn, determines the three-dimensional shape of the protein.

B1.8 The relationship between the structure, properties and functions of carbohydrates

Carbohydrates are important energy sources, for example, sucrose ('sugar', refined from sugar cane), lactose (found in milk) or maltose (used as an energy source for yeast in the brewing industry). Polysaccharides are polymers formed by condensation reactions of many monosaccharide molecules (monomers). As they are such large molecules, they are usually insoluble. This makes them suitable to carry out storage functions (glycogen and starch) and support functions (cellulose).

Type of carbohydrate	Examples	Notes
Monosaccharides	Glucose 	Monosaccharides contain just one sugar molecule and include glucose, fructose and galactose.
Disaccharides	Maltose 	Disaccharides are formed by condensation reactions between two monosaccharides – the bond formed is known as a glycosidic bond. Common disaccharides include maltose (containing two glucose molecules), sucrose (containing glucose and fructose) and lactose (containing glucose and galactose).
Polysaccharides	Amylose Amylopectin Starch Glycogen Cellulose (fibre)	Polysaccharides are polymers of monosaccharide monomers. The most common are starch (straight chain amylose or branched chain amylopectin), glycogen and cellulose, all of which are polymers of glucose.

Test yourself

1 What are the monomers in:
   a proteins
   b polysaccharides?
2 Name the type of reaction in which polypeptides or polysaccharides are formed.
3 What property of polysaccharides makes them good for storage and support?

Key term

Diffusion: is the movement of a substance from a high concentration to a low concentration. For instance, if you drop a crystal of copper sulfate into a beaker of water and watch the blue colour spread then you can see diffusion occur.

B1.9 The relationship between the structure, properties and functions of lipids

Lipids are the only group of large biological molecules that are not polymers. They all contain carbon, hydrogen and oxygen and are usually insoluble in water. More complex lipids can also contain phosphorous and nitrogen.

Lipids consist of two or three fatty acids, which are joined to a molecule of glycerol in condensation reactions. The fatty acids have long hydrocarbon chains, which explains why lipids are generally insoluble.

The main groups of lipids are:
▶ **triglycerides** (for example, fats and oils). These are used mostly as energy stores as well as for insulation (under the skin) and protection (around delicate organs such as the kidneys)
▶ **phospholipids**. These are found in plasma membranes and provide flexibility and help control what can move into and out of cells.

The role of phospholipids in membranes was covered in section B1.2.

Exchange and transport mechanisms

All organisms exchange substances with their surroundings. Single-cell organisms that respire aerobically need to absorb oxygen and get rid of carbon dioxide. They also need to absorb nutrients and get rid of waste products. This happens by the process of simple **diffusion**.

Large multicellular organisms cannot rely on simple diffusion. The distances are too great and it would take too long, as illustrated in Figure 11.11.

Single-celled organism (*Paramecium caudatum*)
Maximum distance for diffusion = 50 μm
Time taken = 8 seconds

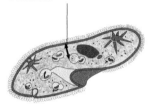

Maximum distance for diffusion = 15 cm
Time taken = 7 hours

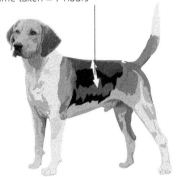

▲ Figure 11.11 The problem of increasing size on the rate of diffusion.

B1.10 How the surface area to volume ratio affects the process of exchange and gives rise to specialised systems

Efficient exchange (of gases, nutrients or waste products) requires three things:
▶ a large surface area for exchange
▶ a short diffusion distance
▶ a high concentration gradient.

The dog in Figure 11.11 has a much longer diffusion distance than the *Paramecium*, although it does have a much larger surface area. However, for efficient exchange the surface area must be large in comparison to the volume. We describe this as having a large surface area to volume ratio. This is illustrated in Figure 11.12. To make it simpler we have used a cube, but the principle is the same for animals with more complex shapes.

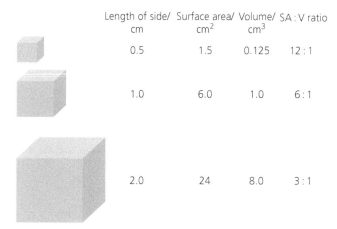

	Length of side/ cm	Surface area/ cm²	Volume/ cm³	SA : V ratio
	0.5	1.5	0.125	12 : 1
	1.0	6.0	1.0	6 : 1
	2.0	24	8.0	3 : 1

▲ Figure 11.12 Decrease of surface area: volume ratio in cubes of increasing size

Where the surface area is small compared to the volume (i.e. a low surface area to volume ratio), organisms cannot rely on simple diffusion across the whole body area like in *Paramecium*. Specialised exchange and transport mechanisms are required to maximise the rate of diffusion, such as lungs or gills and the heart and circulatory system.

Gills and lungs are adapted to make the diffusion distance as short as possible; between the air or water and the blood. For example, the walls of the alveoli in the lungs are only a single cell thick. There are also mechanisms, such as breathing, that maintain a high concentration gradient. Breathing brings air into the lungs that has a higher concentration of oxygen than within the blood. This ensures that oxygen diffuses into the blood.

Temperature and the rate of the chemical reactions needed to support life (**metabolic rate**) also play a part. The heart rate of a mouse is about 500–600 beats per minute, whereas an elephant's heart rate is about 25–30 beats per minute. Why is this?

Think about the following:
Heat loss also depends on surface area to volume ratio.
Mammals, such as mice and elephants, generate body heat from respiration.

More active animals have a higher metabolic rate and therefore require more substances such as glucose for aerobic respiration. This increases their need for specialised exchange and transport mechanisms. In terms of heat loss, the mouse has a larger surface area to volume ratio and therefore has an increased rate of heat loss. Because of this, they need to generate more heat from respiration, increasing their need for glucose and oxygen within their cells, explaining their higher heart rate.

B1.11 The principles of cellular exchange and the transport mechanisms which exist to facilitate this exchange

The lipid component of cell membranes consists of a double layer known as the phospholipid bilayer. As discussed in B1.2, phospholipids consist of a hydrophilic head group and a hydrophobic tail. This is shown in Figure 11.13 and is a very stable structure. Either side of the membrane is in an aqueous medium (meaning the main solvent is water). This means the tails arrange themselves on the centre, due to being hydrophobic. As a result, the hydrophilic heads arrange themselves on the outside, creating the phospholipid bilayer.

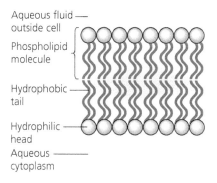

▲ Figure 11.13 Structure of the phospholipid bilayer

The phospholipid bilayer acts as a barrier to diffusion of many substances and only allows small, non-polar molecules through. Polar substances (see section B1.42, page 213) such as water, glucose, amino acids or inorganic ions (e.g. Na^+, K^+ or Cl^-) cannot diffuse through the hydrophobic core of the bilayer. This means that they require specialised transport systems involving proteins and glycoproteins (proteins with sugar molecules attached) embedded in the phospholipid bilayer. We can see an example of an adaptation to increase the efficiency of exchange if we look at the small intestine. The intestine is effectively a tube, and the central space (where digestion occurs) is called the lumen. The epithelial cells that line the small intestine are specialised cells; the plasma membrane facing the lumen is folded into many tiny finger-like projections called microvilli. This greatly increases the surface area for absorption of the products of digestion. Figure 11.14 illustrates the **fluid mosaic model** of the plasma membrane.

> **Key term**
>
> *Fluid mosaic model:* describes the structure of the plasma membrane and how its components are arranged. The proteins, lipids and carbohydrates that are found in the plasma membrane vary in shape, size and location which creates the mosaic pattern. Due to the relatively weak forces between phospholipids, the membrane can be considered to be fluid as these components can move throughout the membrane.

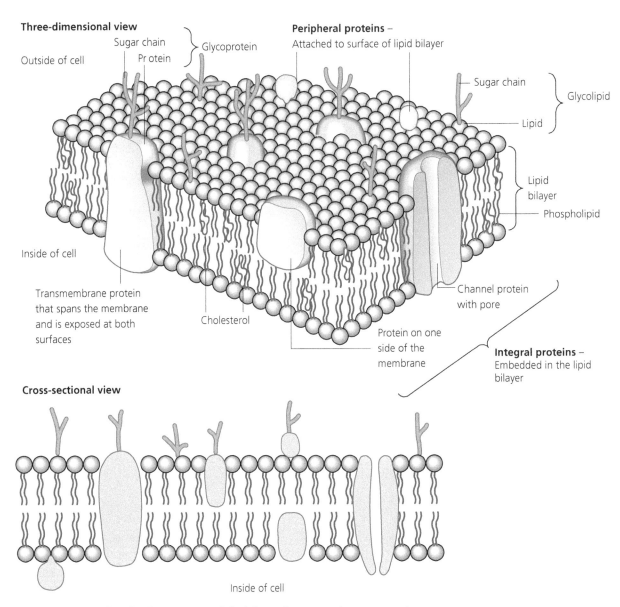

▲ Figure 11.14 The fluid mosaic model of the eukaryotic plasma membrane

Mechanisms: passive, active and co-transport

Simple diffusion, facilitated diffusion and osmosis are all passive processes, meaning that they do not require energy. Therefore, movement is always from high concentration to low concentration, sometimes described as down a diffusion gradient.

Simple diffusion

The phospholipid bilayer with its hydrophobic core can be a barrier to diffusion of polar substances; we say that it is *partially permeable*. So, small, non-polar molecules (e.g. carbon dioxide, oxygen, steroid hormones, lipids or fat-soluble vitamins) can move into the phospholipid bilayer and diffuse across the membrane.

Facilitated diffusion

Polar molecules, such as water, glucose and ions such as Na^+ or Cl^- cannot diffuse across the membrane,

so they rely on **facilitated diffusion**. This is where diffusion is assisted by proteins in the membrane. Ions and small polar molecules (including water) are transported by **channel proteins** (Figure 11.15) that act like pores in the membrane. Some of these can open and close, and these are called **gated channels**. Ion channels are usually specific for particular ions such as Na^+ or Ca^{2+}.

Larger polar molecules, such as glucose or amino acids, use **carrier proteins** (Figure 11.16). They are specific to the substance being transported which binds to the carrier protein in a similar way to a substrate binding to an enzyme (see section B2.5, page 256). The carrier protein then changes shape, which transfers the substance to the other side of the membrane.

The rules of diffusion still apply to both channel proteins and carrier proteins. Substances move only from a high concentration to low concentration.

Channel proteins help the diffusion of ions. Some ion channels have gates that open and close.

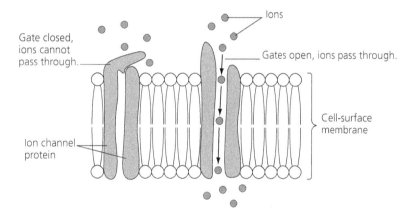

▲ Figure 11.15 Facilitated diffusion through a channel protein. The ones shown here are gated (they can open and close) while others are open all the time

Diffusing molecules bind to a carrier protein. The protein changes shape and takes the molecules through the membrane.

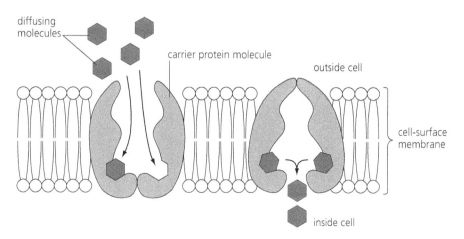

▲ Figure 11.16 Facilitated diffusion by a carrier protein

Osmosis

This is a particular type of facilitated diffusion where water moves across a partially permeable membrane from a high concentration of water molecules to a low concentration of water molecules. Do not be confused by this – a dilute solution (or pure water) will have a high concentration of water molecules, whereas a concentrated solution (e.g. a high concentration of glucose) will have a low concentration of water molecules. Therefore, water moves by osmosis from pure water or a dilute solution to a more concentrated solution.

Active transport

Substances do not move by themselves from low to high concentration (i.e. up a concentration gradient). However, **active transport** is a process that uses energy to move substances against a concentration gradient. This involves carrier proteins that use ATP as a source of energy. The mechanism is like that shown in Figure 11.16 with the addition of ATP as an energy source. These types of active transport proteins are often called pumps.

Co-transport mechanisms

The absorption of glucose from the gut is an example of a co-transport mechanism. Epithelial cells lining the small intestine have carrier proteins that only transport a glucose molecule together with a sodium ion. A sodium ion pump in the plasma membrane pumps sodium ions out of the epithelial cells into the blood capillaries which lowers the concentration of sodium ions inside the epithelial cells. This creates a sodium ion concentration gradient from the inside of the small intestine into the epithelial cells. It is this concentration gradient that causes sodium ions to diffuse into epithelial cells via a co-transport protein which brings with them glucose molecules. This helps ensure all the glucose is absorbed from the small intestine and is known as co-transport. Once inside the epithelial cell, the glucose diffuses down a concentration gradient into the blood. However, as glucose is a polar molecule, this requires a carrier protein too. The process is illustrated in Figure 11.17.

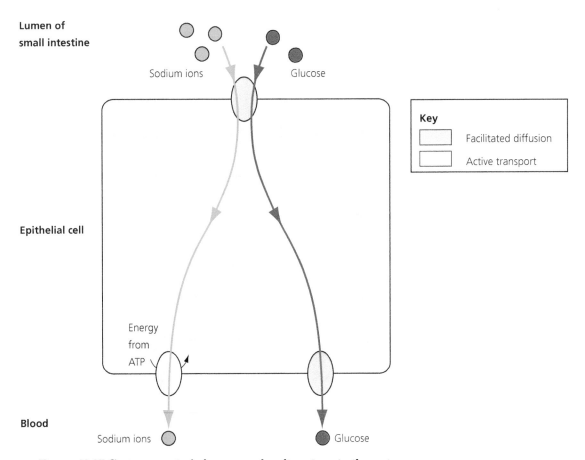

▲ Figure 11.17 Co-transport of glucose and sodium ions in the gut

B1.12 The advantages of having specialised cells in relation to the rate of transport across internal and external membranes

We saw in section B1.10 that multicellular organisms above a certain size can no longer rely on diffusion for exchange of substances with their environments. This usually means that specialised exchange organs have evolved, such as:

▶ gills or lungs for gas exchange
▶ the gut for absorption of nutrients
▶ kidneys for excretion of nitrogenous waste (urea).

Each of these organs have specialised cells adapted to maximise the efficiency of transfer of specific substances. Many of these cells are described as **epithelial** or **endothelial** and they are well adapted to their function.

Epithelial and endothelial tissues are usually single layers of cells that line or coat organs. The easiest way to decide whether a particular cell or tissue is epithelial or endothelial is to consider if it is in contact with the outside world. Of course, the place where it interacts with the 'outside world' can be contained within the body – just think about the gut, which is really just a long tube from mouth to anus.

Think about the following and decide whether they are examples of epithelial or endothelial:

    skin cells
    cells lining the mouth
    cells lining the trachea and lungs
    cells lining the gut
    cells lining the kidney tubules and bladder
    cells lining blood vessels (arteries, veins and capillaries).

Hint: only one of these is an example of endothelial cells!

Research

Choose an organ from the following list and research ways in which specialised cells in that organ are adapted to maximise the rate of transport across membranes:

▷ blood vessels
▷ lungs
▷ gut.

Test yourself

1 Explain why large multicellular organisms require specialised exchange and transport mechanisms.
2 What are the three features of a good exchange surface?
3 Describe the fluid mosaic model of cell membranes.
4 Explain why polar molecules require special transport mechanisms to cross cell membranes.
5 State two differences between active transport and facilitated diffusion.

Genetics

Genetics is the study of inheritance. We resemble our parents, but do not look exactly like either of them. **Genes** are passed from parent to offspring and these interactions control the appearance of those offspring.

The laws of genetics were worked out in the nineteenth century by Gregor Mendel. However, the role of DNA as the genetic material was only established for certain in the middle of the twentieth century.

We can study genetics using Mendel's laws that have been built on by subsequent researchers. We can work out the rules and how to apply them, and in this way, genetics is a type of logic puzzle. More recent approaches involve trying to work out what is happening at the molecular level. That requires an understanding of DNA, RNA and the synthesis of proteins. This section will help you gain a good understanding of this modern approach to genetics.

Key term

Gene: a sequence of bases in DNA that codes for (contains the information to make) a polypeptide, or, in some cases, functional RNA (this is involved in regulating how genes are expressed).

B1.13 The purpose of DNA and RNA as the carrying molecules of genetic information and the role they play in the mechanism of inheritance

Our understanding of genetics is based on some key points. We now know:
▶ Genes consist of **DNA** (deoxyribonucleic acid) and so we can say that DNA holds the genetic information.
▶ Genes control production of proteins by transferring the genetic information from DNA via **RNA** (ribonucleic acid) to the ribosomes where proteins are synthesised.
▶ Proteins are what determine the characteristics of an organism.

To fully understand genetics, we need to understand:
▶ how DNA stores the genetic information
▶ how DNA is replicated to pass on that genetic information to future generations
▶ how DNA and RNA are involved in the production of proteins.

B1.14 The relationship between the structure of DNA and RNA and their role in the mechanism of inheritance

As with all biological molecules, there is a strong relationship between the structure of nucleic acids (DNA and RNA) and their function. Both molecules are polynucleotides – that is polymers of nucleotides (the monomers) – in the same way that a polypeptide is a polymer of amino acids.

Each nucleotide contains a **pentose** (5-carbon sugar), a nitrogen-containing **organic base** and a phosphate group.

▶ In RNA (ribonucleic acid) the pentose is ribose, in DNA (deoxyribonucleic acid) the pentose is deoxyribose.
▶ In RNA the organic bases are adenine (A), cytosine (C), guanine (G) or uracil (U), while in DNA the organic bases are adenine (A), cytosine (C), guanine (G) or thymine (T).

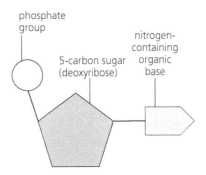

▲ Figure 11.18 General structure of a DNA nucleotide

The nucleotides are joined together in long chains by phosphodiester bonds between the pentose sugars; this is often described as a **sugar-phosphate** backbone. Phosphodiester bonds are formed in a condensation reaction like the one that forms peptide bonds (see Figure 11.10 in section B1.7).

The DNA molecule is a double helix where two very long polynucleotide chains are wound around each other and held together by hydrogen bonds between **complementary** base pairs: A pairs with T and C pairs with G. Complementary base pairing is central to how DNA stores and passes on genetic information, as well as to how genes control the synthesis of proteins.

RNA is a much shorter single-stranded polynucleotide chain.

The structure of DNA and RNA is shown in Figure 11.19.

The sequence of bases (sometimes called the **base sequence**) of DNA is how the genetic information is stored and passed on to future generations.

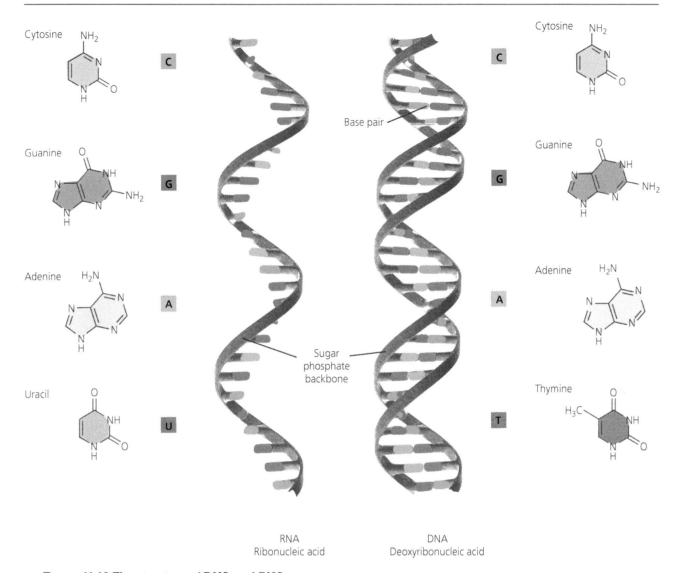

Cytosine

C

Guanine

G

Adenine

A

Uracil

U

Base pair

Sugar phosphate backbone

Cytosine

C

Guanine

G

Adenine

A

Thymine

T

RNA
Ribonucleic acid

DNA
Deoxyribonucleic acid

▲ Figure 11.19 The structure of DNA and RNA

B1.15 The function of complementary base pairing in forming the helical structure of DNA

Complementary base pairing holds the two strands of DNA together in the double helix structure. This makes the DNA molecule very stable, which is important for the molecule that contains and passes on the genetic information. Complementary base pairing is also the basis for how DNA is replicated.

B1.16 The process and stages of semi-conservative replication of DNA

The stages of DNA replication are as follows:
▶ The DNA double helix is progressively unwound. This involves an enzyme (**helicase)** that breaks the hydrogen bonds between the bases, allowing the strands to separate.
▶ Each strand now has unpaired bases.
▶ The strands each act as templates to assemble new strands. DNA nucleotides bind to the unpaired bases through complementary base-pairing.
▶ The enzyme **DNA polymerase** catalyses (speeds up) the formation of the phosphodiester bonds between the nucleotides.

This process is shown in Figure 11.20.

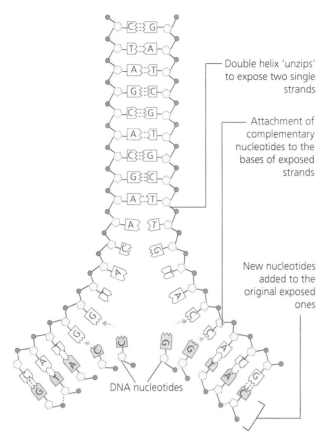

Double helix 'unzips' to expose two single strands

Attachment of complementary nucleotides to the bases of exposed strands

New nucleotides added to the original exposed ones

DNA nucleotides

▲ Figure 11.20 Semi-conservative replication of DNA

Key terms

Semi-conservative replication: when DNA replicates two new double helix molecules are formed, but each one consists of one of the original strands and one newly synthesised strand.

Mutation: a spontaneous change in the sequence of bases in DNA. This can occur in a number of ways. When a mutation occurs within a coding region of DNA a new **allele** can be formed.

Allele: a variant of a gene.

Genetics: the study of how single genes, or a small group of genes, function and how they affect the appearance and functioning of the organism.

Genomics: the study of how all the genes in an organism interact, as well as the role of non-coding sequences of DNA.

Genome: the entire genetic material of an organism. This includes DNA that does not code for proteins as well as the coding DNA (genes).

B1.17 How this semi-conservative replication process ensures genetic continuity between generations of cells

The product of **semi-conservative replication** of DNA is two molecules of DNA, both identical to the original. This is how the genetic information, contained in the DNA, can be passed from one generation of cells to the next. Because the two new molecules are identical to each other and to the original, the new generation of cells will be identical to the previous generation.

B1.18 The link between the semi-conservative replication process and variation

We saw in section B1.17 how semi-conservative replication ensures genetic continuity between generations of cells. However, the process is not always 100 per cent accurate. Sometimes the 'wrong' base is inserted by DNA polymerase. This random event is one source of **mutation**. A mutation is a spontaneous change in the sequence of bases in DNA, although this very rarely results in the formation of new **alleles**, partly because a lot of the DNA does not code for proteins. However, when a mutation occurs within the coding region of a gene, a new allele can sometimes be formed leading to the formation of a new characteristic. This is the source of genetic variation.

Genetic variation is the reason that we do not all look the same and is the basis of natural selection and evolution.

B1.19 The difference between genetics and genomics

We saw at the start of this section how the laws of **genetics** were worked out by a nineteenth-century monk working with pea plants. In contrast, **genomics** requires a great deal of technology to analyse and understand the **genomes**, particularly DNA sequencing and bioinformatics.

Both genetics and genomics are important in medicine. Genetics allows us to understand how inherited disorders like haemophilia, sickle cell anaemia or Huntingdon's disease are passed on. This understanding helps us assess the risk of children inheriting such conditions from their parents. Genomics is being used to investigate the link between all the genes we carry and the development of a wide range of diseases and conditions, from obesity and diabetes to heart disease and cancer.

> **Practice points**
>
> The terms 'genetics' and 'genomics' are similar and easily confused. Use of precise language is essential in science, so it is a good idea to develop good practice at an early stage that will stand you in good stead throughout your career.

> **Test yourself**
>
> 1 What is a gene?
> 2 State two differences between DNA and RNA.
> 3 What is meant by 'complementary base pairing'?
> 4 Give the names of two enzymes involved in DNA replication.

Microbiology

B1.20 The classification and characteristics (size of cell, type of cell, presence of organelles) of the following micro-organisms

Micro-organisms are often thought of simply as pathogens. This may be because that is how we see them having the greatest impact on our lives. However, that is only part of the story.

> Think of all the ways in which micro-organisms are of benefit or even essential. You could include:
>   foods and food production
>   production of medicines, such as antibiotics
>   agriculture
>   production of chemicals and clean-up of chemical contamination.

> **Practice points**
>
> See sections B1.62 and B1.63 (pages 237 and 238) for more about SI units and conversion between units. These are two units that you will encounter when studying micro-organisms.
>
> **Micrometre** (μm) is the most commonly used measure of size when studying micro-organisms and is 10^{-6}m (one millionth of a metre) or 10^{-3}mm (one thousandth of a millimetre).
>
> **Nanometre** (nm) is 10^{-9}m (one billionth of a metre), 10^{-6}mm (one millionth of a millimetre) or $10^{-3}\mu$m (one thousandth of a micrometre).

Bacteria

Bacteria are typically 1–2 μm (micrometres) long (i.e. about 1/1000 to 2/1000 of a millimetre) and usually roughly cylindrical, although other shapes, such as rods and spirals, do occur. They do not have membrane-bound organelles (see sections B1.3 and B1.6) and so are prokaryotes. See Figure 11.8 for a diagram of a typical bacterium.

Fungi

As well as the more familiar mushrooms and toadstools, many fungi are microscopic – the yeasts, including those used in fermentation to produce ethanol, are single cell organisms. Yeast cells are bigger than bacteria – in the range of 4–12 μm.

Fungi are eukaryotes, meaning they have chromosomes contained within a nucleus and other membrane-bound organelles. Multicellular fungi are composed of microscopic threads or hyphae that grow over or through their food source. Figure 11.21 shows a diagram of the mould fungus, *Penicillium chrysogenum* (the original source of the antibiotic penicillin.

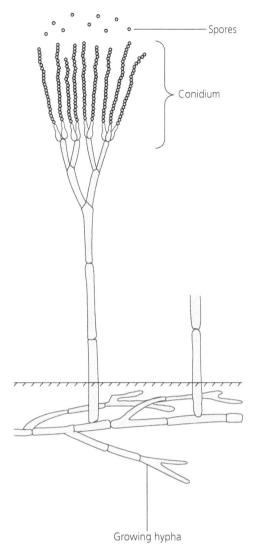

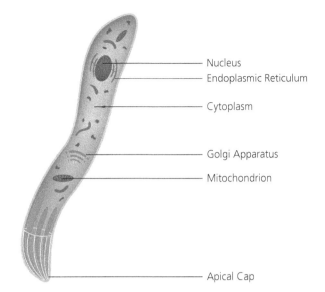

▲ Figure 11.22 Diagram of *Plasmodium*, a microscopic eukaryotic parasite

Viruses

As we saw in section B1.1, viruses are acellular (they are not made of cells) and do not contain organelles in the way that prokaryotes and eukaryotes do. They consist of genetic material (DNA or RNA) surrounded by a protein coat. Sometimes, the protein coat is itself surrounded by an envelope of lipid bilayer and glycoproteins that originated from the cell in which the virus replicated. Most viruses are very small – in the range 20–200 nm, although some giant viruses are as large as 1 µm – about the size of a bacterium and come in many shapes as well as sizes (Figure 11.23).

▲ Figure 11.21 Structure of the mould fungus, *Penicillium chrysogenum*

A honey fungus measuring 2.4 miles (3.8 km) across in the Blue Mountains in Oregon, USA is thought to be the largest living organism on Earth. That is certainly not microscopic! The structures that we recognise as mushrooms or toadstools are actually fruiting bodies, formed from very compact hyphae, that release spores.

Parasites

A parasite is an organism that lives on or in another organism at the expense of that organism. This includes multicellular organisms such as parasitic plants (e.g. mistletoe) and flatworms (e.g. tapeworms). Microscopic parasites are single-celled eukaryotic organisms such as *Plasmodium* (Figure 11.22), the parasite that causes malaria, or *Phytophthora infestans*, a parasite of plants that causes potato blight. Microscopic parasites can be different sizes, but usually in the range 1–10 µm.

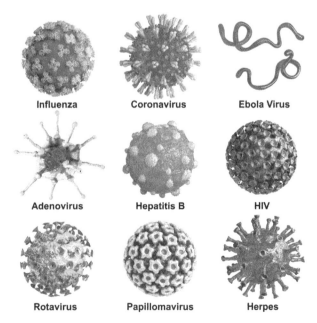

▲ Figure 11.23 A small selection of different virus structures

Are viruses alive? It is a question that has been asked many times and there is no simple answer. What we *can* say is that viruses are **acellular** – they are not made up of cells – and need to infect other cells to reproduce. Think about whether the first principle of cell theory (section B1.1) means that viruses cannot be classed as living organisms. If we consider viruses as alive, is the first principle wrong? Nothing in biology is ever simple!

B1.21 The benefits of using different types of microscopes when investigating micro-organisms

As the name suggests, micro-organisms cannot be seen with the naked eye. Therefore, we need to use some form of microscope to view and study them. The type of microscope used will depend on the size of the micro-organism we want to study – and, as some microscopes are more expensive, what we have access to.

Key terms

Magnification: how much bigger the image is than the actual object we are viewing. It should not be confused with **resolution**.

Resolution: the ability of a microscope to distinguish between two adjacent points. The resolution of a microscope is the smallest distance between two points that can be seen as separate. A high-resolution microscope can show a clearer image.

You have probably encountered resolution when using a camera phone. An old phone will probably have quite a low resolution. If you take a photo you may be able to enlarge the image to the same size as one taken with the latest high-resolution camera phone, but the modern phone will give a much clearer, sharper image.

(a)

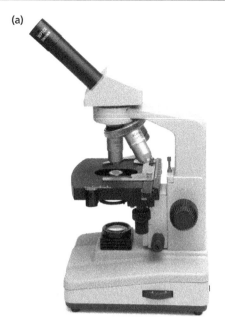

(b)

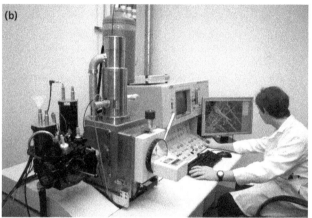

▲ Figure 11.24 (a) a light microscope and (b) an electron microscope

The principle of light and electron microscopes is the same: lenses are used to magnify the image. The difference is that, unlike light microscopes, electron microscopes use a beam of electrons to obtain an image. Whereas light microscopes use glass lenses for **magnification**, electron microscopes use magnets as lenses.

Light microscopes

The good news about light microscopes is that they are relatively inexpensive. They are also relatively easy to use, although they must be used with care to avoid damage. Thin sections of plant and animal tissues are usually prepared, but light microscopes can also be used to examine living micro-organisms as long as they are not too small, like most viruses. Electron microscopes cannot use living material as they have to operate in a vacuum.

Health and safety

Modern light microscopes usually contain a built-in light source. Halogen bulbs are likely to become very hot during use, so make sure you do not touch the bulb or try to disassemble the microscope.

When using a high power objective lens it is easy to drive the lens through the slide as you focus. This is likely to damage the lens as well as creating a hazard from the broken glass of the microscope slide. To avoid this, always lower the stage when changing lens. Ensure the stage is lowered and raised carefully when using the highest power lens. Once in view, only adjust the focus using the fine focus knob to prevent damage to the lens or slide.

Scanning electron microscopes

In a **scanning electron microscope** (**SEM**), the beam of electrons is scanned across the surface of the sample. The electrons bounce off the surface and a computer is used to build up a 3D image of the surface of the sample, showing more surface detail than is possible with a light microscope.

Transmission electron microscopes

In a **transmission electron microscope** (**TEM**), the electrons pass through the sample in the same way as light rays pass through a sample in the light microscope. This means a TEM shows a 2D image of the sample. Very thin sections are required as electrons cannot penetrate materials very deeply. This allows a TEM to reveal details of virus particles or cell organelles that would not be visible with the light microscope. A series of thin sections can be used to take multiple 2D images that can be assembled into a 3D image.

The table compares the approximate magnification and **resolution** of these three types of microscope.

Instrument	Maximum magnification	Maximum resolution
Light microscope	x 1500	200 nm
SEM	x 100000	10 nm
TEM	x 500000–1000000	0.2 nm

Figure 11.25 illustrates what can be seen with the human eye, a good quality light microscope and a transmission electron microscope.

Practice points

Units
The following units are commonly used in microscopy:

μm (micrometre) = 10^{-6} m

nm (nanometre) = 10^{-9} m

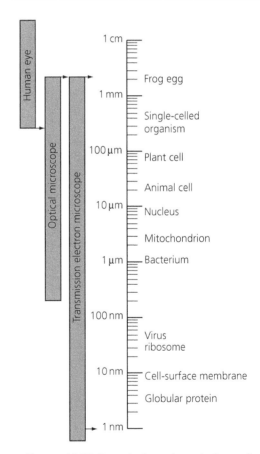

▲ Figure 11.25 A scale from 1 nm to 1 cm showing what can be seen with the naked eye, a light microscope and a transmission electron microscope. A log scale has been used because of the large range of measurements

B1.22 How to calculate magnification from the size of the image and the size of the object

Magnification is simply how much bigger the image is than the object. We can express that mathematically as:

$$magnification = \frac{size\ of\ image}{size\ of\ object}$$

It looks simple – and it is – but it can cause confusion. When calculating magnification, make sure you use the same units for the size of the image and the size of the object. We would normally measure the size of an image (e.g. a photomicrograph) in cm or mm, but microscopic objects are usually measured in μm. It is better to measure the size of the image in mm. When you calculate the magnification, make sure both sizes are expressed in the same units – either mm or μm, as shown in the example. There is more about converting between units in section B1.63.

A photomicrograph of a cell shows a mitochondrion that you know to be 1.5 μm long. You measure the image of the mitochondrion and find that it is 112.5 mm long. What is the magnification?

Method 1: convert the measurement of the image from mm to μm = 112 500 μm by multiplying by 1000 so 112.5 mm.

Now divide the size of the image by the size of the object to calculate the magnification:

$$magnification = \frac{112500}{1.5} = \times\,75000$$

Method 2: convert the size of the object from μm to mm by dividing by 1000 so 1.5 μm = 0.0015 mm

Now, divide the size of the image by the size of the object to calculate the magnification:

$$magnification = \frac{112.5}{0.0015} = \times\,75000$$

As you see, both methods give the same result. Remember that magnification should be quite a large number and always greater than 1. If your answer is not, then you have made a mistake!

B1.23 The uses of differential staining techniques

Staining uses dyes to help make objects clearer under the light microscope by increasing contrast and making it easier to see cells. Unstained cells are almost transparent, which makes it very difficult to see them. Staining makes micro-organisms stand out against the background.

Differential staining adds another dimension by allowing us to distinguish different cell types, including specific types of bacteria or even parasites. There are a number of common ways this is done.

> ### Health and safety
>
> If you are involved in preparing stained sections, make sure that you are aware of any hazards associated with the chemicals used.

Gram staining

The **Gram stain** differentiates bacteria by detecting the peptidoglycan present in the cell wall of Gram-positive bacteria. Crystal violet is used to stain the peptidoglycan and then iodine is added to fix the stain permanently to the peptidoglycan molecules. Stained bacteria appear dark blue or violet. Gram-negative bacteria only have a very thin peptidoglycan layer – they have an outer membrane of lipopolysaccharides – and so the stain can be washed out of the cells. A counterstain of fuchsin or safranin is then used to stain all bacteria red/pink. As Gram-positive bacteria have already been stained dark blue/violet, only the Gram-negative bacteria appear red/pink.

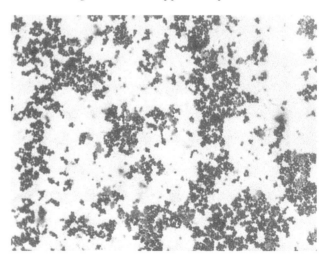

▲ Figure 11.26 Gram staining showing Gram positive (purple) and Gram negative (pink) bacteria

Giemsa staining

The Giemsa stain contains a mixture of Azure B, methylene blue and eosin stains. The methylene blue stains the chromosomes and nucleus dark purple while Azure B and eosin stain the cytoplasm pale blue or pink.

Giemsa stain can be used for:
▶ identification of specific bacteria such as *Chlamydia trachomatis*; these are stained blue-mauve to dark purple
▶ identification of *Plasmodium vivax* and *Plasmodium falciparum*, the malarial parasites; these are stained with a red or pink nucleus and blue cytoplasm
▶ identification of blood diseases such as anaemia and leukaemia; the different blood cells stain differently with Giemsa stain and any abnormalities can be identified (Figure 11.27).

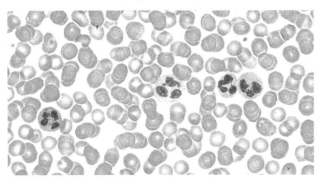

▲ Figure 11.27 A peripheral blood smear stained with Giemsa stain. Red blood cells do not have nuclei and so are stained pink, while the white blood cells have prominent nuclei that stain purple. This allows the different types of blood cell to be identified

Haematoxylin and eosin staining

Also known as H&E staining, this is the most widely used stain in medical diagnosis. The haematoxylin stains cell nuclei blue while the eosin stains the cytoplasm pink.

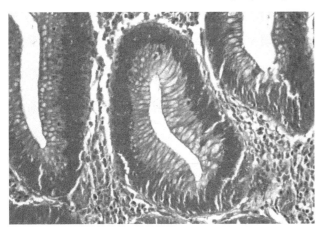

▲ Figure 11.28 Human breast cancer tissue section stained with H&E

Test yourself

1 What is meant by the term pathogen?
2 For each the following pathogens, state whether they are prokaryotes or eukaryotes:
 a The *Plasmodium* parasite that causes malaria.
 b The tuberculosis bacterium.
 c The yeast that causes athlete's foot.
3 Explain the difference between the terms 'magnification' and 'resolution'.
4 Explain why light microscopes usually require thin sections of tissue.
5 Explain why electron microscopes cannot be used to examine living tissue.
6 Give two differences between transmission and scanning electron microscopes.
7 What type of stain would you use for the following:
 a Diagnosis of leukaemia, where there are abnormally large numbers of white blood cells
 b Examination of a food sample to detect food poisoning bacteria
 c Diagnosis of skin cancer from a skin biopsy
 d Diagnosis of a blood-borne parasitic infection.

Immunology

Immunology is the study of the immune system, which is an important part of the body's response to infection. We will see in section B1.30 that the immune system is just one way in which the body defends itself against disease. First, we need to consider the causes of infectious diseases.

B1.24 The nature of infection

Infection describes a microorganism replicating inside the body, resulting in disease. Some organisms, including all viruses, some bacteria and some parasites, infect body cells. Others replicate in organs such as the gut, in the blood or the spaces between cells.

B1.25 Pathogens (causative agents) of infection and examples of resulting diseases

Pathogen	Example of disease	Notes
Bacteria	• chlamydia • gonorrhoea • tuberculosis	Bacterial infections are treated by antibiotics, but bacteria are becoming increasingly resistant to antibiotics.
Viruses	• common cold • mumps • measles	SARS-CoV-2, the coronavirus that causes COVID-19, has recently become the best-known virus.
Fungi	• yeast infection (thrush)	Other fungal skin infections include toenail fungus and athlete's foot.
Prions	• Creutzfeldt-Jakob disease (CJD)	Prions are non-living pathogenic proteins. The mutant form of prion protein, when ingested, can cause normal prion proteins to change shape. This causes damage to the nervous system and eventual death.
Protists	• malaria	Don't confuse the pathogen (*Plasmodium*, a protist) with the *Anopheles* mosquito that transmits the pathogen.
Parasites	• toxoplasmosis	Toxoplasmosis is caused by *Toxoplasma gondii*, a parasitic protoctist. Many multicellular parasites can also cause infections, particularly in developing countries.

You may notice that there is an overlap between protoctists and parasites; malaria and toxoplasmosis are both caused by parasitic protoctists. However, some parasite diseases are due to infection by multicellular parasites, such as tapeworms.

Key term

STI or **sexually transmitted infection:** caused by a pathogen that is passed from person to person during sexual contact.

B1.26 The different ways in which pathogens (causative agents) may enter the body

During the COVID-19 pandemic everyone paid much greater attention to hand hygiene, mask wearing, social distancing and improved ventilation of indoor spaces. Although many people died from COVID-19, one effect of these precautions was that there were far fewer deaths from seasonal flu. This illustrates the importance of understanding the ways in which infections are transmitted.

Direct transmission

▶ Physical contact with an infected person (for example, skin-to-skin contact) or contaminated surface (for example, door handles and other hard surfaces).
▶ Sharing of needles can result in transmission of blood-borne pathogens.
▶ Pathogens such as HIV or hepatitis C virus can be spread by transfusion with contaminated blood or blood products. Unprotected sexual contact can lead to **STIs**.

Airborne transmission

The pathogen is carried by dust or droplets in the air. Some droplets (aerosols) can exist in the air for many hours and inhaling infected droplets can lead to infection. COVID-19 and tuberculosis are spread in this way.

Indirect transmission

Vehicle transmission occurs when infected food or water are ingested (eaten or drunk). Faecal-oral transmission is the result of poor hand hygiene and is a significant cause of food poisoning. Another example of vehicle transmission is from infected blood on inanimate objects such as clothing or bedding.

Another form of indirect transmission is being bitten by an infected **vector** (the organism that transfers the pathogen from host to host). Insect bites can introduce pathogens into the body. The best-known example is the malaria protoctist (*Plasmodium*), for which the vector is the *Anopheles* mosquito. There are many others, including Lyme disease (caused by a bacterium) and Zika fever (caused by a virus).

B1.27 How infectious diseases can spread among populations and communities

Understanding how infection spreads from person to person helps our understanding of how infectious diseases spread among populations. Once we understand that, we can consider ways in which the spread can be prevented, or at least minimised.

Inadequate sanitation includes:

▶ a lack of access to clean water for washing. Clean water is also unlikely to carry water-borne diseases

▶ inadequate sewage disposal, which increases the risk of faecal-oral transmission of a wide range of pathogens, including parasites that have evolved alongside human populations.

Dense populations lead to overcrowding in households as well as a lack of social distancing outside the home. These will both increase the rate of transmission by direct, airborne and indirect routes.

Inadequate healthcare infrastructure, such as inadequate hospitals or clinics, or a lack of doctors or nurses, increases the risk of disease spreading unnoticed as well as making it harder to treat and prevent further spread.

Ignorance can be deadly. Lack of accessible health promotion information means that people are less likely to take necessary precautions to prevent spread of infection. They may also be more resistant to prevention measures, such as vaccination.

It is worth noting that while all these factors are more prevalent in countries with developing economies, they are associated with areas of deprivation worldwide.

Case study

Between 1846 and 1860, a worldwide pandemic of cholera was responsible for over a million deaths worldwide. Cholera was thought to be caused by particles of decaying matter in the air ('miasma'). During 1854 there was a severe outbreak near Broad Street in the Soho district of London. A physician, John Snow, had been studying cholera for several years by this time. He mapped all the outbreaks of cholera in the district and showed that they were concentrated around a public water pump on Broad Street. Snow did not understand that cholera was caused by a bacterium, but the evidence he collected showed that in this outbreak it was transmitted by infected water from the Broad Street pump and not miasma. The authorities were unwilling to accept the results of Snow's work, although the pump handle was temporarily removed to prevent its use.

Snow also investigated the quality of water provided by different water companies. Individual houses in the same area received their water supply from different companies. Snow showed that there was a higher incidence of cholera in those households receiving their water from two large companies who extracted it directly from the River Thames. At that time, the river was heavily contaminated with raw sewage. Other households obtained their water from smaller companies who provided cleaner, better filtered water. These households had much lower incidence of cholera.

John Snow's work was a good example of the scientific method, including his use of statistical analysis. His work led, eventually, to great improvement in sanitation through the installation of more efficient sewage handling and provision of clean water.

▶ Think about the nineteenth-century cholera outbreak. Can you see any parallels with the COVID-19 pandemic?

▶ John Snow investigated a range of different factors in order to see a pattern. Can you think of factors that should be considered when studying the COVID-19 pandemic?

▶ Do you think that we have learned the lessons of John Snow's work?

B1.28 The definition of an antigen and an antibody

Key terms

Antigen: a substance that is recognised by the immune system as self (the body's own cells) or non-self (foreign cells and pathogens) and stimulates an immune response. Antigens are found on pathogens but also on the surfaces of all body cells.

Antibody: a blood protein that is produced in response to a specific antigen. An antibody binds specifically to an antigen in a similar way to an enzyme binding specifically to its substrate.

B1.29 The link between antigens and the initiation of the body's response to invasion by a foreign substance

We can think of **antigens** as chemical markers rather like ID cards. They are usually proteins or **glycoproteins** (proteins with sugar molecules attached) on the surface of pathogens or body cells.

Some antigens on the surface of body cells allow the immune system to distinguish between the body's own cells ('self' antigens) and foreign cells, including pathogens ('non-self' antigens).

The response to invasion by a foreign substance involves several stages that we can think of as defence mechanisms. The immune response is part of that process but is not the only part.

B1.30 The stages and cells involved in the body's response to an antigen

The stages of defence against non-self antigens, for example, on pathogens, is shown in Figure 11.29.

Physical and chemical barriers

The first line of defence is to keep pathogens out. The skin plays a significant part as an external barrier. Mucous membranes are also important. These line the gut, airways and reproductive system. Goblet cells produce thick, sticky mucous that helps to trap bacteria and other pathogens. Antimicrobial proteins and peptides also help to destroy pathogens and can also be involved in stimulation of the immune system.

Lysozyme is an enzyme that hydrolyses bonds in the cell wall components of some bacteria. This weakens the cell walls, meaning that the bacteria swell and burst. Lysozyme is present in tears, helping to protect the surface of the eyes, as well as in breast milk providing protection to infants while their immune systems are developing.

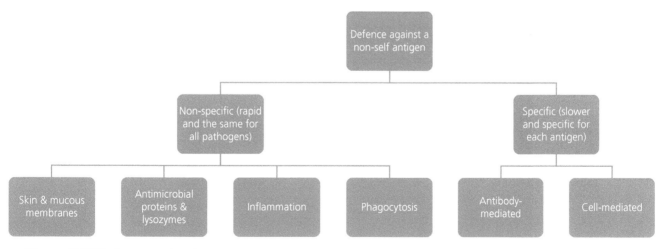

▲ Figure 11.29 Defence mechanisms

Inflammation

Inflammation is a response to injury or infection where the area becomes hot, red and swollen as a result of increased blood flow. Mast cells respond to tissue damage (caused by injury or infection) by secreting histamine. This **cell-signalling** compound stimulates a range of responses, including:

- increased blood flow in capillaries;
- capillaries begin to leak more, allowing fluid to enter the tissues resulting in swelling;
- **phagocytes** leave the blood and enter the tissues where they can engulf foreign material.

Histamine also stimulates cells to release cytokines, including interleukins that lead to more inflammation. Cytokines also lead to the promotion of **phagocytosis**.

Key terms

Innate immunity: the non-specific mechanisms, present from birth, that protect against a wide range of pathogens. These mechanisms, including inflammation, phagocytosis and antimicrobial proteins, work quickly, but are not always very effective.

Inflammation: a local response to injury and infection.

Cell-signalling: the process by which cells communicate with each other, usually by release of chemicals such as histamine, cytokines and interleukins.

Phagocytes: produced in the bone marrow and circulate in the blood. Some leave the blood and are present in the tissues.

Phagocytosis: the process of a phagocyte engulfing a pathogen or other foreign material.

Lymphocytes: small white blood cells. B lymphocytes, or B cells, are responsible for antibody production. Different types of T lymphocytes, or T cells, play different roles in the immune response.

Phagocytosis

Chemicals released by pathogens into the blood attract phagocytes. Receptors on the surface of phagocytes bind to antigens that are present on the surface of most pathogens, and this leads to the phagocyte engulfing and digesting the pathogen (Figure 11.30).

Some types of phagocyte known as macrophages do not completely digest the pathogen. Instead, antigens from the partially digested pathogen are processed and then appear on the plasma membrane of the macrophage. These are then known as antigen-presenting cells (APCs), as they 'present' the antigens to **lymphocytes** (T cells). This process of antigen presentation initiates the immune response. This is the slower, more specific and more effective stage of defence against infection.

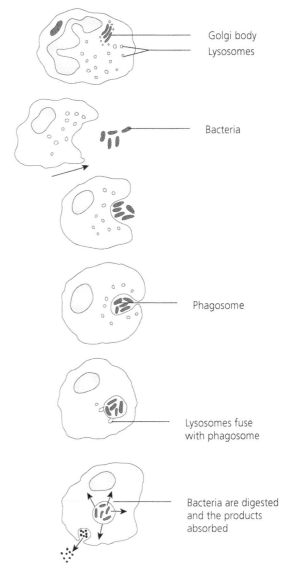

Golgi body
Lysosomes

Bacteria

Phagosome

Lysosomes fuse with phagosome

Bacteria are digested and the products absorbed

▲ Figure 11.30 The stages in phagocytosis

The role of T cells

The two main groups of T cells are T helper cells (T_H cells) and cytotoxic T cells (T_C cells) also known as T killer cells (T_K cells).

T_H cells have a type of cell-surface receptor known as CD4. There are many different shapes of CD4 receptor corresponding to the millions of antigen shapes that we might encounter. When a T_H cell encounters an APC, the CD4 receptors on the T_H cell may be complementary to the antigen, i.e. the shapes match. If so, the following events happen:
- The T_H cell binds, via its CD4 receptor, to the APC.
- This activates the T_H cell.
- The activated T_H cells divide by mitosis to form a clone of active T_H cells and memory cells.
- Activated T_H cells are then able to activate T_w and B cells.

This is shown in Figure 11.31.

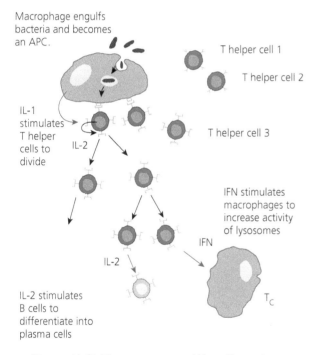

▲ Figure 11.31 The activation of T_H cells in the immune response. Three different T_H cells are shown, but only T_H cell 3 has a CD4 receptor complementary to the antigens on the APC

When body cells are infected by pathogens such as viruses, they also process pathogen antigens and present them on their cell surfaces, becoming APCs. T_C cells also have cell-surface receptors known as TCRs. Like CD4 receptors on T_H cells, there are many different shapes of TCR, complementary to different antigens.

If a T_C cell encounters an APC with complementary antigens the following series of events occurs:
- The T_C cell becomes activated. This process also involves T_H cells.
- Once activated, the T_C cells will divide by mitosis to form a clone of activated T_C cells and memory cells.
- The T_C cells will bind to the surface of other infected cells and destroy them.

It might seem drastic to destroy the body's own cells, but infected cells will usually end up dead anyway and the action of T_C cells prevents pathogens such as viruses replicating inside the infected cells. This process is shown in Figure 11.32.

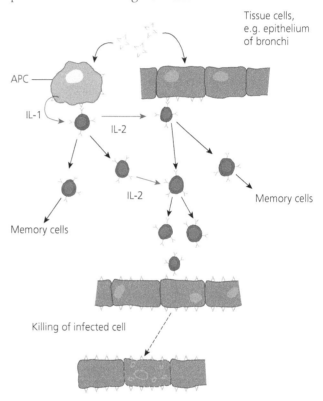

▲ Figure 11.32 Activation of T_C cells by infected host cells leads to killing of other infected cells

The role of B cells

B cells also have cell surface receptors, the B cell receptor or BCR. Again, there are many different shapes of BCR. When a B cell encounters antigens complementary to (i.e. which match) its BCR, the following events take place:
- The antigen binds to the BCR.
- B cell takes in the BCR and antigen.
- The B cell processes the antigen and so becomes an APC.

At this stage, activated T_H cells become involved.

▶ Any activated T_H cells with complementary CD4 receptors will then bind to the antigens on the APC.

▶ The T_H cell then secretes cytokines (such as IL-1 and IL-2).

▶ The cytokines activate the B cell.

▶ The activated B cell divides to form a clone of activated B cells and memory cells.

▶ The activated B cells differentiate to form plasma cells that produce large quantities of antibodies.

This process is shown in Figure 11.33.

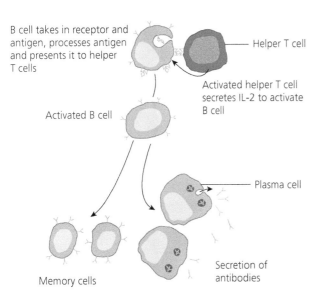

▲ Figure 11.33 Selection and activation of B cells with a BCR complementary to pathogen antigens

Antibodies bind to antigens on the surface of pathogens. Antibodies protect against pathogens in several ways, including:

▶ binding to toxins produced by bacteria and making them harmless;

▶ cross-linking pathogens so that they are too large to spread or infect cells;

▶ signalling to phagocytes to engulf the pathogens;

▶ binding to pathogen proteins that the pathogens use to enter body cells, for example, the spike protein on the surface of the SARS-CoV-2 virus.

B1.31 The differences between cell-mediated immunity and antibody-mediated immunity

In the previous section, we looked at the parts that T cells and B cells play in the immune response. It is useful to separate the two parts of the immune response. T_H cells play a key role in both of these parts, but there are important differences between the way in which the immune system protects the body from pathogens.

In the **cell-mediated response**, T cells destroy pathogens by destroying infected body cells. This means the pathogens cannot replicate and infect more body cells. Antibodies are not involved in the cell-mediated immune response.

In the **antibody-mediated** response, B cells produce antibodies, and it is the antibodies that lead to destruction of pathogens. Some antibodies are known as antitoxins, because they bind to and neutralise toxins produced by pathogens.

B1.32 The role of T and B memory cells in the secondary immune response

If we are infected by a pathogen, it takes the body about 10–17 days to produce antibodies. This process, known as the primary response, was described in section B1.31 and explains why we get ill with an infection. Once antibodies (and active T_C cells) are produced, the pathogen is removed and we get better. Plasma cells do not live long in the blood and the antibodies they produce are gradually broken down.

If, after some time, we are infected with the same pathogen, then antibodies are produced more rapidly and in much larger quantities. This is known as the **secondary immune response** and is illustrated in Figure 11.34.

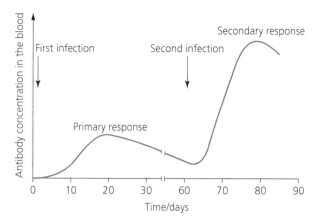

▲ Figure 11.34 The change in the concentration of antibodies during the primary and secondary immune responses

We saw how memory T cells and memory B cells were produced during the primary immune response. These remain in the body for a long time. When they encounter the pathogen for a second time, they trigger a stronger and quicker immune response by multiplying much more rapidly to form clones of plasma cells and T_C cells. The plasma cells produce high concentrations of antibodies in just a few days. In this way, the secondary immune response can clear pathogens from the body before we even show symptoms of the disease.

The secondary immune response is the basis of how vaccines work. A vaccine will stimulate the primary immune response without needing exposure to the pathogen, This means that memory T and B cells are produced. If a vaccinated person is later infected with the pathogen, a rapid secondary response will prevent them from becoming ill, or will reduce the severity of the illness.

Test yourself

1 Give two examples of diseases caused by each of the following types of pathogen:
  a Bacteria   b Viruses
2 Explain the difference between a pathogen and a vector.
3 Give two examples of each of the following methods of disease transmission:
  a Direct   b Airborne   c Indirect
4 Describe two factors that cause infections to spread within populations.
5 What is an antigen?
6 Give two differences between the specific and non-specific immune responses.
7 What is meant by an antigen presenting cell (APC)?
8 Give two differences between the primary and secondary immune responses.

Project practice

You are working in a lab that analyses ingredients for use in food manufacture. It is important that these meet required standards of identity, purity and safety. Choose one of the following areas.
▶ DNA analysis of meat to confirm species (lamb, beef, pork, etc.)
▶ Microbiological analysis of ingredients for contamination with pathogens.
▶ Immunological techniques for confirming identity and purity of ingredients.

You then need to carry out the following:
1 Research a strategy.
  a Carry out a literature review.
  b Justify why you have chosen specific sources and not others.
2 Plan a project using the sources that you selected in your literature review.
  a Set out the techniques you would use in your chosen form of analysis.
  b Include all appropriate risk assessments.
  c Identify the data that you would need to collect and how you would record the data.

3 Analyse the data.
  a You will normally be presented with the data you need, as there will not be time to actually carry out the investigation.
  b Produce a report of your analysis; think about what statistical tests you might need to apply.
4 Present your outcomes and conclusions in the form of a scientific poster showing:
  a the techniques being used
  b the strengths and weaknesses of your chosen technique
  c your conclusions about the technique you have chosen.
5 Group discussion covering topics such as:
  a the need for food analysis
  b the practicality of different techniques
  c do these techniques help to reassure consumers?
6 Reflection – write a reflective evaluation of your work.

Assessment practice

1   For each of the following organelles, state whether they are found in eukaryotes, prokaryotes, or both:
   a   Nuclei
   b   Plasma membrane
   c   Ribosomes
   d   DNA associated with proteins
   e   Plasmids

2   Which of the following statements is true?
   A   Lipids are polymers of fatty acids and glycerol.
   B   Polysaccharides are highly soluble molecules which makes them suitable as energy stores.
   C   Proteins are used as a form of storage molecule.
   D   Proteins, polysaccharides and lipids are all formed by condensation reactions.

3   A short section of DNA has the following base sequence.
   A G C T T A G C T
   Give the base sequence of the complementary strand of DNA.

4   Explain how semi-conservative replication ensures genetic continuity from one generation of cells to the next.

5   A student was using a light microscope to study a stained section of animal tissue. Explain what type of stain is likely to have been used in preparing the section.

6   A class was studying micrographs of animal cells.
   a   One micrograph showed a mitochondrion that was labelled as being $1.5\,\mu m$ long. A student measured the micrograph and found that the image of the mitochondrion was $11.3\,cm$ long. Calculate the magnification of the microscope.
   b   Another micrograph was labelled 'x500 000'. The student measured the thickness of the plasma membrane on the micrograph and found it was $2\,mm$ wide. Calculate the actual width of the plasma membrane in nm.
   c   Explain what type of microscope will have been used to create the micrographs.

7   Antibiotics can be used to treat bacterial infections because they do not harm eukaryotic cells.
   a   Explain why antibiotics cannot be used to treat malaria.
   b   Explain why malaria can be controlled using insecticides.

8   During the COVID-19 pandemic, the UK government's advice was built around the slogan 'Hands, face, space'. This was later amended, adding 'fresh air'. Evaluate the use of this slogan.

9   Which of the following are involved in antibody-mediated immunity?
   A   B cells, T helper cells, plasma cells, phagocytes
   B   B cells, T killer cells, plasma cells
   C   Phagocytes, T killer cells, plasma cells
   D   Phagocytes, T killer cells, T helper cells

10   During the early stages of development of a COVID-19 vaccine, the concentration of antibodies in the blood of volunteers was measured in the weeks after vaccination. One vaccine produced significantly more antibodies than another type of vaccine. However, both were found to be similarly effective in preventing COVID-19 infection. Suggest an explanation for this finding.

B1.33–B1.44: Core science concepts: Chemistry

Introduction

Chemistry is the study of substances, their properties and how they combine to make other substances. All substances are made up of atoms, so we need to understand the structure of atoms and the arrangement of electrons within atoms. This is important because electrons are responsible for the bonds between atoms and chemical reactions involve making and breaking bonds.

This chapter also covers basic concepts of acids and bases, the factors that affect the rates of chemical reactions and some methods we can use to analyse substances.

Learning outcomes

The core knowledge outcomes that you must understand and learn:

Materials and chemical properties
B1.33 the relationship between the atomic structure and physical and chemical properties of metals

B1.34 how the arrangement of electrons is linked to the way in which elements are situated within groups in the periodic table

B1.35 the correct names for sub-atomic particles and their position in an atom – protons, electrons and neutrons

Acids/bases and chemical change
B1.36 the physical properties of acids

B1.37 the concept of strong and weak acids (as distinct from dilute and concentrated solutions)

B1.38 how to determine the name of the salt produced in acid-base reactions

Rates of reaction and energy changes
B1.39 the principles of collision theory

B1.40 the effect of temperature on rates of reaction

B1.41 the definition of a catalyst and the role of catalysts in a reaction

Chemical analysis of substances
B1.42 the principles of tests and techniques that are used to separate, detect and identify chemical composition

B1.43 the tests that could be used to quantify components in a mixture

B1.44 the principle of titration.

Practice points

Standard form

Throughout the coming chapters, and elsewhere in this book, you will see numbers written in what is known as **standard form**.

For example, the number 3200 can be written as 3.2×10^3.

10^3 is one thousand ($10 \times 10 \times 10$; try it on your calculator). So we are writing a number which we might say as three thousand two hundred as 3.2 thousands, hence 3.2×10^3.

3200 is not a very large number, so you might ask, 'why bother?' But what about the following number?

3200000000000000000000000

That takes too much space to write. But it also has the complication of working out millions, billions and so on. It is much easier to write it as 3.2×10^{21}.

The same is true of very small numbers:

5.5×10^{-11} is much easier to work with than 0.000000000055. Can you be sure to keep track of all those zeroes?

Numbers in standard form always have two parts. The number before the '×' sign is always greater than or equal to 1 but less than 10. The number after the '×' sign is always a power of 10. Here are some examples:

$1.25 \times 10^3 = 1250$

$3.5 \times 10^{-4} = 0.00035$

You can count decimal places to convert between standard form and 'ordinary' numbers.

$$\overset{\text{3 jumps}}{1.25 \times 10^3 = 1\,250}$$

$$\underset{\text{4 jumps}}{3.5 \times 10^{-4} = 0.00035}$$

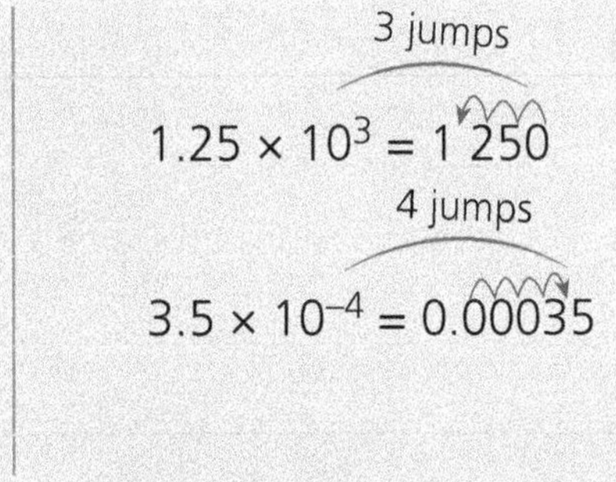

▲ Figure 12.1 Converting between standard form and 'ordinary' numbers

There are two ways to work out if the power of 10 is positive or negative:

▶ If the decimal point jumps to the left, it is a positive power of 10. If it jumps to the right, it is a negative power of 10.
▶ If the number is greater than 10, then the power is positive. If the number is less than 1, the power is negative.

You will find more detail about this topic with further examples in section B1.64.

Materials and chemical properties

Materials science is the study of the properties of solid materials and how they are determined by the chemical and physical composition of the material. Materials science is of great importance in the modern world. It inhabits the space between chemistry, physics, biology and engineering. Therefore, in chemistry, we must study both the physical and chemical properties of substances and understand how they are related to get a full understanding of the substances we are studying.

B1.33 The relationship between the atomic structure and physical and chemical properties of metals

Key terms

Ions: atoms that have lost electrons (positive ions) or gained electrons (negative ions).

Delocalised electrons: 'free' electrons that are not associated with any single atom.

Physical properties

Metals are usually solids at room temperature (mercury is the only metal that is liquid at room temperature), with particles packed closely together in a repeating three-dimensional grid arrangement called a lattice. The atoms in the lattice have lost their outer shell electrons, so the structure consists of metal **ions** and a sea or cloud of **delocalised electrons**. The attraction between the positive ions and the negative electrons holds the particles together in the lattice. This is known as **metallic bonding** and is shown in Figure 12.2. Metallic bonding explains the physical properties of metals.

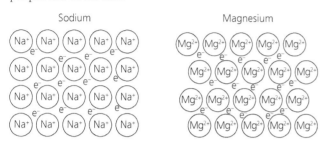

▲ Figure 12.2 A two-dimensional diagram of metallic bonding in sodium and magnesium

Conductivity (electrical and thermal)

The delocalised electrons are free to move throughout the lattice. This means that they can carry an electric current (a flow of charge), so metals are good electrical conductors.

The delocalised electrons can move and vibrate and so transfer thermal energy from one to another through the metal, making metals good conductors of heat.

Malleability/ductility

Metals are **malleable** (they can be hammered into shape) and **ductile** (they can be drawn into wires). This is because of the layered structure. As you can see in Figure 12.3, if a force is applied to a metal, the layers can slide without disrupting the bonding.

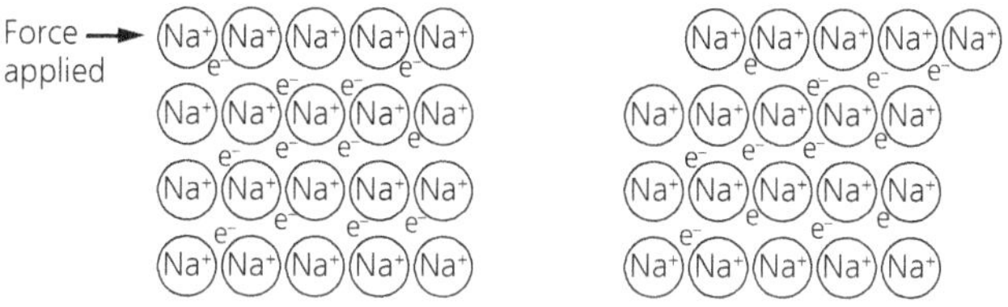

▲ Figure 12.3 Applying a force to part of a metal causes the layers to slide over each other without disrupting the bonding

Strength

The close packing of the particles in a metal explains their high density. You can see from Figure 12.2 that the Mg^{2+} ions have twice the charge of the Na^+ ions and there are also twice as many delocalised electrons. This is because magnesium has 2 electrons in its outer shell, so forms 2+ ions, whereas sodium has only 1 electron in its outer shell, so forms 1+ ions (see section B1.34). This means that magnesium has 2+ ions and twice as many delocalised electrons compared with the 1+ ions and fewer delocalised electrons in sodium. Therefore, there is a greater electrostatic attraction between the positive ions and delocalised electrons in magnesium. As a result, more energy is needed to overcome these forces in magnesium and so magnesium is stronger and has a higher melting point than sodium. The transition metals have even more delocalised electrons, which explains their greater strength and density – for instance, iron or tungsten. These are very hard compared with sodium which is soft enough to be cut with a knife and spread with a spatula.

Chemical properties

See section B1.34 for more information about how elements are divided into groups and blocks.

Group 1

Group 1 metals are all highly reactive. They react with water to form hydrogen gas and the metal hydroxide, for example, sodium:

$$2Na(s) + 2H_2O(l) \rightarrow 2NaOH(aq) + H_2(g)$$

They react with oxygen to form solid white oxides, for example:

$$4Na(s) + O_2 \rightarrow 2Na_2O(s)$$

Group 1 metals all have a single electron in their outer shells and so they form ionic compounds containing 1+ ions (Na^+, K^+, etc.). The outer electrons are further from the nucleus as you move down the group, from lithium to rubidium. This means there is a weaker force of attraction between the negative outer shell electron and the positive nucleus (because of the increased distance and increased shielding from the other electron shells). That means the outer electron is lost more easily in rubidium, making it more reactive.

The reactivity of group 1 metals increases down the group from lithium (least reactive) to rubidium (most reactive).

Transition metals

Transition metals are much less reactive with oxygen and acids compared with either group 1 or group 2 metals. Iron will rust (in other words, form iron oxide), but only when in contact with air containing water vapour – there is no reaction with dry air or oxygen-free water. Other transition metals react even more slowly, which makes them resistant to **corrosion**.

> ### Key term
>
> **Corrosion:** the process where metals react with substances in the air to form oxides, carbonates, hydroxides or other compounds.

The reaction of transition metals with acids is variable – it depends on the metal and the acid.

For example, iron will react with hydrochloric acid to produce iron chloride and with sulfuric acid to produce iron sulfate. In both cases hydrogen is also produced and the iron forms Fe^{2+} ions.

The reaction with nitric acid is a little different because concentrated nitric acid is a stronger oxidising agent than either sulfuric, hydrochloric, or dilute nitric acids. This means that the Fe is oxidised to Fe^{3+} rather than Fe^{2+}. Also, the nitrate (NO_3^-) in nitric acid is reduced to nitrogen monoxide (NO).

Copper is one of the least reactive transition metals and does not react with dilute hydrochloric or sulfuric acids. However, nitric acid and concentrated sulfuric acids are strong oxidising agents, and will both react with copper:

Dilute nitric acid reacts to form copper nitrate and nitrogen monoxide.

Concentrated sulfuric acid reacts to form copper sulfate and sulfur dioxide.

Notice that hydrogen is not produced in either case.

The table compares some properties of transition metals with group 1 metals.

Property	Group 1 metals	Transition metals
Melting points	Low	High
Density	Low – lithium, sodium and potassium are less dense than water	High
Hardness and strength	Soft and weak (can be cut with a knife)	Hard and strong
Reactivity with:		
Oxygen	High to very high (need to be stored under a protective layer of oil)	Slow to very slow
Chlorine	React vigorously to produce solid metal chlorides	Less reactive. Iron wool heated strongly will react with chlorine to produce iron (II) chloride
Water	Vigorous to very vigorous	Relatively unreactive

> ### Reflect
>
> Think about metallic bonding – positive ions surrounded by a sea of delocalised electrons.
>
> Can you use this to explain the difference in physical properties between the group 1 metals and the transition metals?

The relationship between the structure and properties of the following materials

Composite materials

As the name suggests, composite materials are made from two or more materials that have different properties. Reinforcing fibres or particles are embedded in a softer matrix that helps to bind the material together. Some common examples include:

- Concrete, used since Roman times, consists of small stones embedded in a matrix of sand and cement.
- Fibreglass has the strength of glass but is not brittle. Relatively rigid glass fibres act as reinforcement in a flexible resin matrix. Fibreglass can be created in a mould to produce a strong but lightweight material.
- Carbon fibre consists of carbon nanotubes embedded in a polymer matrix. It can combine great strength in a lightweight material. Although it is more expensive than metals such as iron and steel, carbon fibre is used increasingly in engineering applications (automotive and aircraft), sports equipment (golf clubs and tennis racquets) and many others.

Ceramics

Ceramics are made from materials such as clay, sand and other minerals that are moulded and then baked to form strong bonds between the atoms in the structure. This makes them hard, strong under compression and chemically unreactive.

- Clay ceramics have been made for thousands of years by moulding clay and heating in a furnace to produce decorative items and tableware.
- Glass is made from sand (a silicate) with either sodium carbonate and limestone to make soda-lime glass or with boron trioxide to make the stronger, heat-resistant borosilicate glass.

Polymers

Polymers are long chain molecules made from repeating monomer units. For example, poly(ethene) is a polymer with many thousands of repeating units based on ethene.

Some polymers have weak forces between the chains – this makes them softer and more flexible – while others have strong bonds holding the chains in a stronger and more rigid structure.

Polymers are chemically unreactive and electrical insulators (as they do not have any delocalised electrons to carry a current). This makes them very useful for storing and packaging items to keep them from changing.

Poly(ethene) is known commonly as **polythene**. Low density poly(ethene) (LDPE) and high density poly(ethene) (HDPE) are examples of **thermosoftening** polymers. Weak forces between the molecules allow the chains to slide over each other. This means they can be heated to soften them, allowing them to be moulded and reformed. Recycling thermosoftening polymers is therefore relatively easy and it is used in various consumer products to take advantage of this fact.

- HDPE is stronger and more rigid because it has a higher density. This is because it contains unbranched chains that can pack together more tightly. HDPE is used for bottles and other containers.
- LDPE is weaker and more flexible because it has a low density. This is because the chains are branched and therefore cannot pack together as easily. LDPE is used for bags, sheets and films.

Thermosetting polymers use heat to create cross-links between the chains. This makes the polymer stronger and less flexible. However, thermosetting polymers cannot be heated and reformed, which means they are less easily recycled.

Research

Make a list of the different types of material covered in this section. Choose at least one of each type.

Then make a list of the different uses of the materials. For each example, try to show how the properties are related to the use.

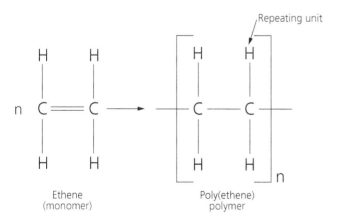

▲ Figure 12.4 Formation of poly(ethene) from ethene. The square brackets with n indicates that the repeating unit is repeated n times (i.e. many times)

The simplest definition of an **atom** is a **nucleus** surrounded by a cloud of **electrons**. The atom is more complex than this, but this model of the atom can be useful in some circumstances.

The **nucleus** is at the centre of the atom. It consists of **protons** and **neutrons** and contains most of the mass.

Elements consist of only one type of atom. **Compounds** consist of two or more types of atoms.

Electrons do not orbit the nucleus like planets orbiting the sun. Instead, they are located in **shells**, numbered from 1 to 7. Each shell is higher energy than the previous.

We can think of each shell being further from the nucleus than the last, and you will often see shells represented like this in diagrams. However, this can be misleading, making it look as if the electrons really are orbiting the nucleus. It is better to think of these 7 shells as being the 7 principal energy levels.

Electrons within each shell are located in various orbitals – s, p, d and f. The **orbitals** are found in different shells as follows:

- Each shell has only one s orbital.
- In shell 2 and above, there are also p orbitals.
- In shell 3, there are also d orbitals.
- In the highest shells, there are also f orbitals.

The modern periodic table is arranged in increasing order of **atomic number**, but the elements are also arranged according to their electronic structure into the s, p, d and f blocks.

The periodic table is also organised into **groups** – the columns in the table (Figure 12.5). Elements in the same group have similar chemical properties. This is because they have the same number of electrons in the outer shell.

The periodic table can be divided into blocks depending on which type of orbital the outer electrons are located in (Figure 12.6):

- Group 1 (the 'alkali metals') have one electron in the outer shell. It is in an s orbital, so these are all in the s block.
- Group 2 (the 'alkaline earths') have two electrons in the outer shell, both in the s sub-shell, so these are also in the s block.
- Groups 3 through 0 have from three to eight electrons in the outer shell. These are all in p orbitals, so these are in the p block.
- The d block includes the transition metals.

> ### Key terms
>
> **Atomic number:** refers to the number of protons in the nucleus.
>
> **Group:** refers to the columns in the periodic table. Elements in each group have the same number of outer shell electrons. Period refers to the rows in the periodic table. Elements in each period have the same number of shells.
>
> **Orbitals:** are where electrons are located. Each orbital can be empty or can contain one or two electrons.

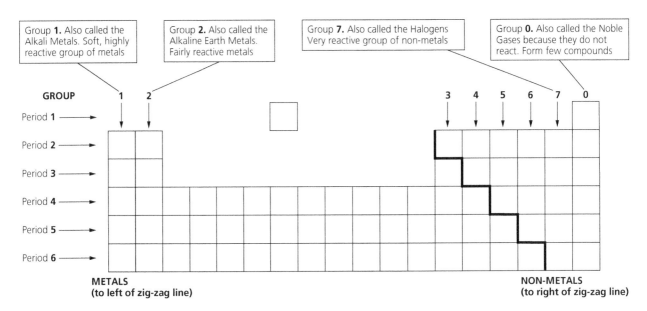

Group **1.** Also called the Alkali Metals. Soft, highly reactive group of metals

Group **2.** Also called the Alkaline Earth Metals. Fairly reactive metals

Group **7.** Also called the Halogens Very reactive group of non-metals

Group **0.** Also called the Noble Gases because they do not react. Form few compounds

METALS (to left of zig-zag line)

NON-METALS (to right of zig-zag line)

▲ Figure 12.5 The periodic table showing the groups and periods

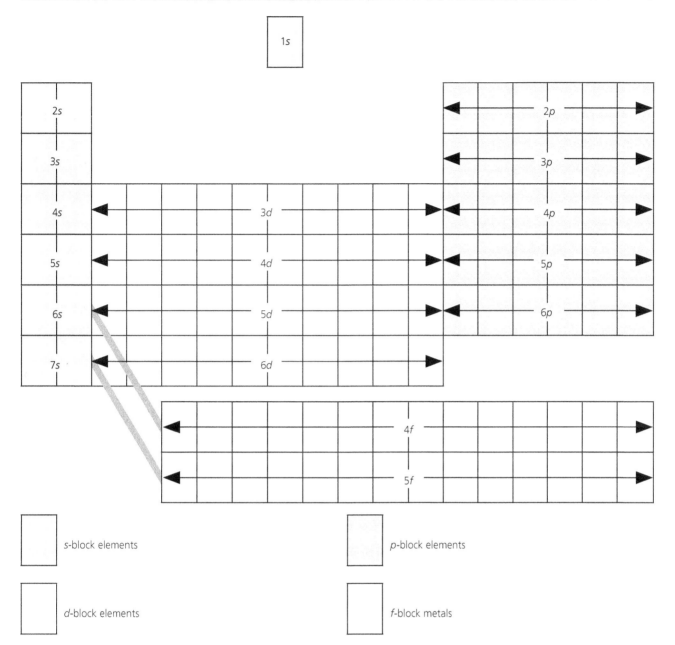

▲ Figure 12.6 The periodic table divided into blocks according to electronic structure

B1.35 Sub-atomic particles – protons, electrons and neutrons

An atom was originally thought of as being the smallest particle that an element could be broken down into. We now know that is not the case. Nuclear physics studies the many types of sub-atomic particle, but in chemistry we are only concerned with three types:

Protons are found in the nucleus and have a charge of +1 and a relative mass of 1.

Neutrons are also found in the nucleus. They have no charge but also have a relative mass of 1.

Electrons are found in orbitals around the nucleus and have a charge of –1. The relative mass of electrons is approximately 1/2000th that of a proton or neutron.

From this, you can see that the majority of the mass of an atom is in the nucleus.

Acids/bases and chemical change

pH is a measure of the hydrogen ion concentration. It is a **logarithmic scale**, usually from 0 to 14, although pH values can go above or below this range. Because it is a logarithmic scale, a change of one pH unit means the hydrogen ion concentration changes by a factor of 10. At room temperature, pure water has a pH of 7, which is considered the neutral point. Below pH 7 the solution is **acidic** and above pH 7 the solution is **alkaline**. **Neutralisation** occurs when acid and **base** react to form water and a salt.

Acid-base reactions are particularly important in chemistry. As well as being used to determine concentrations in titration (see section B1.44), they are used for preparation of a wide range of salts.

B1.36 The physical properties of acids

Acids release hydrogen (H^+) ions – we say that they are H^+ **donors**. This explains the properties of acids:

- They are an irritant (cause inflammation of the skin) and often corrosive.
- They react with bases in a neutralisation reaction to produce a salt and water (see B1.38).
- They react with most metals to form hydrogen gas (H_2).
- Because they have a high concentration of H^+ ions, they have a pH value less than 7.

B1.37 The concept of strong and weak acids

It will be useful to look at sections B1.62 and B1.63 on units before going further if you are at all unsure about use of the different units used for concentration in this section. Anyone who has tried drinking orange squash without diluting it will be familiar with the idea of concentrated and dilute solutions. To use a more chemical example, if we prepare a solution containing 0.1 mol hydrochloric acid in $1\,dm^3$ water, the hydrochloric acid will dissociate completely into its ions:

$$HCl(aq) \rightarrow H^+(aq) + Cl^-(aq)$$

In this equation, '(aq)' means **aqueous**, a solution in water. The concentration of H^+ will be $0.1\,mol/dm^3$ and the solution will have a pH = 1.00. This process is usually described as **dissociation** of the acid (into its constituent ions), but you will also come across the term **ionisation** to describe the same thing.

We can do the same with nitric acid or sulfuric acid:

$$HNO_3(aq) \rightarrow H^+(aq) + NO_3^-(aq)$$

$$H_2SO_4(aq) \rightarrow 2H^+(aq) + SO_4^{2-}(aq)$$

The only difference with sulfuric acid is that the concentration of H^+ will be $2\,mol/dm^3$ because 1 mol of sulfuric acid releases 2 mol of H^+.

In contrast, if we have a $1\,mol/dm^3$ solution of ethanoic acid and measure the pH, it will be 2.88 because the H^+ concentration is only $0.0013\,mol/dm^3$. This is because ethanoic acid is a weak acid, which means that it only partially **dissociates** in aqueous solution. We can represent this as the following reversible reaction:

$$CH_3COOH(aq) \leftrightharpoons H^+(aq) + CH_3COO^-(aq)$$

A **reversible reaction**, also known as an equilibrium reaction, is represented by the double arrow $\leftrightharpoons$ and can move in either direction. In any **equilibrium**, there will be a mixture of all the components (reactants and products) in varying proportions. In the case of weak acids, such as ethanoic acid, the position of the equilibrium lies well to the left, i.e. only a small amount will dissociate – most will remain as CH_3COOH.

Once we understand the difference between strong/weak and concentrated/dilute, it should be clear that we can have a dilute solution of a strong acid and a concentrated solution of a weak acid.

You will also see from the above that when we have solutions of the same concentration, e.g. $0.1\,mol/dm^3$, the pH of the strong acid (pH = 1.00) will be lower than that of the weak acid (pH = 2.88). If you consider the H^+ concentration you will see that ethanoic acid (the weak acid) has a H^+ concentration almost one thousand times lower than the hydrochloric acid.

This illustrates another important feature of pH. For each one unit decrease in pH, the H^+ concentration increases by a factor of 10.

B1.38 How to determine the name of the salt produced in acid-base reactions

Bases, such as sodium or potassium hydroxide, react with acids, such as hydrochloric acid, in a neutralisation reaction to produce a salt plus water. For example, with hydrochloric acid and sodium hydroxide:

$$NaOH(aq) + HCl(aq) \rightarrow NaCl(aq) + H_2O(l)$$

Group 2 hydroxides (hydroxides of Group 2 elements) react in the same way. The salt formed always takes the name of the metal (positive ion) together with the name corresponding to the acid used. Ammonia solution (ammonium hydroxide) reacts in the same way to form ammonium salts.

The table shows the names of salts formed from some common acids.

> ### Practice points
>
> The table shows the modern (systematic) names of the chemicals. You will probably encounter older, non-systematic names as well, such as 'acetate' for 'ethanoate'. There are many other examples, although we will not be covering those chemicals here.

Acid	Formula of acid	Type of salt formed	Formula of anion	Example
Hydrochloric	HCl	Chlorides	Cl^-	Sodium chloride, NaCl
Sulfuric	H_2SO_4	Sulfates	SO_4^{2-}	Sodium sulfate, Na_2SO_4
Nitric	HNO3	Nitrates	NO_3^-	Sodium nitrate, $NaNO_3$
Phosphoric	H_3PO_4	Phosphates	PO_4^{3-}	Sodium phosphate, Na_3PO_4
Ethanoic	CH_3COOH	Ethanoates	CH_3COO^-	Sodium ethanoate, CH_3COONa

> ## Test yourself
>
> 1 What is meant by:
>   a an acid   b a base?
> 2 Explain the difference between a concentrated solution of a weak acid and a dilute solution of a concentrated acid.
> 3 Name the salts produced in the following acid-base reactions:
>   a Potassium hydroxide and nitric acid.
>   b Calcium hydroxide and phosphoric acid.
>   c Ammonia solution and sulfuric acid.

Rates of reaction and energy changes

We need to understand the factors that affect the rate of chemical reactions. That understanding is important if we are running a chemical factory producing ammonia for use in artificial fertiliser or preparing a sample of a medicine such as aspirin in the laboratory.

B1.39 The principles of collision theory

Our understanding of rates of reaction and energy change in chemical reactions is based on **collision theory**. This was originally worked out for reactions between gases, but the principles also apply to reactions in solution. We can summarise this theory in three statements:

1. Molecules must collide in order to react.
2. Molecules must have sufficient energy when they collide. Chemical reactions involve breaking chemical bonds (which requires energy) before new bonds are made (which releases energy). The energy required to break bonds is known as the activation energy. If molecules that have less energy than the activation energy collide, they will just bounce off each other without reacting.
3. Molecules must be in the correct spatial orientation when they collide. The bonds being broken and reformed will be in specific positions in space. This means molecules must be aligned correctly when they collide – see Figure 12.7.

B1.40 The effect of temperature on rates of reaction

As temperature increases, the kinetic energy of the molecules increases. This means they move faster. If they move faster, they are more likely to collide. So increasing temperature increases the rate of reaction in two ways:

- The probability of collision increases.
- The proportion of molecules with sufficient energy to react also increases.

On the other hand, decreasing temperature decreases the kinetic energy of molecules. Therefore, molecules move slower, decreasing the probability of successful collisions and hence decreases the rate of reaction.

B1.41 The definition of a catalyst and the role of catalysts in a reaction

Catalysts increase the rate of chemical reactions. They do this because they provide an alternative reaction pathway that has a lower activation energy. This means that more molecules will have sufficient energy to react and this will increase the rate of reaction. In short, more reactants will exceed the activation energy and so can successfully collide.

Transition metals are often used as catalysts. In some reactions between gases the solid transition metal catalyst provides a surface for the reaction to take place. In other cases, the transition metal takes part in the reaction. Acids are common catalysts in organic chemistry. In both cases, the catalyst will participate in the reaction but will be reformed at the end of the reaction. This means that the catalyst is not **permanently** changed. It is used, but not used up. In all cases, remember the catalyst provides an alternative reaction pathway that has a lower activation energy.

Test yourself

1. What are the three statements of collision theory?
2. Explain why temperature increases the rate of reaction.
3. Explain how a catalyst increases the rate of reaction.

Unsuccessful collision as the OH⁻ approaches the I$^{\delta-}$

Successful collision as the OH⁻ approaches the C$^{\delta+}$

O   C

I   H

▲ Figure 12.7 Effect of orientation on reaction outcome

Chemical analysis of substances

Our knowledge of chemistry is built upon the ability to analyse substances. We need to understand the composition of substances and how this changes when they undergo chemical reactions.

B1.42 The principles of tests and techniques that are used to separate, detect and identify chemical composition

Separation is central to most forms of analysis. As well as separating the components of a mixture, we sometimes need to **quantify** them (work out how much of them there is).

Examples of analysis based on separation of the components of a mixture include:
- detecting additives in foodstuffs
- analysing urine samples to detect use of performance-enhancing drugs in sport (doping)
- determining the purity of pharmaceutical raw materials.

List as many types of substance you might need to analyse as you can. For each one, decide whether it would be enough to know what is present (**qualitative analysis**) or whether you would need to know how much of each substance is present (**quantitative analysis**).

Key terms

Chromatography: the separation of the components of a mixture dissolved in a liquid or gas (the mobile phase) carrying it through a structure holding the stationary phase.

Adsorption: when a substance (e.g. a gas, liquid or solute) binds to or attaches to another, usually solid.

Adsorbent: often used to describe the stationary phase in chromatography because substances become adsorbed to it during separation.

In all types of **chromatography**, separation depends on substances in a mixture having a different **affinity** or attraction towards two phases. The **stationary phase** is fixed while the **mobile phase** (a liquid or gas)

is able to move. It is important, in this context, that we use the term **adsorption** for the process of binding to a material in the stationary phase. (This is not the same as **absorption**, which is a term sometimes used, incorrectly, in this context. In absorption, one substance is taken in or **absorbed** by another, rather like a sponge absorbing water.)

Thin layer chromatography

Thin layer chromatography (TLC) is used to separate non-volatile mixtures such as amino acids, pharmaceuticals or dyestuffs. These are substances that cannot be easily vaporised. Separation is based on their solubility in the mobile phase (solvent) or affinity for (attraction to) the stationary phase (on a coated plate). The stationary phase is a thin layer of **adsorbent** such as silica gel, alumina or powdered cellulose on a flat, **inert** (non-reactive) support, such as glass or (more usually) plastic.

TLC can be used to detect the number of components in a mixture. The stages are:
- A pencil line (the **origin**) is drawn about 1 cm from one short edge of the TLC plate.
- The sample or samples are applied in solution at various points along the origin and left to dry.
- The plate is placed vertically in a container with a shallow layer of solvent, so that the origin is above the level of the solvent (Figure 12.8).
- The container is covered or sealed to prevent evaporation. The solvent will be drawn up the paper by **capillary action** (the process where a liquid is drawn into narrow spaces like those between the fibres in filter paper or a kitchen towel).
- When the solvent reaches the origin, substances in the sample will dissolve and begin to move.
- Substances with greater affinity for the stationary phase will move more slowly than substances with a greater affinity for the solvent and so the mixture will become separated.
- After the solvent has moved far enough up the paper (usually almost to the top), the plate is removed from the container.
- The position that the solvent has reached (the **solvent front**) is marked.

Efficient separation requires choice of solvent (mobile phase) so that the different components of the mixture will have different solubility in the mobile phase. If all components are equally soluble, they will all move the same distance. If they are insoluble, they will remain at the origin. Therefore, different solvents or mixtures of solvents with a range of **polarities** (ability to mix or dissolve) are used depending upon the substances being separated.

In the same way, the nature of the stationary phase will determine how far a particular substance moves. If a substance has a high affinity for the stationary phase, it will not move far up the plate. If it has a low affinity for the stationary phase, it will move further.

TLC is often used to separate the different coloured dyes in a mixture or to analyse the different coloured pigments in a plant extract. In other cases, the components of the mixture are colourless. This means we need to make the spots visible in some way. A common method is to use a dye that will bind to the chemicals in our mixture:

▷ Ninhydrin will stain amino acids purple. If a TLC plate is used to separate a mixture of amino acids, it can be dried and then sprayed with ninhydrin solution. The amino acids spots are then stained purple.

▷ Iodine vapour will stain many chemicals brown.

Another method to make the spots visible is to use a TLC plate that contains an inert fluorescent dye. After the separation is done and the plate is dried it can be illuminated with UV light. The spots will appear dark on a bright background.

If we analyse a substance by TLC and see just a single spot, it suggests the substance is pure.

However, it is possible that two substances move the same distance. To be sure a sample is pure, we need to repeat the analysis in a different system, for example, with a different solvent mixture or different adsorbent.

As well as separating the components of a mixture, paper chromatography and TLC can be used to identify the components. One way is to run pure samples (**standards**) alongside the mixture. The distance travelled by a component of the mixture can then be compared to the distance travelled by one of the standards.

Another method is to calculate the **retention factor** or **R_f value**. R_f values can be used in TLC to identify unknowns based on standard published literature values.

Column chromatography

Column chromatography uses similar stationary phases to TLC but a much wider range is available, particularly those used in the separation of biologicals. In all cases, the sample is applied to the top of the column and **eluted** with a suitable mobile phase, the **eluent**. The stationary phase runs the entire length of the column. The substances then travel down the column based on their affinity for the eluent. Substances with a higher affinity for this mobile phase, will be separated and eluted earlier from the bottom of the column. The advantage of column chromatography is that the **eluate** can be collected in small amounts

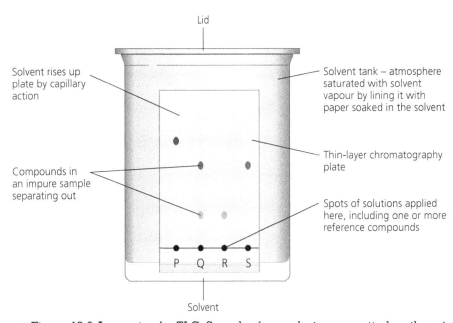

▲ Figure 12.8 Apparatus for TLC. Samples for analysis are spotted on the origin at P, Q, R and S

(**fractions**) as it emerges from the bottom of the column. Different substances will be in the different fractions because they elute from the column at different times. This allows column chromatography to be used for purification of a single chemical compound as well as for analysis.

> ### Key terms
>
> **Elution:** to wash out. In column chromatography this means 'washing out' a substance that has become adsorbed to the column (stationary phase).
>
> **Eluent:** the solvent (mobile phase) used to wash substances out of a column.
>
> **Eluate:** the mobile phase, containing dissolved substances, as it emerges from a column.

Gas chromatography

Gas chromatography (GC) is used to separate and analyse **volatile** compounds (ones that can be vaporised). GC uses an inert carrier gas as the mobile phase. The stationary phase can be a thin layer of high boiling point liquid on an inert solid support packed into a column. Substances that have a higher affinity for the stationary phase, will interact with it more, meaning it will take longer for it to emerge from the column. Substances with a lower affinity for the stationary phase, will then emerge sooner and be detected first. More recently, capillary GC uses a polymer lining a very fine capillary column as the stationary phase. The column will be coiled inside an oven to maintain the relatively high temperature needed (Figure 12.9).

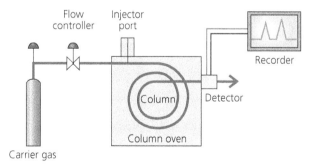

▲ Figure 12.9 Apparatus for GC

The sample is injected into the gas stream and the components of the mixture being separated will interact with the stationary phase to different degrees and so will emerge from the column at different times. The substances emerging from the column are detected, usually by a **flame ionisation detector (FID)** or a **thermal conductivity detector (TCD)**. These work with a wide range of substances, although other detection methods are available for specific applications.

The time taken between injection and detection of a particular component is known as the **retention time**. We saw how R_f values can be used in TLC to identify unknowns based on standard published literature values. The same is true of retention time in GC. However, conditions used for the analysis must be identical to those used when determining the standard values. This includes use of the same column (not just the same stationary phase), mobile phase, temperature, etc.

Another way to confirm the identity of a substance in a mixture is to add a purified sample of that substance (a standard) to the sample when it is injected onto the column (GC or HPLC, see below). If the substance in the mixture **co-elutes** with the standard (i.e. has the same retention time), it is strong evidence of identity.

High performance liquid chromatography

High performance liquid chromatography (HPLC) is a type of column chromatography that uses very small particles and high pressures to achieve better separation. The very fine particles make it difficult to force the solvent through and so HPLC requires powerful pumps and pressure-resistant columns to be fitted. The sample being analysed or purified cannot simply be applied to the top of the column – the system is sealed and under pressure – so the sample needs to be introduced by injection through a valve or port. The advantages of using HPLC for separation are much greater speed and higher resolution (ability to separate two very similar substances). HPLC has become one of the standard methods of separation in analytical laboratories. It can also be scaled up to operate as a purification method on a much larger scale, handling grams or even kilograms of substance.

A more recent development of HPLC, **ultra-high performance liquid chromatography** or **UPLC**, uses capillary columns to separate mixtures before analysis by mass spectrometry.

Mass spectrometry

Mass spectrometry (MS) can identify the amount and type of compound, which makes it very useful in identifying unknown compounds. The sample is ionised by removing electrons. The mass spectrometer then measures the ratio of mass to charge (m/z) of the positive ions produced. This can be done in various ways:

- By measuring how far the ions were deflected by a magnetic field; ions with greater mass are deflected less (Figure 12.10).
- More recent methods such as **time of flight (TOF)** measure how long it takes for ions to reach the detector; ions with greater mass are slower and take longer.

MS provides the molar mass (M_r) of the compound, but additional information about the structure of the compound can be obtained from the way in which the ion breaks up (fragments) in the mass spectrometer.

By feeding the output of a GC column into a mass spectrometer, it is possible to identify the compounds being separated much more accurately than simply by retention time. This technique, known as GC-MS, is used widely because of the compact nature of the equipment, its speed and relatively low cost. Applications (uses in the real world) include airport screening for drugs and explosives, fire forensics (investigating the causes of a fire) and space exploration. Probes containing miniaturised GC-MS have been sent to Mars, Venus and Titan.

Research

A similar approach to GC-MS can be taken with liquid chromatography. In LC-MS, a very small capillary column is used for separation of complex mixtures such as proteins and peptides. Other applications involve analysis of drug metabolites in pharmacology and anti-doping laboratories. An internet search for 'applications of LC-MS' will help you learn more.

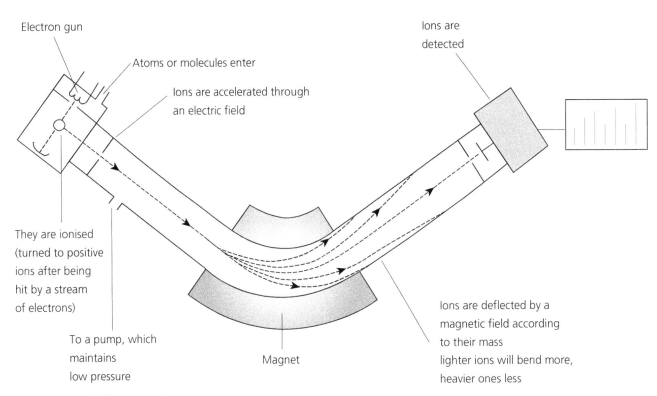

Electron gun

Atoms or molecules enter

Ions are accelerated through an electric field

Ions are detected

They are ionised (turned to positive ions after being hit by a stream of electrons)

To a pump, which maintains low pressure

Magnet

Ions are deflected by a magnetic field according to their mass lighter ions will bend more, heavier ones less

▲ Figure 12.10 Diagram of a mass spectrometer

B1.43 The tests that could be used to quantify components in a mixture

GC, HPLC and MS are all powerful analytical techniques that separate the components of a mixture. This means they can be used to determine whether a sample is pure or is contaminated with other substances. To be really useful, we need to be able to calculate the percentage of each component, so that we can work out the percentage purity.

In both GC and HPLC, substances emerging from the column pass through a detector. The output from the detector gives a type of graph called a chromatography trace where the area under the peaks is proportional to the amount of the substance. This data is captured by a computer and analysed to show the percentage of each component.

It is also possible to use a standard of known content to convert the relative values we get from measuring peak area into absolute values, i.e. actual masses, in µg or mg.

In mass spectrometry, the output gives the relative abundance of each component. It is possible to determine the absolute quantity of each component using standards made with heavy isotopes.

B1.44 The principle of titration

Neutralisation occurs when an equal number of **moles** of H^+ (from an acid) and OH^- (from a base) react together to form water. If a strong acid, such as hydrochloric acid, reacts with a strong base, such as sodium hydroxide, in exactly equal proportions, the mixture will be neither acid nor base but neutral (pH 7.0). At this point an **indicator** will change colour. We use this in the process of **titration** to calculate the concentration of a solution of acid or base.

In titration we have one unknown – the number of moles in the **analyte**. We titrate this against the **standard solution**. Titration involves adding the standard solution from a **burette** to a known volume of the analyte in a conical flask until we reach the **end point** – when the indicator changes colour. This should be the **equivalence point**, where we have equal moles of acid and base (H^+ and OH^-). Because we know the concentration of the standard solution, we can use the **titre** to calculate how many moles were required to

reach the equivalence point. This will tell us the number of moles of analyte in the flask and therefore the concentration of the analyte.

Key terms

Mole: an amount of substance. This helps us to work out the reacting proportions in any reaction. We can also use the mole to work out reacting masses or volumes. The abbreviation is **mol**.

Indicator: a substance that changes from one colour to another or from coloured to colourless depending on whether it is in acidic or basic solution.

Analyte: the solution of unknown concentration in a titration.

Standard solution: the solution of known concentration in a titration.

Burette: a long glass tube that has a tap at the bottom and is marked in $0.1cm^3$ divisions. It is used to deliver an accurate volume of liquid (the **titre**) to reach the **end point**.

End point: the point in a titration where the indicator changes colour.

Equivalence point: the point of neutralisation where the number of moles of acid and base are equal. This should ideally be the same as the **end point**.

Titre: the volume of standard solution needed to neutralise the analyte (i.e. to reach the end point of the titration).

Test yourself

1 Explain the difference between the terms adsorption and absorption.
2 Explain what is meant by the mobile phase in chromatography.
3 Give two examples of a stationary phase.
4 Explain the similarities and differences between column chromatography and HPLC.
5 State the meaning of the following terms:
   a titre
   b end point
   c equivalence point.

Project practice

You have been asked to prepare visuals for a school outreach display that your company is supporting.

Choose one of the following themes and prepare a poster or PowerPoint illustrating how this is relevant to your workplace. Use resources from the internet where necessary.

▶ 'The right material for the job' – the ways in which different types of material are suited to particular applications in science or healthcare.

▶ 'Not just a corrosive liquid' – the importance of strong and weak acids in industry, foodstuffs and analysis.

▶ 'Time is money' – the importance of rates of reaction in industrial processes.

▶ 'What's in my lunch?' – the use of chemical analysis to test the quality and purity of the food and drink we consume.

Your presentation should include at least one graph, chart or diagram to help illustrate the relevance of your theme to your workplace.

Assessment practice

1 Explain the reasons for the properties and the uses of the following metals:

  a Sodium is soft and low density, but iron is hard and high density.

  b Sodium reacts vigorously with cold water, but iron only reacts very slowly with water.

  c Copper is used to make electric wiring.

  d Copper is used to make pipes in central heating systems.

2 The table shows a range of applications with the corresponding material. For each combination, select the appropriate explanation for why the material is suited to the application.

A Lightweight, can be formed into wires (ductile) good electrical conductor

B Lightweight, strong

C Strong, can be moulded into shape (malleable)

D Inert, strong and can be moulded into shape

E Transparent, inert, heat-resistant

F Strong, inert, very heat resistant

G Strong, can be formed into shape, non-conductor of electricity

Application	Material	Explanation (letter)
Body panel for budget-priced car	Steel	
Body panel for high performance sportscar	Carbon fibre composite	
Bottle for storage of corrosive liquids	High density polyethene (HDPE)	
Crucible for heating solids at high temperatures	Clay ceramic	
Flask for use in reflux where the contents are heated for a prolonged period	Borosilicate glass	
Insulator for use in high-voltage power lines	Clay ceramic	
Cable for use in high-voltage overhead power lines	Aluminium	

3 Use the periodic table to complete the following table.

Element	Group	Period
Sodium		
Magnesium		
Iron		
Nitrogen		
Neon		

4 Classify the following as to whether they are (i) strong or weak and (ii) concentrated or dilute:
  a a 5 mol/dm^3 solution of ethanoic acid
  b a 0.001 mol/dm^3 solution of hydrochloric acid
  c a 10 mol/dm^3 solution of sulfuric acid.

5 Name the salts that would be produced by mixing each of the following pairs of acid and base:
  a nitric acid and calcium hydroxide
  b sulfuric acid and barium nitrate
  c hydrochloric acid and ammonia
  d ethanoic acid and potassium hydroxide.

6 One method of preparing a salt involves adding an excess of insoluble substance such as a metal, metal oxide, metal hydroxide or metal carbonate to an acid. Suggest how you could prepare a solution of copper sulfate from insoluble copper oxide and sulfuric acid. Explain the advantage of this method over mixing an acid and a soluble base.

7 Collision theory states that particles with sufficient energy must collide in order to react.
  a Explain why increasing the concentration of reactants in solution will increase the rate of reaction.
  b Suggest why increasing the temperature has a bigger effect on rate of reaction than increasing the concentration.

8 You have been asked to analyse the amino acids in a food supplement to show that it contains the amino acid lysine.
  a Outline how you would use TLC with ninhydrin spray and a pure sample of lysine to do this.
  b You find that the food supplement contains two spots. The impurity spot moves a shorter distance than lysine. What can you tell about the properties of this impurity?
  c Describe how you could use chromatography to obtain a sample of the impurity for further analysis.

9 You have been asked to analyse a sample of petrol to see if it is contaminated. You are provided with a pure sample of the potential contaminant.
  a Explain how you could use GC to show that the contaminant was present in the sample.
  b Explain how you could use MS coupled to GC to identify an unknown contaminant in the sample.

10 Describe how you would determine the percentage purity of the following:
  a crude pellets of sodium hydroxide
  b a volatile compound for use in food flavouring using gas chromatography.

B1.45–B1.64: Core science concepts: Physics

Introduction

Physics is the study of atoms, particles, energy, forces, mechanics and waves – in fact, the whole physical universe.

There is overlap with chemistry in places – atomic structure, for example. The overlap with biology is less obvious, but biophysics is an important part of modern biology. It uses physical methods to understand the interactions between biological molecules and within biological systems. Also, physics is based very firmly on a foundation of mathematics. In fact, mechanics is often an option in mathematics courses and the difference between theoretical physics and applied mathematics can be hard to spot! The field of medical physics is another important area of overlap, where physics contributes to treatment and diagnosis.

Whichever way you look at it, a good understanding of physics is important for work in many areas of science, healthcare and engineering.

Learning outcomes

The core knowledge outcomes that you must understand and learn:

Electricity
B1.45 the definitions of, and how to calculate, charge and current using $Q = It$

B1.46 the definitions of, and how to calculate, current, potential difference and resistance, using Ohm's law $V = IR$

B1.47 how to calculate total resistance of multiple fixed resistors in a series and parallel circuit

B1.48 the difference between alternating and direct current

B1.49 the properties of mains electricity in the United Kingdom

Magnetism and electromagnetism
B1.50 magnetism and magnetic poles

B1.51 magnetic fields

B1.52 the uses of electromagnetism and electromagnets

Waves
B1.53 the definition of a wave

B1.54 the relationship between frequency, wavelength and speed using the wave equation $v = f\lambda$

B1.55 the properties of longitudinal and transverse waves

B1.56 the uses of different types of waves

Particles and radiation
B1.57 the types and properties of ionising radiation

B1.58 the definitions of half-life and count-rate

B1.59 the main types of radioactive decay in relation to unstable nuclei

B1.60 how radiation interacts with matter

B1.61 the applications of radioactivity within the health and science sector

Units
B1.62 the use of the international system of units (SI)

B1.63 how to convert between units

B1.64 the importance of using significant figures and science notation.

Electricity

Electricity in the home, laboratory or industry is usually **alternating current (AC)**. However, low-voltage lighting circuits or anything that you plug into a 5V USB power supply (like a phone charger) will use **direct current** or **DC**.

We now know that an electric **current** is caused by a flow of electrons, i.e. from negative to positive poles of a battery or power supply. However, electricity and electric currents were discovered long before electrons. Investigators realised that something was flowing – which they called **current** – but did not know what it was. They decided that it must flow from positive to negative.

Although we understand electricity better now, we still follow this convention. You will sometimes see this described as **conventional current**, to distinguish it from flow of electrons. There has been so much work done assuming that current flows from positive to negative that we would have to rewrite too many formulae and textbooks!

Why are electrons negative and protons positive? What do we mean by positive and negative?

Those are philosophical questions that we could spend hours thinking about. But what is important is the fact that electrons and protons have *opposite* charges, and we just call one 'negative' and the other 'positive'.

B1.45 The definitions of, and how to calculate, charge and current using $Q = It$

The size of the **current** is the rate of flow of **charge**. That means that when current (I) flows past a given point in a circuit for a length of time (t), we can calculate the charge (Q) that has flowed past this point using the formula:

$$Q = I \times t$$

Where Q is in **coulombs**, I is in amps and t is in seconds. This can be rearranged to give:

$$I = \frac{Q}{t}$$

In other words, current equals charge divided by time. This is the amount of coulombs of charge per second, which is the rate of flow of charge, which is the definition of current.

To give an example, if a battery charger passes a current of 3.0A through a rechargeable cell (battery) for a period of two hours, the total charge transferred to the cell can be calculated (don't forget to convert hours to seconds!):

$$Q = I \times t = 3.0 \times (2 \times 60 \times 60) = 21\,600\,\text{C}$$

> ### Key terms
>
> *Charge:* a fundamental property of many subatomic particles. Electrons, by convention, have a negative charge. The unit of charge is the **coulomb (C)** and the symbol is Q.
>
> *Coulomb (C):* the unit of charge. **Current** is the rate of flow of charge past a given point in a circuit, i.e. how fast it flows past. The unit of current is the ampere (A), often shortened to 'amp', and the symbol is I.

B1.46 The definitions of, and how to calculate, current, potential difference and resistance, using Ohm's law $V = IR$

The relationship between **current**, **potential difference** and **resistance** is one of the most useful concepts in electronics:

$$V = I \times R$$

Potential difference (**voltage**) is the driving force that causes charge to flow round a circuit. It is defined as the electrical work done per unit of charge flowing through components in the circuit. The unit of potential difference is the volt (V) and the symbol is also V. Potential difference is often abbreviated to 'p.d.'.

Resistance is the property of any component of a circuit that slows the flow of current. The unit of resistance is the ohm (Ω) and the symbol is R.

To calculate the current that is flowing in a circuit with resistance $3\,\Omega$ and a potential difference of 12V, we simply rearrange the formula and substitute the values:

$$I = \frac{V}{R} = \frac{12}{3} = 4\,\text{A}$$

221

B1.47 How to calculate total resistance of multiple fixed resistors in a series and parallel circuit

For resistors connected in **series**, the total potential difference (p.d.) of the supply is shared between the components (see Figure 13.1) and the current through each component is the same (see Figure 13.2).

Key terms

Series: circuits where the components are connected in line, end to end between the positive and negative terminals of the power supply. See Figure 13.1 for an example of a series circuit.

Parallel circuits: where the components are each connected separately to the positive and negative terminals of the power supply. See Figure 13.3 for an example of a parallel circuit.

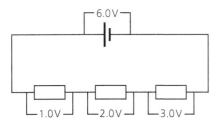

▲ Figure 13.1 Circuit diagram showing p.d. across resistors connected in series

▲ Figure 13.2 Circuit diagram showing current in series circuits

When resistors are connected in series, the total resistance is equal to the sum of the individual resistors, so the total resistance of a circuit with n resistors is given by:

$$R_{total} = R_1 + R_2 + R_n$$

In **parallel circuits**, the p.d. across each route is the same (see Figure 13.3) and the total current through the whole circuit is the sum of the currents passing through each of these possible routes (see Figure 13.4).

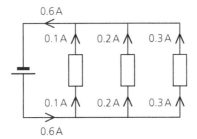

▲ Figure 13.3 Circuit diagram showing sharing of p.d.s in parallel circuits

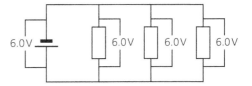

▲ Figure 13.4 Circuit diagram showing sharing of current in parallel circuits

When single resistors are connected in parallel, the potential difference across each will be the same, but the total resistance will be less than the sum of all the resistances. The total resistance is given by:

$$\frac{1}{R_{total}} = \frac{1}{R_1} + \frac{1}{R_2} + \frac{1}{R_n}$$

So, if three resistors of $5\,\Omega$, $10\,\Omega$ and $15\,\Omega$ are connected in parallel to a power supply we can calculate the total resistance of the circuit by substituting the numbers in the equation:

$$\frac{1}{R_{total}} = \frac{1}{R_1} + \frac{1}{R_2} + \frac{1}{R_3} = \frac{1}{5} + \frac{1}{10} + \frac{1}{15} = 0.37$$

Therefore, rearrange this to find R_{total}:

$$R_{total} = \frac{1}{0.37} = 2.7\,\Omega$$

B1.48 The difference between alternating and direct current

In **direct current**, DC, the **conventional current** constantly flows in one direction from the positive terminal to the negative terminal – although we now know that it is electrons flowing from negative to positive.

Alternating current, AC, as used in the mains electric supply in the UK and most other countries, has current that constantly changes direction alternating potential difference where the positive and negative ends alternating.

B1.49 The properties of mains electricity in the United Kingdom

By using alternating current, it is possible to transmit electricity over long distances at high voltages – between 275 kV and 400 kV (1 kV = 1000 V). This reduces the current and so less electrical energy is **dissipated** (scientific term for wasted) as heat.

The high voltage used for transmission can then be transformed to a lower voltage of 230 V for supplying to residences and businesses, and higher voltages can be supplied to businesses with a greater demand for energy.

The frequency of mains electricity in Europe is 50 Hz (see page 228), which means there are 50 cycles per second. This is the frequency at which electricity is generated.

Test yourself

1 What do the following terms mean?
   a current
   b potential difference (voltage)
   c resistance.
2 1.2C of charge flows through a light bulb in a time of 30s. Calculate the current flowing through the bulb.
3 A 9V battery is connected to a 4Ω resistor. Calculate the current flowing through the resistor.
4 Two more resistors of 5Ω and 6Ω are added in parallel with the 4Ω resistor.
   a Calculate the total resistance.
   b What voltage battery would you need to use to maintain the same current as in question 3?

Magnetism and electromagnetism

We are all familiar with **magnets** and **magnetism**. The earth's **magnetic field** causes a compass needle to point towards the (magnetic) north pole. Magnetic fields are invisible, but we can see their effects all around us. A fridge magnet holds notes, family photos or children's artwork on a fridge door. Magnetic clasps are used on handbags and briefcases. Magnets have many other uses and **electromagnets** are particularly useful in science and engineering.

Key terms

Magnet: a material or object that produces a magnetic field.

Magnetism: the force experienced by some types of metals in the earth's magnetic field or in a magnetic field of a magnet. It is also defined as the attractive or repulsive force produced by a moving electric charge.

Magnetic field: a region where magnetic materials experience a force.

Electromagnet: produced when a current flows through a coil of wire.

B1.50 Magnetism and magnetic poles

All magnets have two **poles**, **north** (or north-seeking) and **south** (or south-seeking). These are determined by whether the pole points towards the earth's magnetic north pole (north-seeking) or south pole (south-seeking). For this reason, we call magnets **dipoles**. However, if you cut a magnet in half, you produce two smaller dipoles – you cannot produce a magnetic **monopole** (just a north or south pole).

The north and south poles of a magnet are where the magnetic forces are strongest because the **magnetic field** is strongest. When magnets are placed close together, they will **attract** or **repel** each other, even if they are not touching. For this reason, we say that magnetism is a **non-contact force**. Other types of non-contact force include:

▶ electrostatic force – opposite charges attract, even if they are not in contact
▶ weight – objects with mass attract each other and the greater the mass, the greater the attraction. A satellite falling out of orbit does not have to be in contact with the earth to feel the attractive force of the earth's gravity.

If we place two north poles or two south poles close together, they will repel each other. If we place north and south poles together, they will attract each other.

There are two types of magnet: **permanent** and **induced**. A permanent magnet produces its own magnetic field. If we bring a **magnetic material** close to a permanent magnet, the magnetic material becomes an induced magnet. It has its own induced poles and magnetic field (Figure 13.5).

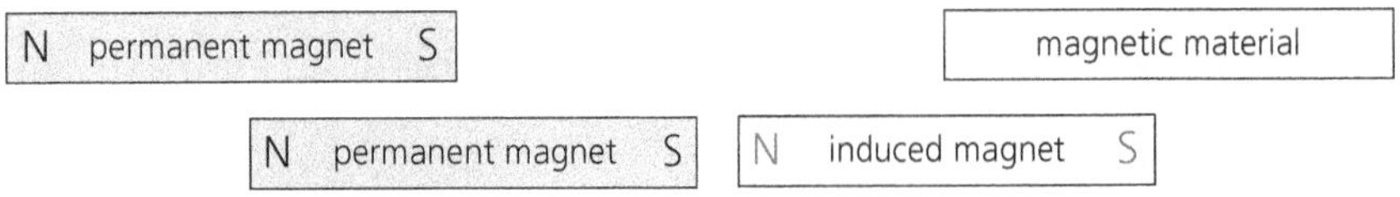

| N | permanent magnet | S |

| magnetic material |

| N | permanent magnet | S | | N | induced magnet | S |

▲ Figure 13.5 A permanent magnet causes magnetic material to become an induced magnet when they are placed close together. **N** and **S** are induced poles

Key terms

Permanent magnet: produces its own magnetic field.

Induced magnet: an object that can become a magnet when it is placed in a magnetic field.

Magnetic materials: such as iron, steel, nickel and cobalt will experience an attractive force when placed in a magnetic field.

Lines of magnetic flux: indicate the direction and strength of a **magnetic field**.

If the induced magnet is removed from the magnetic field of the permanent magnet, it will quickly lose most or all of its magnetism.

B1.51 Magnetic fields

We cannot see a magnetic field, but we can see its effect using iron filings (Figure 13.6). The small pieces of iron act as temporary magnets and line up along the magnetic field lines, also called **lines of magnetic flux**. This means that we can see the shape of the magnetic field.

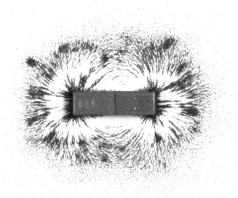

▲ Figure 13.6 Iron filings reveal the invisible lines of flux around a bar magnet

The iron filings allow us to see the shape of the field, but they do not show the direction of the lines of flux. We can plot the magnetic field lines around a bar magnet using a **compass needle**. The north pole of the magnet in the compass needle

will always point towards the south pole of the bar magnet and away from the north pole. We can use this to plot the flux lines of the field around a bar magnet (Figure 13.7). If we move the compass far enough away from the bar magnet, it will point to the earth's north pole.

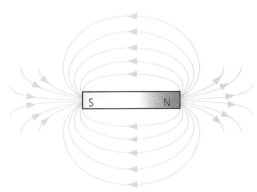

Field lines around a bar magnet

▲ Figure 13.7 Flux lines for a bar magnet. The lines are closest together near the poles, showing where the field is strongest

We follow the convention that the flux lines always go from north to south, indicated by the arrows on the lines in Figure 13.7. You can see how the flux lines are closest together near the two poles. This is where the magnetic field is strongest. Another way to describe the strength of the magnetic field is the **magnetic flux density**, which is defined as the number of magnetic flux lines that pass through an area of $1\,m^2$.

The north pole of the magnet in a compass needle points towards the south pole of a bar magnet. But the north pole of the compass needle points towards the geographic north pole of the earth.

Does this mean that the earth generates its own magnetic field? If so, does it mean that the south pole of this 'magnet' is at the geographic north pole?

If you place a compass next to a wire carrying an electric current, you will see the compass needle move as you move the compass around the wire.

This must mean that the current flowing through the wire is generating a magnetic field; you are using the compass needle to trace the magnetic flux lines around the wire.

If you reverse the flow of current in the wire you will notice that the compass needle now points in the opposite direction. We can therefore tell that the direction of the magnetic field has reversed (Figure 13.8).

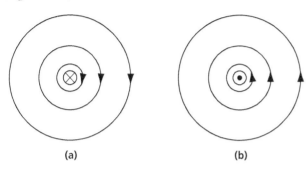

▲ Figure 13.8 The magnetic flux lines of a wire carrying a current. Symbols inside the wire indicate the direction of the current. In (a) the current is flowing into the page – away from you – and in (b) the current is flowing out of the page – towards you

From this we can conclude that moving charges (the electric current) generate a magnetic field. This phenomenon is of great importance in physics, engineering, electronics and many areas of our everyday lives.

Notice that the magnetic flux lines get further apart as the distance from the wire increases. This must mean that the strength of the field decreases as distance from the wire increases.

If we increase the size of the current flowing through the wire, we will see that the magnetic flux lines get closer together. Therefore, the greater the current, the stronger the magnetic field.

B1.52 The uses of electromagnetism and electromagnets

The magnetic field generated by a current flowing through a wire can be used to make magnets that can be switched on or off – these are **electromagnets**.

A **solenoid** is created by wrapping wire into a coil. This increases the strength (flux density) of the magnetic field when a current passes through the coil of wire (Figure 13.9). All the magnetic flux lines around each loop of wire line up with each other. This results in lots of flux lines all pointing in the same direction, very close to each other.

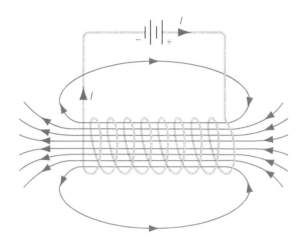

▲ Figure 13.9 The magnetic field generated by a solenoid

Note how close together the magnetic flux lines are in the centre of the solenoid. This means that this is where the magnetic field is strongest. Outside the coil, the magnetic field is just like the one around a bar magnet (Figure 13.7).

You can increase the strength of the magnetic field of the solenoid even more by placing an iron bar in the centre of the coil. This iron core becomes an induced magnet when the current flows – we have made an electromagnet. When the current is switched off, the magnetic field disappears.

A **portative** electromagnet is designed to hold material in place. An example is the type of electromagnet used for lifting materials made of iron or steel in a scrapyard.

A **tractive** electromagnet is one that applies a force and moves another object. An example is a simple solenoid where the coil surrounds a plunger. When the current is switched on, the solenoid applies a force to the plunger making it move. This can be used in many different ways, such as in valves or switches.

Electromagnetic induction

Key terms

Ammeter: measures the flow of current; always connected in series. **Voltmeters** measure the potential difference in a circuit and are always connected in parallel.

Electromotive force (emf): like **potential difference**, except that it refers to power supplies such as cells (batteries), generators or mains power supplies. These transfer other forms of energy, such as light energy or kinetic energy, into electrical energy. The unit of emf is also the volt (V).

Generator effect: when a **potential difference** (voltage) is induced in a wire that experiences a change in magnetic field.

Electromagnetic induction can be demonstrated using the apparatus shown in Figures 13.10 and 13.11.

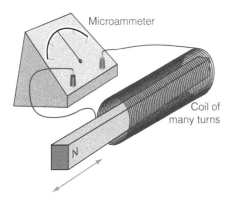

▲ Figure 13.10 Using a moving magnet to produce a potential difference

Figure 13.10 shows that if you move the bar magnet into the coil, you will see the needle on the micro **ammeter** flick in one direction and then in the opposite direction when the magnet is pulled out. The reading will be zero when the magnet is stationary. As the magnet moves, lines of magnetic flux are being crossed by the wires in the coil. Therefore, moving the magnet into the coil induces a potential difference. Providing the circuit is complete, this induces a current resulting in the reading on the microammeter. Moving the magnet back out causes the induced potential difference to change direction and therefore the current flow changes direction. The effect is the same if you move the coil and keep the magnet still.

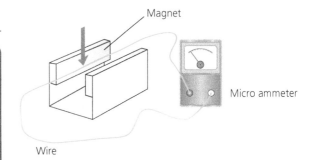

▲ Figure 13.11 Moving a wire into a magnetic field induces a potential difference

Figure 13.11 shows that, as the length of wire moves between the two magnets, the micro ammeter will flick in one direction when the wire moves down and in the opposite direction when the wire moves up. In this case, a potential difference is induced in the wire because there is a moving electric charge (the electrons in the wire) **perpendicular** (at a 90-degree angle) to the magnetic field and they experience a force. This causes them to move towards one end of the wire, creating a potential difference across the wire.

The principle of electromagnetic induction is the basis of the **generator effect** and is used in dynamos and alternators to generate an electric current (Figure 13.12).

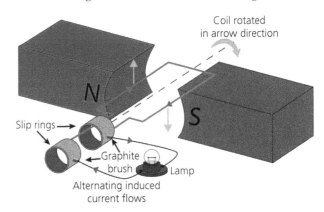

▲ Figure 13.12 A simple alternator

The motor effect

When a current passes through a wire placed in a magnetic field, it moves because a force acts on it. When a current passes through a wire, we know it induces a magnetic field. This results in the wire exerting a magnetic force onto the permanent magnet. When the current-carrying wire is placed in between the two poles of the magnet the magnetic fluxes combine to form what is known as a **catapult field**. This field exerts a force on the wire causing it to move. Figure 13.13a shows the lines of magnetic flux between the poles of two magnets and Figure 13.13b shows the lines of magnetic flux around the current-carrying wire.

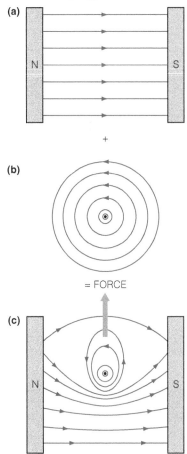

▲ Figure 13.13 (a) A uniform magnetic field between two poles of a magnet; (b) the field around a current-carrying wire; (c) the catapult field. The force is from the strong field (greater flux density) to the weak field (lower flux density)

You can use Fleming's left-hand rule (Figure 13.14) to work out the direction of the force on the wire.

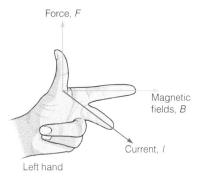

▲ Figure 13.14 Fleming's left-hand rule. Using your left hand, point your first finger in the direction of the field, your second finger in the direction of the current and your thumb will point in the direction of the force (motion)

As the name suggests, the **motor effect** is the basis of electric motors (Figure 13.15).

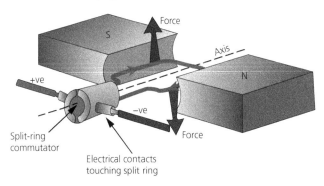

▲ Figure 13.15 A DC electric motor. The current flowing in the wires means they experience a force perpendicular to the magnetic field. The split-ring commutator reverses the current every half turn to ensure the motor keeps turning in the same direction

Key term

Motor effect: when a current-carrying wire is placed between magnetic poles. The magnetic field around the wire interacts with the magnetic field it is placed in. This causes the wire and magnet to exert a force on each other and can cause the wire to move.

Look at Figure 13.15 and use Fleming's left-hand rule to work out the direction of the magnetic field. Is that the direction you expect? It is always worth checking your answer!

Induction heating

Induction heating has many applications. Perhaps the most familiar is the 'induction hob' used to heat saucepans in domestic and commercial kitchens. An electronic oscillator passes a high-frequency alternating current through an electromagnet. This generates a rapidly alternating magnetic field. If a pan made of iron or steel is on the hob then currents, known as **eddy currents**, are generated inside the conductor (the pan). These currents generate heat which will then heat the contents of the pan.

Induction heating has the advantage that the heat is generated inside the object itself, so it is rapid. Also, it does not rely on conduction from an external heat source, such as a naked flame in a gas hob or a hot element in a traditional electric hob. This means that the hob becomes hot only because of being in contact with the hot pan and cools quickly once the pan is removed. It is therefore safer in many ways than using naked flame or a traditional electric hob, since heating does not occur without the conductor.

There are many other applications of induction heating, particularly in manufacturing, including:
▶ welding and brazing
▶ induction furnaces to produce molten metals
▶ cap sealing of containers in food and pharmaceutical industries
▶ heat treatment (e.g. hardening) of metals.

Research

There are many uses for permanent and temporary magnetic materials such as iron, steel, cobalt and nickel.

There are also many applications of electromagnets in electric and electromechanical devices:
▶ transformers
▶ motors, generators, alternators
▶ loudspeakers and microphones
▶ induction heating
▶ MRI machines
▶ cranes and sorting or separation equipment used in recycling
▶ relays.

Select a number of these applications and see if you can find others. For each, research the application and how our understanding of the topics covered in this section has allowed advances in technology. In particular, think about:
▶ the type of magnetic material used – permanent or temporary
▶ how the type of material relates to its function
▶ is this an example of the generator effect, motor effect or magnetic induction?
▶ for electromagnets:
  – does the material affect the properties of the electromagnet?
  – is the electromagnet tractive, portative or neither?

Test yourself

1 Explain the difference between a permanent magnet and an induced magnet.
2 What is the convention for drawing flux lines of a magnetic field?
3 How do flux lines indicate the strength of a magnetic field?
4 Describe how you would make an electromagnet.
5 Explain the difference between portative and tractive electromagnets.

Waves

We learn about waves from an early age – a toddler dropping a pebble into a still pond and watching the ripples is observing waves. Studying waves in water can help us understand a lot about how waves behave. However, watching waves crashing onto a shore, or even just watching the tide come in on the beach, can give us a misleading impression of what a wave is.

Key terms

Amplitude: the maximum displacement of any point from the equilibrium position (Figure 13.16).

Frequency: in Hz, the number of complete waves that pass a given point in one second (Figure 13.17). One complete wave is a **cycle**, so a frequency of 1 Hz corresponds to one cycle per second. This is a very low frequency, so you will often see frequency measured in kHz (kilohertz, 10^3 Hz), MHz (megahertz, 10^6 Hz) or GHz (gigahertz, 10^9 Hz).

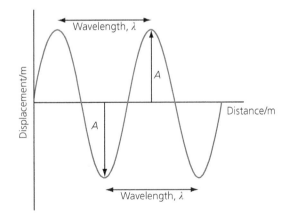

▲ Figure 13.16 Wave terminology. A represents the amplitude. This is an example of a transverse wave

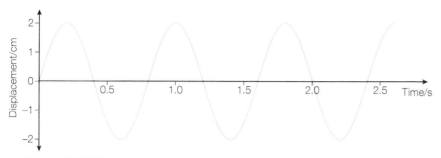

▲ Figure 13.17 If you measure from one peak to the next peak you can calculate that the **frequency** is 1.25 Hz. You will get the same answer if you measure from trough to trough

B1.53 The definition of a wave

Waves transfer **energy** from one energy store to another, they do not transfer **matter** (e.g. objects or water, for example). For example, electromagnetic waves transfer energy from the nuclear store of energy in the Sun to the thermal store of energy on Earth. This is why waves crashing onto the shore can be misleading. When waves travel through a medium (e.g. the ocean) the particles in the medium (i.e. water molecules) **oscillate** (move up and down). However, the particles stay in the same place, only the energy is transferred. When the wave hits the beach, things change, because now the particles of water are interacting with the particles of sand, and the wave is changed into a movement of the water – and anything floating in it, which is why things get washed up on the beach by the waves.

B1.54 The relationship between frequency, wavelength and speed using the wave equation $v = f\lambda$

The speed of a wave is a measure of how fast energy is transferred. More simply, it is the speed the wave is moving at. The relationship between wave speed, frequency and **wavelength** is the same for all types of wave and is given by the wave equation:

$$v = f\lambda$$

where:

v = wave speed in m/s

f = frequency in Hz

λ = wavelength in m.

For example, radio waves travel at the speed of light (approx. 3.0×10^8 m/s). A radio station transmits at a frequency of 92.3 MHz, calculate the wavelength of the radio waves.

$92.3\,\text{MHz} = 92.3 \times 10^6\,\text{Hz} = 9.23 \times 10^7\,\text{Hz}$

Rearrange the equation above to give:

$$\lambda = \frac{v}{f}$$

So

$$\lambda = \frac{30 \times 10^8}{9.23^7} = 3.25\,\text{m}$$

> ### Key term
>
> **Wavelength:** the distance between the same point in successive cycles. For example, the distance from the peak of one wave to the peak of the next wave (Figure 13.16). The standard unit of wavelength is the metre, m. However, **electromagnetic** waves can have wavelengths from 10^4 m to 10^{-15} m, so you will often see wavelengths expressed in units such as **nanometres**, nm. (1nm = 1×10^{-9} m – see section B1.63, page 238 for more about units.)

B1.55 The properties of longitudinal and transverse waves

There are two types of wave, **transverse** and **longitudinal**.

Most waves are **transverse**, where the waves transfer energy in a direction at right angles to the direction in which the particles are vibrating. This means the oscillations are **perpendicular** (at right angles) to the direction of energy transfer (Figure 13.16). The particles move away from and towards the horizontal line – this is represented by the arrow 'A' (for amplitude) – so the energy transfer is in the forward direction even though the particles just move from side to side, or up and down, depending on how you look at the wave.

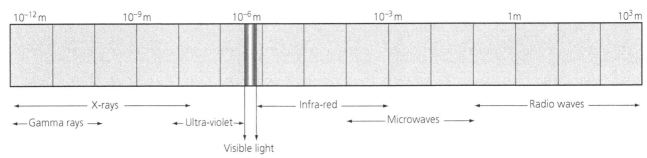

▲ Figure 13.18 The electromagnetic spectrum

Key terms

Electromagnetic waves: include gamma rays, X-rays, visible light, microwaves and radio waves. Their energy is carried by oscillating electric and magnetic fields.

Electromagnetic spectrum: describes all the different types of electromagnetic waves. The properties of the different types of waves vary considerably, so we usually consider the spectrum as seven groups, with slight overlaps (Figure 13.18).

Examples of transverse waves include:

▶ all **electromagnetic waves**, e.g. light, radio waves, X-rays (Figure 13.18)
▶ ripples in a pond or waves in the ocean
▶ a wave on a string, e.g. a violin or guitar string that is made to vibrate when it is bowed or plucked.

Electromagnetic waves are a little more complicated because the wave represents changes in the electric and magnetic fields, but the principles are the same.

Longitudinal waves transfer energy in the same direction in which the particles are vibrating. This means that the oscillations are **parallel** to the direction of energy transfer. These oscillations create **compressions** (where the particles get closer together) and **rarefactions** (where the particles move further apart) (Figure 13.19).

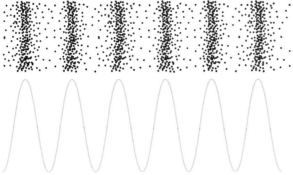

Trace shown on an oscilloscope

▲ Figure 13.19 Changes in air pressure at a microphone diaphragm, measured over a period of time

Examples of longitudinal waves include:

▶ sound waves moving in air
▶ **ultrasound** (very high-frequency sound waves) that move through materials, for example:
  – metals, when used to detect cracks in components or structures
  – the body, when used in medical imaging such as scans in pregnancy
▶ **shock waves**, such as some types of **seismic** waves (produced by earthquakes).

Reflect

You can use a metal spring such as a Slinky to demonstrate both transverse and longitudinal waves (Figure 13.20).

  If you lay the spring out on a surface, hold it at one end and wiggle it from side to side, you will see a transverse wave move along the length of the spring.

  If you push the end of the spring quickly back and forward, you will see a longitudinal wave move along the length of the Slinky.

Does a longitudinal wave in the Slinky explain why, when you speak, the sound waves travel through the air, but they do not create a vacuum in your mouth?

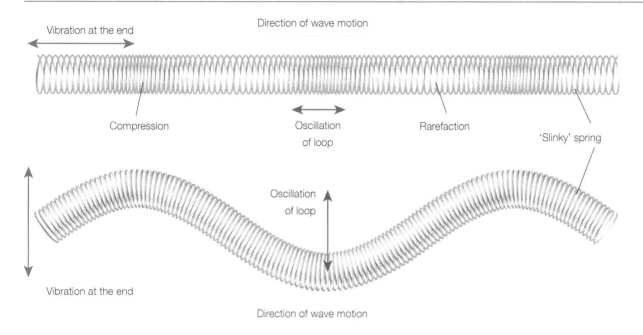

▲ Figure 13.20 Longitudinal (top) and transverse (bottom) waves in a Slinky spring

B1.56 The uses of different types of waves

Waves transfer energy, but they can also transfer information. The simplest example is signalling, using beacon fires to send warnings or flashlights to send signals using Morse code – light waves are being used to transfer information.

Communication

Radio waves are probably a better example of the use of waves in communication. Since the pioneering experiments of Marconi, we now use radio waves for many different types of communication. You can probably think of many more examples than these:
▶ TV and radio broadcasts (although not satellite TV – see below)
▶ wireless broadband (WiFi).
▶ Bluetooth® and other wireless technologies used in computing and control systems.

Microwaves have a slightly shorter wavelength/ higher frequency than radio waves but can also be used for short-range radio communication. A major use of microwaves now is in satellite communication, including satellite TV. This uses microwaves of a wavelength that is not absorbed by water molecules in the atmosphere.

Medical uses

There are numerous examples of the use of waves of different types in science and medicine:
▶ X-rays for imaging are used in different ways, from a simple X-ray taken to show a broken bone to the much more complex CT (computerised tomography or CAT) scan. A CT scan can build up a three-dimensional image of the whole body or part of the body.
▶ Gamma rays are used for cancer treatment, to kill cancer cells.
▶ Gamma rays can also be used to sterilise medical instruments and equipment.
▶ Ultrasound is used in scanning, particularly to create images of soft tissues that don't show up well on X-rays for imaging, such as the heart, liver, kidneys, gallbladder and major blood vessels. This can be useful in diagnosis of some types of cancer.
▶ Ultrasound can also be used for cleaning. The high-frequency sound waves help to dislodge dirt from objects as diverse as jewellery, electronic components and medical and laboratory equipment.

Food processing

We saw in section B1.52 how magnetic induction can be used for heating in both domestic and commercial food processing. Waves are also important in food processing:

▶ Infrared heating uses electromagnetic radiation with a longer wavelength than visible light:
  – Heat lamps to keep food hot before serving, for instance, in canteens.
  – Infrared ovens, particularly small ones, heat up more rapidly than conventional electric or gas ovens.
▶ Microwave heating can be used for cooking and reheating food. This uses microwaves of a different wavelength to satellite communication. In microwave heating, energy from the microwaves is absorbed by water molecules in the food. This transfers the energy to the water molecules, causing them to heat up. The energy is then transferred to the rest of the food by heating, which quickly cooks the food.
▶ Gamma radiation can be used in food preservation (see section B1.61).

You are at a festival listening to a band and realise that the lead guitar produces high-frequency sounds, while the bass guitar produces low-frequency sounds. Also, you can tell that the drum kit is similar – the hi-hat cymbals produce high-frequency sounds and the bass drum produces low-frequency sounds. In spite of that, you hear all the different sounds at the right times – this band has rehearsed!

What does that tell you about the relationship between frequency and speed of sound?

Could you devise an experiment to test your hypothesis?

Test yourself

1 What is meant by the frequency of a wave?
2 What are the units of frequency used for waves?
3 A student was studying waves in a swimming pool of 25m in length. They created a wave that had a frequency of 5Hz and wavelength of 0.05m. Calculate how long, in seconds, the wave would take to travel the length of the pool.
4 Give two examples of transverse waves and one example of longitudinal waves.
5 Give two examples of the use of microwaves.

Particles and radiation

This is an area where chemistry and physics overlap. You will have learnt about the structure of the atom in section B1.33 and about sub-atomic particles in section B1.35 in the chapter on chemistry. Knowledge of these sub-atomic particles helps us understand the basis of radiation and radioactive decay.

B1.57 The types and properties of ionising radiation

Key terms

Ionising radiation: any form of radiation that interacts with matter, resulting in **ionisation** of that matter.

Ionisation: the formation of charged particles from neutral molecules or atoms by adding or removing electrons.

There are three types of **ionising radiation** that you need to know about: alpha (α), beta (β) and gamma (γ). The first two are types of particle, whereas gamma radiation is a form of electromagnetic radiation, mentioned in sections B1.55 and B1.56. It helps us understand the properties of these types of radiation if we understand their nature and origin.

Alpha radiation

Alpha particles are helium **nuclei** (plural of nucleus), meaning that they consist of two **protons** and two **neutrons** which give them a positive charge (+2). On the atomic scale, alpha particles are relatively large compared to beta particles (they have a relatively large mass). This means that they can easily remove electrons from atoms when they collide with them. This makes alpha radiation highly ionising.

However, because of their size, alpha particles do not penetrate materials very far – they are more likely to collide with atoms or nuclei of atoms. Alpha radiation can travel just 1–2 cm in air and can be absorbed by a sheet of paper.

Beta radiation

A beta particle is a fast-moving **electron**, therefore they have almost no mass (about 1800 times less than a proton) and a negative charge (–1). Beta particles are less ionising than alpha particles and can penetrate materials a moderate amount.

Beta radiation can travel approximately 15 cm in air and can be absorbed by a sheet of aluminium about 5 mm thick.

Gamma radiation

Unlike alpha and beta radiation, gamma radiation is a form of electromagnetic radiation – like X-rays or radio waves (see section B1.55). Gamma rays tend to pass through atoms rather than be absorbed by them. This makes them only weakly ionising, but with high penetrating power. Gamma rays have a range of many kilometres of air but can be absorbed by thick sheets of lead or several metres of concrete.

B1.58 The definitions of half-life and count-rate

Key terms

Radioactive decay: the random process that occurs when an unstable nucleus loses energy by giving out **alpha** or **beta** particles or **gamma** radiation.

Activity: the rate at which a radioactive source decays. The unit of activity is the **becquerel** (Bq) where 1 Bq = 1 decay per second. **Count-rate** is the number of radioactive decays recorded each second. You can see that activity and count-rate can be used interchangeably.

Half-life: the time taken for half the unstable nuclei in a sample to decay.

Radioactive decay is detected and measured by using a **Geiger-Muller tube** and counter – usually referred to as a **GM tube** or **Geiger counter**. This measures the radiation such as alpha particles, beta particles or gamma rays that are emitted when a radioactive source decays.

The rate at which a radioactive source decays can be plotted in a graph of activity against time or number of nuclei against time (Figure 13.21).

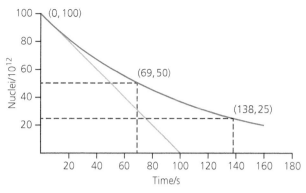

▲ Figure 13.21 A graph of radioactive decay showing how the number of nuclei decreases with time

There is a lot of useful information in Figure 13.21. First, we can see that the rate of radioactive decay (the **activity**) decreases with time. The tangent (blue line) is drawn at time = 0. If we measure the slope, we can calculate the rate of decay. Try drawing tangents at other times, such as 60 s and 120 s. You will see that the slope of the line is a maximum at time = 0 and then decreases. This shows that the rate of decay is decreasing. This is called an **exponential decay**.

Look at how the number of nuclei change with time. We start with 100×10^{12} nuclei but after 69 s there are 50×10^{12} nuclei remaining – the number has reduced by half. After another 69 s (at 138 s) the number has reduced by half again to 25×10^{12} nuclei. This is characteristic of exponential decay: the time it takes for the number of nuclei to halve is constant. We call this the **half-life**, sometimes written as $T_{1/2}$.

We can make the same calculation from a graph where the y-axis (vertical axis) is **activity** rather than number of nuclei.

B1.59 The main types of radioactive decay in relation to unstable nuclei

There are two important facts about radioactive decay that you need to know:
1 Radioactive decay is **spontaneous**. Radioactive nuclei are **unstable**, which is why they decay. However, there is nothing that triggers decay.
2 Radioactive decay is **random**. We cannot predict which nucleus in a radioactive source will decay nor can we predict when it will decay.

As we saw in section B1.58, we can measure the half-life of a radioactive source and then use this to make predictions about the behaviour of the radioactive source, even though the decay is a random process.

Alpha decay

We saw in section B1.57 that an alpha particle consists of two neutrons and two protons and is equivalent to a helium nucleus. It is represented by the symbol $^4_2$He in nuclear equations. The top number represents the mass number – the sum of the protons and the neutrons. The bottom number represents the atomic number (see section B1.34), i.e. the number of protons. You generally will not see atomic and mass numbers featured in chemical equations as they do not change in a chemical reaction. However, because nuclear equations can involve transformation of one element into another, it is useful to include these to show the changes that have taken place.

An alpha particle is formed when an unstable nucleus loses two neutrons and two protons and becomes more stable in the process. An example is the decay of uranium-238 into thorium-234, represented by the nuclear equation:

$$^{238}_{92}U \rightarrow \, ^{234}_{90}Th + \, ^4_2He$$

In this case, the uranium-238 has lost 2 neutrons and 2 protons, so the atomic number (number of protons) decreases by 2, forming thorium and the mass number decreases by 4, so the isotope of thorium is thorium-234. We can see that the mass and atomic numbers all balance, because the alpha particle (helium nucleus) has 2 protons and 2 neutrons, giving it an atomic number of 2 and a mass number of 4.

Beta decay

A beta particle is a high-speed electron. It is represented in nuclear equations by the symbol $^{\,\,0}_{-1}$e. This electron is formed and then ejected at high speed from the nucleus when a neutron turns into a proton. This means that the number of protons increases by 1, therefore the atomic number increases by 1 although the mass of the nucleus remains the same. An example is the decay of carbon-14 into nitrogen-14, represented by the nuclear equation:

$$^{14}_{6}C \rightarrow \, ^{14}_{7}N + \, ^{\,\,0}_{-1}e$$

Note that in both alpha and beta decay, a new element will be formed as the nucleus produced will always have a different number of protons. The atomic number determines which element it is, so this means the atomic number changes and therefore the element changes.

Gamma radiation

Gamma radiation is a form of electromagnetic radiation from the nucleus produced when excess energy is lost from the nucleus. When a nucleus decays by alpha, beta or other types of emission, the protons and neutrons in the nucleus are often left in an excited state. The protons and neutrons then return to a lower energy level and the difference in energy is emitted as gamma radiation. Unlike alpha and beta decay, there is no change to the atomic structure and so no new elements are formed.

Health and safety

Some radioactive sources have a short half-life and others have a long half-life. A source with a short half-life will usually decay rapidly – i.e. the nuclei will be very unstable and the source will have a high activity. This means it will emit a high level of radioactivity, which could be dangerous. However, it will become safe relatively quickly as the unstable nuclei decay quickly.

A long half-life means that the source will have a low activity and will emit a relatively low level of radioactivity. However, it could still be dangerous because it will go on emitting radiation for a long time – even for millions of years.

We also need to consider the type of radiation emitted. For example, an alpha source is the most dangerous via contamination; an alpha source absorbed into the body will do damage to any tissues it is in contact with. On the other hand, a beta source is more dangerous by irradiation as beta particles are more penetrating.

B1.60 How radiation interacts with matter

Radiation interacts with matter in two ways: **ionisation** and **excitation**.

Ionisation

Ionisation occurs when electrons are removed from atoms or molecules to produce positive ions. All forms of radioactive decay produce radiation that can cause ionisation – hence the term **ionising radiation**.

We saw in section B1.57 that alpha radiation is the mostly highly ionising while gamma radiation is the least ionising. Beta radiation is in between the two in its ability to cause ionisation.

Excitation

Excitation occurs when radiation transfers energy to atoms or molecules. Excitation involves moving an electron to a higher energy level (**shell** or **orbital**). If enough energy is transferred to the electron, it will be removed from the atom and ionisation will have occurred. Gamma radiation can cause excitation. In fact, other types of electromagnetic radiation such as X-rays and visible light can cause excitation of electrons. This is why some types of ultraviolet (UV) light can cause sunburn or, in the worst cases, skin cancer. The UV light causes excitation of electrons in the DNA molecules found in skin cells, leading to ionisation which can result in cancer-causing mutations.

B1.61 The applications of radioactivity within the health and science sector

Health and safety

Before we look at applications of radioactivity, it is worth considering the biological effects of radiation – how radiation interacts with biological materials, including the human body.

The human body is made up of many complex molecules. Removal or addition of an electron changes a molecule chemically – this means that it may behave differently in any interaction with other molecules. As you have learned from Biology, interaction between molecules is central to how the body works.

UV light has sufficient energy to ionise biological molecules. Alpha, beta and gamma radiation all have energy millions of times greater than that of UV light. This means that these types of ionising radiation can be highly dangerous because they can change the chemistry of the body. The function of enzymes can be changed, cells can be damaged and the DNA in cells can be damaged, which can lead to cancer.

Case study

Because radiation can be so damaging, we must exercise great care when handling radioactive sources.

Hasini is working as a technician in a college science laboratory. She is responsible for the safe storage and demonstration of radioactive sources.

Hasini decides that, for each source, she must consider the following:
▶ the activity of the source
▶ the type of radiation (see section B1.57).

Based on these two considerations, Hasini can decide what protection is required.
▶ Alpha radiation is highly damaging but has low penetration, so can be relatively easily screened. This makes it relatively harmless outside the body, but if a source emitting alpha radiation enters the body, it can be highly damaging.
▶ Beta and gamma radiation are less ionising, but they have much greater penetration and may require screening with thick sheets of lead.

The table shows the radioactive sources kept in the college laboratory. For each source, the type of radiation emitted is shown.

Source	Radiation emitted
cobalt-60	pure gamma
strontium-90	pure beta
americium-241	alpha and some gamma

There are strict regulations covering working with radioactive sources, including those described in Chapter A3 (page 35).
▶ What precautions should Hasini use for storing each source?
▶ What precautions should she take when using each source?

Radioactive tracers

A major use of radioactive tracers is in medical diagnosis (next section).

In industry, radioactive tracers can be used to detect the presence of materials such as dust, cellulose fibres, glass fragments or organic materials as these all adsorb radioactive tracers from solution.

Another use of radioactive tracers is in monitoring hydraulic fracturing ('fracking') used to extract oil and gas from rocks.

Medical diagnostic applications

Radioactive **isotopes** (**radioisotopes**) can be injected or swallowed and their course around the body followed using an external detector. One example is iodine-123. This is taken up by the thyroid gland, just like the non-radioactive isotope (iodine-127). The radiation given out can be detected and used to show whether the thyroid gland is absorbing iodine correctly.

Radioactive tracers are usually gamma emitters as these are weakly ionising but highly penetrating. This means that they are less likely to cause damage to the body but can be easily detected from outside the body. Radioisotopes used as tracers also need to have short half-lives so that their activity disappears soon after the procedure is completed.

Computerised tomography uses X-rays to build up a three-dimensional image of the body – known as a CT scan. More recently, gamma radiation detectors have been used to detect gamma emission of a radioactive tracer from many angles. A computer can then build up an image from the points of emission. This technique, known as **single photon emission computerised tomography (SPECT)** is now the main scanning technology used to diagnose a range of medical conditions.

Iodine-123 is an example of a **diagnostic radiopharmaceutical**. Various radioisotopes can be attached to biologically active substances, such as amino acids, hormones, therapeutic drugs, etc. These can be used together with SPECT to examine a wide range of processes:
▶ blood flow to the brain
▶ functioning of the liver, lungs, heart or kidneys
▶ assess bone growth.

Food preservation

Sufficiently high doses of radiation will kill micro-organisms in food. Some of these may be harmful (**pathogens**), so **irradiation** can kill the organisms that cause food poisoning, for example.

Other micro-organisms cause food to 'go off' – these are known as **spoilage** micro-organisms. Because irradiation will kill these spoilage micro-organisms as well, it can help to preserve food or extend the shelf-life of fresh or prepared foods.

Key terms

Isotopes: atoms of the same element (they have the same number of protons) but with different number of neutrons.

Radioisotopes: unstable isotopes that undergo radioactive decay emitting radioactivity as they do so.

Radiopharmaceuticals: therapeutic drugs (medicines) that incorporate radioisotopes. Some radiopharmaceuticals are used to treat disease whereas **diagnostic radiopharmaceuticals** are used to help in diagnosis of disease.

Irradiation: the process of exposing an object to radiation. It does not make the object radioactive.

Research

The Food Standards Agency is responsible for ensuring the safety of the food that we buy and eat. Its website has a lot of information about the use of irradiation to preserve food.

Visit the website: www.food.gov.uk

Search for 'irradiated food', then research the following:
▶ Is irradiated food safe to consume?
▶ How does irradiation change food?
▶ What types of food can be irradiated and sold?
▶ How can you tell that a food has been irradiated?

Dating deceased organisms

Radiocarbon dating measures the age of any object that contains organic material. This includes dead organisms as well as anything made from dead organisms – such as a wooden table or pair of leather shoes.

Carbon-14 (^{14}C) is the radioactive isotope of carbon. The isotope is constantly made in the atmosphere from nitrogen using energy absorbed from **cosmic radiation**. This is the reverse of the reaction shown in section B1.59 illustrating beta-decay.

The carbon-14 then combines with atmospheric oxygen to produce carbon dioxide. This is taken up by plants and used in **photosynthesis** to produce sugars and other organic compounds. Plants are eaten by animals and so the carbon-14-containing organic compounds are incorporated into animal tissues as well. Both plants and animals respire and emit carbon dioxide. This means that the level of carbon-14 remains constant while the organism is alive.

Once the organism dies, it no longer exchanges carbon-14 with the environment and the carbon-14 undergoes radioactive decay. We can calculate how long ago the organism died by measuring the proportion of carbon-14 in a sample of bone, skin, fur, wood, etc.

The half-life of carbon-14 is about 5730 years. That means we can only date samples from organisms that died less than about 50 000 years ago. After that long, there will not be enough carbon-14 left to measure accurately.

Key terms

Radiocarbon dating: the process that uses a radioactive isotope of carbon, **carbon-14**, to determine the age of an object containing organic material.

Cosmic radiation: (also called cosmic rays) originates from the sun as well as from outside the solar system. It is a mixture of high energy protons (hydrogen nuclei), alpha particles (helium nuclei) and beta particles (electrons).

Photosynthesis: the process that plants use to make complex organic compounds such as sugars from carbon dioxide and water using light energy.

Test yourself

1 Explain how gamma radiation is different to alpha or beta radiation.
2 Nobelium-254, $^{254}_{102}$No, undergoes radioactive alpha decay to form an isotope of the element fermium, Fm.
   a What is meant by isotopes of an element?
   b Write a balanced equation to show the radioactive (alpha) decay of nobelium-254.
3 A radioactive source had an activity of 500000 Bq and a half-life of 2.5 years. Calculate the activity remaining after 10 years.
4 The only foods that can be legally irradiated for sale in the UK are dried herbs and spices. Explain what type of radiation would be most appropriate for irradiation of herbs and spices.

Units

You will have encountered units already in this chapter – for length (e.g. wavelength) as well as various electrical measurements (charge, current, voltage). There have been many different types of unit used over the years. The **imperial** system uses fluid ounces, pints and gallons for volume and feet, inches and miles for distance. The **metric** system uses millilitres and litres for volume and millimetres, metres and kilometres for distance. Getting measurements from these two systems mixed up can be confusing or worse. In 1999, NASA lost the $125 million Mars orbiter because of confusion between imperial and metric units.

B1.62 The use of the international system of units (SI)

The international system of units is known as SI, from the French *Système international*. It is a modern form of the metric system but extends to units used for all kinds of measurement, particularly in science and engineering.

The metric system originally used **standards** for several units, such as the standard **kilogram** and standard **metre**, both of which were made of platinum–iridium alloy and kept in Paris. These are no longer sufficiently accurate for modern measurements and have been replaced by definitions based on **defining constants**, such as the speed of light, the Planck constant and the Avogadro constant.

There are seven SI base units, as shown in the table.

Unit name	Unit symbol	Quantity name
ampere	A	electric current
candela	cd	luminous intensity
kelvin	K	temperature
kilogram	kg	mass
metre	m	length
mole	mol	amount of substance
second	s	time

There are many other SI units that are derived from these base units. Some that you may have come across include:

- hertz (Hz) for frequency
- volt (v) for electrical potential difference
- coulomb (C) for electric charge
- joule (J) for energy, work or heat
- watt (W) for power.

B1.63 How to convert between units

The metre and kilogram are both SI base units. That does not make them useful for measuring very short lengths or small masses. If you look at a ruler, the smallest division is usually the **millimetre** (mm). Paracetamol tablets are usually 500 **milligrams** (mg). The **prefix** 'milli' represents one thousandth, so we can work out the following conversions:

Length – metres (m) and millimetres (mm)

$1\,m = 1000\,mm$ or $10^3\,mm$

$1\,mm = 0.001\,m$ or $10^{-3}\,m$

Mass – grams (g) and milligrams (mg)

$1\,g = 1000\,mg$ or $10^3\,mg$

$1\,mg = 0.001\,g$ or $10^{-3}\,g$

Prefixes

You can see that millimetres use the prefix 'milli' to represent one thousandth. There are other prefixes used in SI units, as shown in the table.

Prefix name	Prefix symbol	Base 10
Tera	T	10^{12}
Giga	G	10^9
Mega	M	10^6
Kilo	K	10^3
Hecto	H	10^2
Deca	Da	10^1
Deci	d	10^{-1}
Centi	c	10^{-2}
Milli	m	10^{-3}
Micro	µ	10^{-6}
Nano	n	10^{-9}
Pico	p	10^{-12}

Many of these prefixes will be familiar, although not all of these examples are SI units:

- centimetre (cm) is common in everyday measurement
- micrometre (µm) is common in microscopy
- nanometre (nm) is commonly used to measure wavelength of visible light
- megabytes (MB), gigabytes (GB) and terabytes (TB) are used in computing for capacity of hard drives, memory modules and data transmission speeds. Unfortunately, these can be misleading because they can be calculated in **binary** as well as in **decimal**. This is the reason that a 400 GB hard drive (decimal) is shown by Microsoft Windows as only 372 GB (binary).

Notice that the base unit of mass is the kilogram, which is named as if the base unit is the gram.

Volume

The SI unit of volume is the cubic metre (m^3). This is too large to be useful in many situations – particularly in biology, chemistry and everyday life – it is much easier to ask for 1 litre of orange juice than asking for '$0.001\,m^3$ please'.

In chemistry, the most common units of volume are the litre (L) and millilitre (mL):

$1\,L = 1000\,mL$

$1\,mL = 10^{-3}\,L$

Using 'l' as the abbreviation for litre can be confusing – it can look like the number 1. That is why the SI symbol for litre is 'L' and so 1 millilitre is 1 mL. You may see 'l' used as the symbol for litre, as in 1 ml (1 cm^3) but it is likely to be printed in italics, as in 1l. Because of this, particularly in chemistry, you are more likely to use $1\,dm^3$ instead of 1 litre.

Practice points

Another unit you will see used quite often is the cubic decimetre or dm^3. This is the same volume as one litre and is preferred now in chemistry. You will often see concentrations expressed in mol/dm^3 or $mol\ dm^{-3}$, for example. This is replacing the older method of mol/L or M (molar), but you will still encounter these units sometimes.

In fact, you might encounter other unfamiliar units in the workplace. Different industries are moving at different rates towards using SI units.

B1.64 The importance of using significant figures and science notation

Powers of 10

You will have encountered several numbers in this chapter and elsewhere using powers of 10. This is very convenient in science (see **Standard form** below) because it allows us to write very large or very small numbers more easily. There is another advantage: when we multiply two numbers shown as powers of 10 we **add** the powers. This is illustrated in the following example:

$100 \times 1000 = 100\,000$

Using powers of 10, 100 becomes 10^2 and 1000 becomes 10^3. This means that we can rewrite the calculation as follows:

$10^2 \times 10^3 = 10^5$

Notice that when we multiply the numbers shown as powers of 10, we simply add the powers to get the answer. This makes life so much easier!

If you look at the table of prefixes in section B1.63 you will see a range of values from 10^{-12} (pico) all the way up to 10^{12} (tera). These are examples of scientific notation where we use powers of 10 rather than writing the number out in full. For example:

1 million can be written as $1\,000\,000$ or 10^6

1 nm is equal to $0.000\,000\,001$ m or 10^{-9} m

It is not difficult to see that using powers of 10 makes it much easier to write (and understand) very large or very small numbers.

Standard form

When we use scientific notation, using powers of 10, we follow a convention known as **standard form**. A number in standard form has two parts. For example, $2\,650\,000$ would be written as:

2.65×10^6

The first part is a number between 1 and 10. The second part is a power of 10. So we should not write that number as 26.5×10^5.

By using numbers in standard form, calculations become less cumbersome. Also, with a little practice, it is easier to write down and understand very large or very small numbers using standard form.

Significant figures

Measure the line in the Reflect box below.

Reflect

Measure the line below with a ruler. What answer do you get?

If the printing process has worked correctly, that line should be 132.2917mm long. Can you measure it that accurately with a ruler?

If the smallest division on your ruler is 1mm, it is possible that you could estimate between the smallest divisions. But how reliable would your measurement be?

Because there is **uncertainty** in any measurement, we must take care not to imply a greater accuracy than is possible. In the case of the measurement of the line (132.2917mm), we would probably round it down to 132.3mm, but is that the **true value**? We cannot estimate 0.3 of the smallest division.

Therefore, we use **significant figures**. These are the digits in a number that we believe are reliable. In the example, the first three digits (1, 3 and 2) are reliable, but the fourth digit, after the decimal point, is not. There is some uncertainty in the third digit (2), but we can be confident that the **true value** is closer to 132 mm than to 133 mm.

This value of 132 mm is quoted to three **significant figures**, the number of digits that we believe are reliable.

If the smallest division on the ruler was only 10 mm, we would have to give a value of 130 mm – this is **two** significant figures.

Here are some more examples of a different measurement (1257.59) to various numbers of significant figures.

Number	Significant figures
1257.59	6
1258	4
1300	2
1000	1

From this example, you can see that **trailing zeros** do not count as significant figures – we call them **placeholders** as they stand for the 1s and 10s.

Similarly, **leading zeros** (usually after the decimal point) do not count towards the number of significant figures – so 0.075 m has two significant figures. We could write it as 75 mm and it would be the same length with the same uncertainty.

Trailing zeros in a **decimal** do count as significant figures. So, a concentration of 0.0500 mol/dm^3 is to **three** significant figures.

One important feature of significant figures is that they help us to reduce the chances of data errors:
▶ making sure data is not recorded to more digits than the measurement allows (e.g. ruler, balance, burette)
▶ in calculations, ensuring that we report the same number of digits as the original measurements allow.

As a rule in **calculations**, we look at the number of significant figures in the items of data. A calculator may give an answer to 10 digits, but using all of them would give **spurious** digits – they would not all be

reliable. We should not report the answer to more significant figures than are in the item of data with the **lowest** number of significant figures.

In a titration (see section B1.44) we might have a standard solution that is 0.0500 mol/dm^3 (three significant figures) and a titre of 24.5 cm^3 (three significant figures). This means that we can calculate the number of mol:

mol = concentration × volume

mol = $0.0500 × 0.0245 = 1.225 × 10^{-3}$ or 0.001225 mol

However, we should only report this answer to three significant figures, or $1.23 × 10^{-3}$ mol.

There are some exceptions to this rule. For example, if we have a set of data and calculate a mean, it can be acceptable to quote the mean to one more significant figure than the values in the data. This is justified because calculating a mean allows us to get closer to the **true value** than any individual data point would be.

Test yourself

1 Give the SI unit for the following measurements:
   a mass
   b length
   c temperature.
2 Which of the following is not an SI unit?
   a candlepower
   b coulomb
   c joule
   d watt.
3 Complete the following table of conversions.

Convert from	Convert to	Answer
1250mm	m	
0.0005kg	g	
1250000J	kJ	
0.005W	mW	
2.5MHz	Hz	

4 Convert the following to numbers in standard form to three significant figures:
   a 1256000
   b 0.000000135023

Project practice

You are working for a company that provides components for use in lateral flow diagnostic test kits, like those used for COVID-19 testing in schools. The kits include swabs used to collect a sample from the nose and throat. These are currently sterilised with ethylene oxide gas. People have been concerned that ethylene oxide is dangerous and that children should not be exposed to the risk.

Exposure to ethylene oxide in large quantities can have serious health consequences. However, you know that there is no ethylene oxide remaining on the swabs after sterilisation and ethylene oxide has been used safely for decades to sterilise medical equipment and devices, including swabs and dressings. There has never been any evidence of harm caused to patients or users.

You have been asked to investigate other sterilisation methods and prepare a report with recommendations for alternatives to ethylene oxide.

You will need to carry out the following:

1 Research the options:
  a Carry out a literature review of sterilisation methods.
  b Investigate sources of safety information.

2 Prepare a plan:
  a Make a shortlist of possible methods.
  b Justify your choice of methods.
  c Prepare risk assessments of the different methods.
3 Analyse the data in comparison with the current method (ethylene oxide) and reach conclusions:
  a comparison of effectiveness of different methods
  b relative costs, including any specialist equipment needed and consumable items, e.g. chemicals
  c possible damage to components caused by the different methods
  d risk factors associated with the different methods, either to people carrying out the tests or to end users/patients.
4 Present your analysis and conclusions in the form of a PowerPoint presentation.
5 Group discussion covering:
  a the relative effectiveness and safety of the different methods
  b would they be more or less effective than the existing method
  c the perception of risk by members of the general public.
6 Write a reflective evaluation of your work.

Assessment practice

1 Karen has an iPhone XR. She has found out that it has a battery with a capacity of approximately 2900 mAh. A charge of 1 mAh is transferred when a current of 1 mA flows for 1 hour.
  a Calculate the total electric charge, in coulombs, in the iPhone XR battery.
  b Karen sees that the charger supplied with the iPhone XR is marked '5 V, 1000 mA'. She also has a charger for an iPad mini that is marked '5 V, 2100 mA'. Karen believes that the iPad charger will charge the iPhone more quickly. Is Karen correct? Calculate the time taken to charge the iPhone using each type of charger. Give your answers in hours, minutes and seconds.
2 The figure shows a circuit with a 9.0 V battery and three resistors.

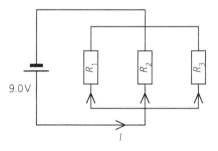

  a Is this an example of a series or parallel circuit?
  b The resistors are marked as follows: R1 = 10 Ω, R2 = 22 Ω, R3 = 47 Ω. Calculate the total resistance in the circuit. Give your answer to an appropriate number of significant figures.
  c Based on your answer to part b, calculate the current flowing from the battery. Give your answer to an appropriate number of significant figures.

3 Describe **two** ways you could use to make a piece of iron magnetic.

4 Electromagnets are used in the recycling industry.

   a What type of electromagnet would be found in a crane used for lifting scrap iron in a scrapyard?

   b Explain how electromagnets can be used to separate scrap iron from scrap copper.

   c A strong magnet will not attract aluminium. However, if a mixture of scrap aluminium and scrap plastic moves along a conveyor belt, eddy currents are created in the aluminium. These eddy currents cause the aluminium to be attracted to the magnet so it can be separated from the plastic. Explain how a current is formed in the aluminium.

5 Which of the following statements about waves is true?

   a Gamma radiation is a type of transverse wave.

   b Light waves transfer energy but sound waves transfer matter.

   c The displacement of particles in a longitudinal wave is perpendicular to the direction of travel.

   d The displacement of particles in a transverse wave is parallel to the direction of travel.

6 Make sketch graphs of transverse waves travelling along a Slinky. The y-axis should show the transverse displacement of the waves and the x-axis should show the distance along the Slinky.

   a Two waves with the same frequency but one wave has double the amplitude of the other.

   b Two waves with the same amplitude. One wave has double the frequency of the other wave.

7 Two radioactive sources were contained in glass bottles stored inside lead-lined containers. Both sources had approximately the same activity marked on the container. The bottles were removed from their containers and the following measurements taken using a Geiger counter:
   – Source A gave a very low reading.
   – Source B gave a very high reading.
   – When a sheet of aluminium was placed between source B and the Geiger counter, the reading fell by 75 per cent.
   – When a sheet of lead was placed between source B and the Geiger counter, the reading fell by 95 per cent.
   Explain what type of radiation was emitted by each source. Justify your answer.

8 A sample of radioactive material has been spilled. Discuss the options that you would consider in order to deal with the situation. You should include consideration of type and level of radiation emitted, half-life, and level or intensity of the radiation.

9 You are investigating sound waves moving in air or in water.

   a In air, the wavelength of the sound is 250 cm and its frequency is 131 Hz. Calculate the speed of the wave. Give your answer in the correct SI units to an appropriate number of significant figures.

   b Sound from the same source travels in water at 1480 m/s but the frequency remains the same. Calculate the new wavelength to the appropriate number of significant figures.

10 You have been asked to produce a wallchart showing the electromagnetic spectrum. You have been given the following table of values for the wavelength of the different types of electromagnetic radiation. Convert these values to metres (m) expressed in standard form so that they can be plotted more easily on the wallchart.

Type of radiation	Wavelength	Wavelength in standard form (m)
Gamma rays	1 pm to 10 pm	
X-rays	5 pm to 10 nm	
Ultra-violet	5 nm to 380 nm	
Visible light	380 nm to 750 nm	
Infra-red	750 nm to 1 mm	
Microwaves	0.1 mm to 10 cm	
Radio waves	10 cm to 2000 m	

B2.1–B2.14: Further science concepts: Biology

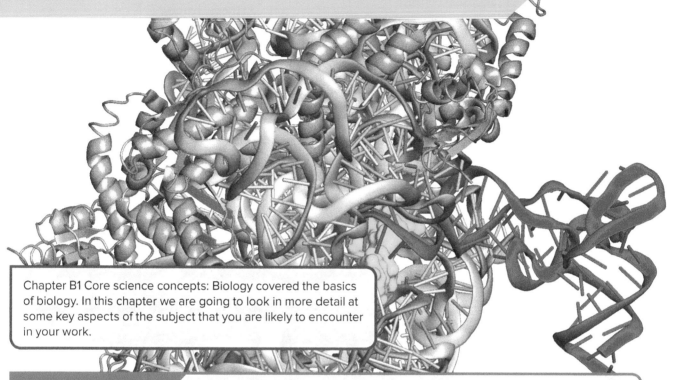

Chapter B1 Core science concepts: Biology covered the basics of biology. In this chapter we are going to look in more detail at some key aspects of the subject that you are likely to encounter in your work.

Learning outcomes

The core knowledge outcomes that you must understand and learn:

Classification of biological molecules

B2.1 the molecular structures and functions of proteins, carbohydrates, lipids and nucleic acids

Enzyme and protein structure

B2.2 the role of DNA bases in the production of amino acid chains, which form proteins

B2.3 how the process of protein synthesis occurs

B2.4 the properties of enzymes that are determined by their tertiary structure

B2.5 how enzymes' mechanism of action allows them to catalyse a wide range of intracellular reactions

Cell cycle

B2.6 the function of both mitosis and meiosis in nuclear division within cells

B2.7 the characteristics of each of the stages of mitosis, including the behaviour of chromosomes and the cellular structure at each stage

B2.8 how the process of meiosis, including phase 1 and phase 2, results in the formation of haploid gametes from diploid cells in the reproductive organs

B2.9 the significance of the differences between mitosis and meiosis

Cellular respiration

B2.10 how respiration results in the breakdown of glucose to produce the energy-carrying molecule adenosine triphosphate (ATP)

B2.11 how ATP provides a source of energy for biological processes

B2.12 the comparative amounts of energy produced by different respiratory substrates (lipids, proteins and carbohydrates)

Pathogens

B2.13 the definition of a pathogen

B2.14 examples of different types of pathogens and the diseases they can cause.

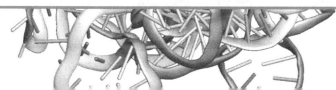

Classification of biological molecules

Chapter B1 Core science concepts: Biology, sections B1.7 to B1.9, covered the relationship between the structure and function of the three main classes of biological molecules:

▶ proteins
▶ carbohydrates
▶ lipids.

We are now going to look at these in more detail and add the fourth major class of biological molecules, the nucleic acids.

B2.1 The molecular structures and functions of proteins, carbohydrates, lipids and nucleic acids

Proteins

In section B1.7 we looked at the basic structure of polypeptides as polymers of amino acids and how proteins consist of one or more polypeptide chains with a recognisable 3D structure. We will now look at how that structure is formed.

First, we need to think a little about bonding. There are three types of bonds involved in determining the structure of the proteins that you need to know about:

▶ **hydrogen bonds**
▶ **ionic bonds**
▶ **disulfide bonds** (a type of **covalent bond**).

Key terms

Hydrogen bond: the attraction between slight positive and negative charges when hydrogen atoms are attached to oxygen or nitrogen atoms.

Ionic bond: the attraction between a positively charged R group of one amino acid and the negatively charged R group of another amino acid.

Disulfide bond: covalent bond between the sulfur atoms of the R groups of two cysteine amino acids.

Covalent bond: forms when atoms share one or more pairs of electrons; the electrons are attracted to the nuclei of both atoms, and this holds the two atoms together. Covalent bonds are very strong.

Hydrogen bonds are formed when hydrogen atoms are covalently bonded to oxygen or nitrogen atoms. When this happens, the hydrogen develops a slight positive charge and the oxygen or nitrogen develops a slight negative charge. Therefore, there is an attractive force between these charges. This can be between different molecules, such as in water, or between different parts of the same molecule, which happens in proteins.

Figure 14.1 shows how hydrogen bonds hold water molecules together.

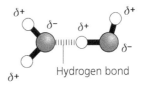

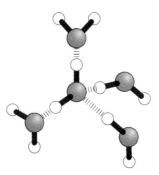

▲ **Figure 14.1 Hydrogen bonds in water**

In proteins, hydrogen bonds form between amino acids in the polypeptide chain and help to hold the structure together. This will be described in more detail below.

Ionic bonds are formed between atoms or groups of atoms that lose electrons to form positive ions (for example, Ca^{2+} or NH_4^+) and those that gain electrons to form negative ions, such as CO_3^{2-} or Cl^-. Because these have opposite charges, strong electrostatic forces hold the ions together, such as in $CaCO_3$ (calcium carbonate) or NH_4Cl (ammonium chloride).

If you look back at Figure 11.9 on page 178, you will see that amino acids all have **R groups**, also called side chains. The 20 different types of amino acids found in proteins have different R groups with a range of chemical structures (Figure 14.2). Some of these contain carboxyl groups (—COOH) that can lose H^+ to form —COO$^-$ while others contain amino groups (—NH$_2$) that can gain H^+ to form —NH$_3^+$. Attraction between these positive and negative charges forms ionic bonds that help to maintain the 3D structure of proteins.

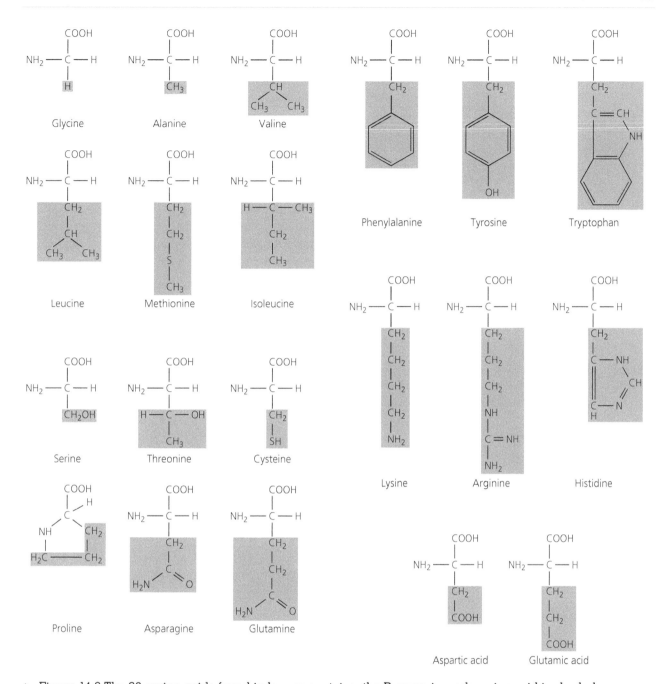

▲ Figure 14.2 The 20 amino acids found in human proteins; the R group in each amino acid is shaded

The other type of bond involved in the 3D structure of proteins is a **covalent** bond formed between the R groups of two cysteines (see Figure 14.2). This bond is known as a **disulfide bond** or **bridge**, because it forms between the sulfur atoms in each cysteine and bridges or holds together different parts of the polypeptide chain (see Figure 14.3 for an example).

▲ Figure 14.3 The primary structure of a protein; this uses the three-letter abbreviations of the names of each of the amino acids; it also shows three disulfide bridges between pairs of cysteines

Figure 14.3 shows the **primary structure** of a protein. This is simply the order of amino acids in a polypeptide chain. The primary structure is important because it determines how the protein folds up into its specific 3D structure. The structure of a protein determines its function.

We can think of **secondary structure** as different types of structural elements that are found in most proteins.

The secondary structure of a protein describes how the polypeptide chain (primary structure) is folded into specific structures, such as α-helices (singular = alpha helix) and β-sheets (pronounced 'beta').

Figure 14.4 shows more detail of an α-helix and β-sheet. Note how hydrogen bonds between parts of different amino acids along the chain hold the structure together in the α-helix or hold two chains together in a β-sheet. Hydrogen bonds in the β-pleated sheet hold two lengths of polypeptide chain together.

The **tertiary structure** of a protein describes its 3D shape. The polypeptide chain folds into elements such as α-helices and β-sheets, connected by sections of the polypeptide chain that do not have any recognisable structure. However, the folded protein will have a definite 3D structure determined by the hydrogen bonds, ionic bonds and, in some proteins, disulfide bridges. The number and location of these bonds (and therefore the 3D shape of the protein) is determined by the R groups in the amino acid monomers.

Globular proteins are folded into a variety of shapes with each individual protein having its own particular shape. This huge diversity of shapes reflects the very wide range of functions performed by these proteins, such as binding, signalling, transport and biological catalysts (enzymes).

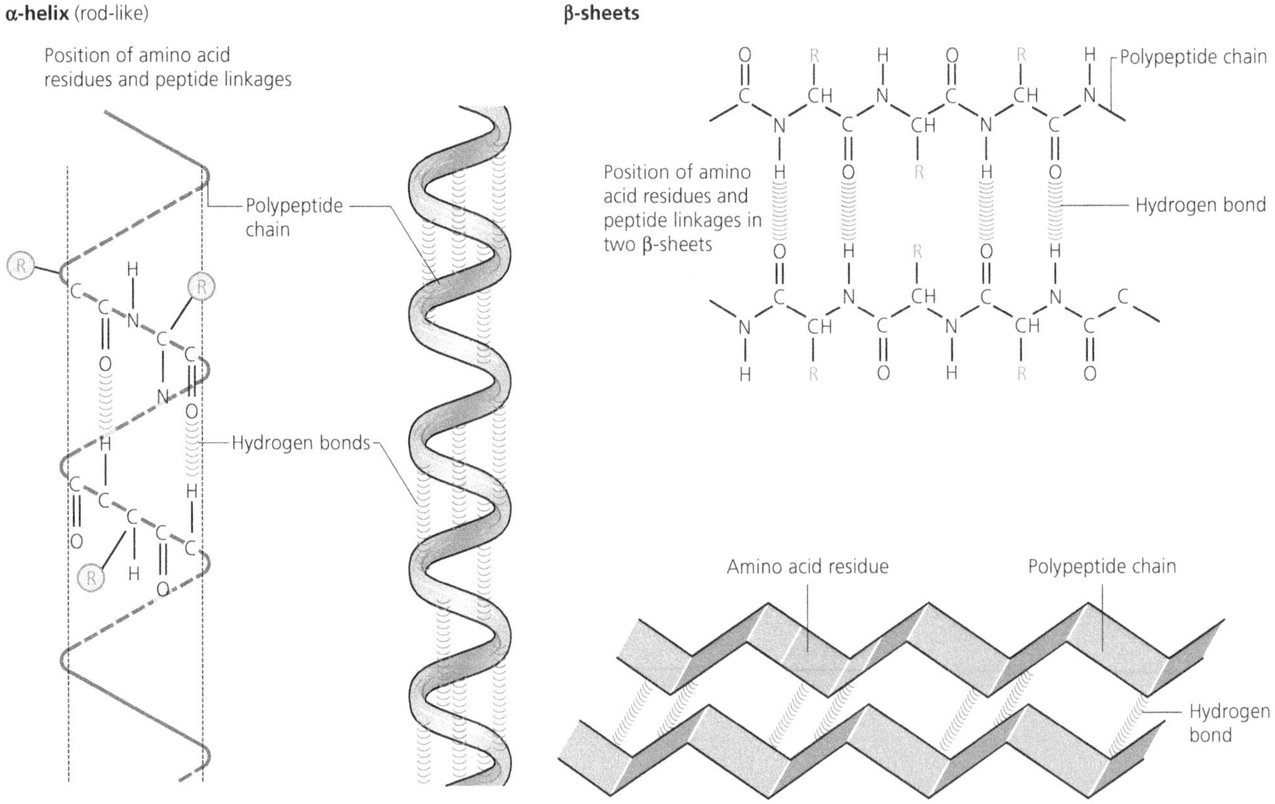

α-helix (rod-like)

Position of amino acid residues and peptide linkages

Polypeptide chain

Hydrogen bonds

β-sheets

Position of amino acid residues and peptide linkages in two β-sheets

Polypeptide chain

Hydrogen bond

Amino acid residue

Polypeptide chain

Hydrogen bond

▲ Figure 14.4 The structure of an α-helix (left) and β-sheet (right) showing how hydrogen bonds hold the polypeptide chain in a specific secondary structure

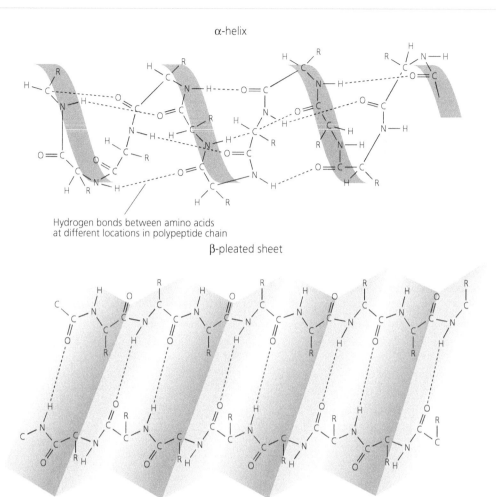

α-helix

Hydrogen bonds between amino acids
at different locations in polypeptide chain

β-pleated sheet

▲ Figure 14.5 The structure of an α-helix and a β-pleated sheet showing how hydrogen bonds hold the polypeptide chain in place

Figure 14.6 is a computer-generated model of a globular protein showing the tertiary structure. The polypeptide chain is represented as a ribbon twisted into α-helices and three strands of the ribbon (polypeptide chain) form a β-sheet.

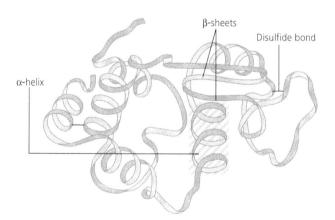

α-helix

β-sheets

Disulfide bond

▲ Figure 14.6 A 3D model of a globular protein showing elements of secondary structure as well as disulfide bonds forming cross-links helping to stabilise the molecule

Practice point

You will come across different conventions when showing protein structure. Sometimes the polypeptide chain is represented by a thin cylinder, like a wire, or by a ribbon. The α-helices might look like coiled springs (as in Figure 14.6) or be shown as tubes (as in Figure 14.7), whilst β-sheets are often represented by broad ribbons with arrowheads.

An internet image search for 'protein structure' will provide many examples.

In Figure 14.6 the secondary structure elements can be seen more clearly together with the sections of polypeptide chain between the elements.

The final level of protein structure is the **quaternary structure**. This refers to the fact that many functional proteins are formed from two or more polypeptide chains. Figure 14.7 shows one example, the haemoglobin protein found in red blood cells that is responsible for transport of oxygen from the lungs to the tissues.

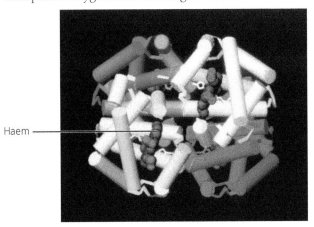

▲ Figure 14.7 Structure of haemoglobin showing the four polypeptide chains, each one in a different colour

Haemoglobin consists of four polypeptide chains, shown in four different colours in Figure 14.7. You can also see the secondary structure in this image – the cylinders represent α-helices. Another feature to note is the **haem group**. There are four of these in haemoglobin, one attached to each polypeptide chain; the haem is where an oxygen molecule binds, allowing haemoglobin to transport oxygen in the blood.

Fibrous proteins are formed of long chains that run parallel to each other, linked by cross bridges to form stable molecules. These act as structural polymers, for example collagen (Figure 14.8), which is found in the tendons and ligaments that hold the bones together

and attach muscles to bones. The structure of collagen means that it is both strong and flexible.

(a)

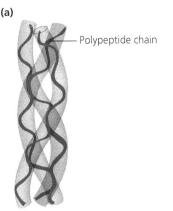

Polypeptide chain

(b)

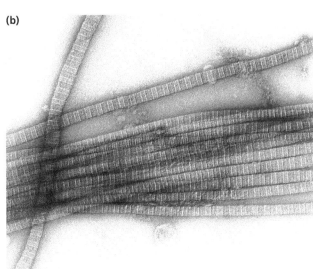

▲ Figure 14.8 Structure of collagen: (a) the three polypeptide chains coiled around each other and (b) an electron micrograph showing collagen fibres

---

Practice point

Summary: How different elements present in R groups contribute to protein structure and function

You will see from Figure 14.2 that the R groups of amino acids contain elements besides carbon and hydrogen and these all play an important part in determining the structure of proteins, and the function of proteins depends on them having the correct tertiary structure.

Hydrogen bonding can involve R groups containing oxygen and/or nitrogen and contributes to holding different parts of the polypeptide chain together forming the tertiary structure.

Ionic bonding can also occur between R groups containing oxygen or nitrogen and also contributes to formation of the tertiary structure.

The amino acid cysteine has an R group that contains sulfur. These can form disulfide bonds that also contribute to holding parts of the polypeptide chain together (see Figure 14.3) as the protein forms its tertiary structure.

As well as these bonds contributing to the formation of the tertiary structure of a protein, some play a direct role in the function of the protein. For example, R groups in the active site of enzymes sometimes form hydrogen bonds or ionic bonds with substrate molecules. This plays an important role in the mechanism of action of enzymes.

Levels of protein structure

Figure 14.9 summarises the four levels of proteins found in globular proteins. Note that the structures are shown in a more simplified way here compared to Figure 14.4.

Fibrous proteins do not fit easily into the four levels shown in Figure 14.9. If you look at Figure 14.8 you will see that collagen consists of helical polypeptides (secondary structure) held together in a triple helix (quaternary structure) but does not have anything that could be described as tertiary structure.

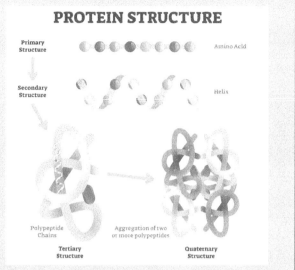

PROTEIN STRUCTURE

▲ Figure 14.9 The four levels of structure in globular proteins; not all proteins have quaternary structure

Carbohydrates

Section B1.8 covered the relationship between the structure, properties and functions of the main types of carbohydrate.

All carbohydrates are composed of carbon, hydrogen and oxygen. The basic units of carbohydrates are monosaccharides. Examples include glucose (Figure 14.10), fructose and galactose.

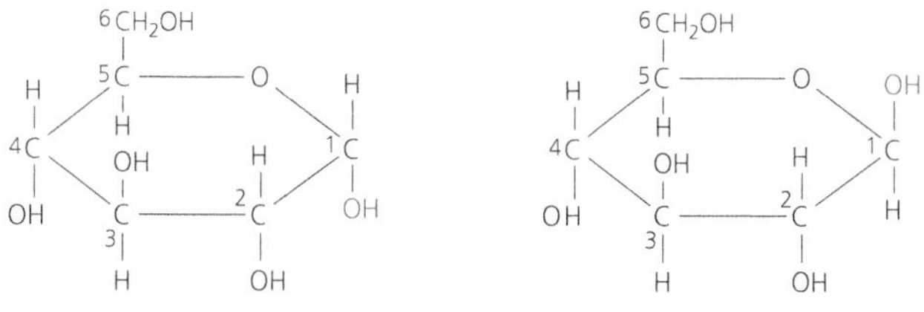

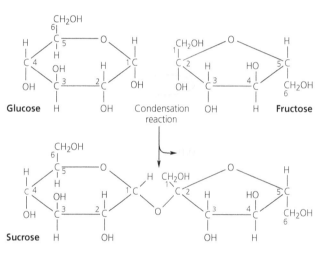

▲ Figure 14.11 Condensation reaction between two monosaccharides, forming a glycosidic bond

▲ Figure 14.10 The structure of α- and β-glucose

You will see from Figure 14.10 that glucose exists as two forms (**isomers**): alpha-glucose (α-glucose) and beta-glucose (β-glucose). The only difference between these isomers is the position of the —OH group on carbon number 1, which is pointing below the ring in α-glucose and above the ring in β-glucose. We will see the importance of this shortly.

Monosaccharides can combine in pairs to form **disaccharides** through a **condensation reaction**, forming what is known as a **glycosidic** bond (Figure 14.11).

Disaccharides can be formed from the same monosaccharide. For example, **maltose** consists of two glucose units. Alternatively, they can be formed from two different monosaccharides. You can see this in Figure 14.11 where **sucrose** (table sugar from sugarcane or sugar beet, which we use in baking and so on) is formed from glucose and fructose.

Polysaccharides, like polypeptides, are polymers in which the monomer (see section B1.8) is a monosaccharide. **Glycogen** and **starch** are both polymers of α-glucose. Starch is used by plants as a food store and exists in two forms (Figure 14.12):
- amylose, a straight-chain polymer
- amylopectin, a branched-chain polymer.

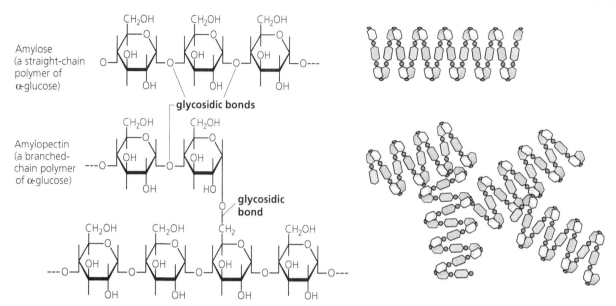

▲ Figure 14.12 The structure of starch showing amylose (top) and amylopectin (bottom)

Glycogen is also used as a food store but is even more branched than amylopectin. Glycogen is found in animals where it is stored mostly in muscle and liver cells.

Both glycogen and starch are insoluble. This makes them well suited to their role as energy stores in the cell because they do not affect the water balance of the cell. Storing the equivalent amount of glucose would cause water to enter the cell by osmosis (see section B1.11, page 181).

Because of the way the —OH on carbon-1 points below the ring, it causes the polysaccharide chain to curl into a helix. This makes it very compact, which is an advantage for a food-storage polysaccharide.

Because of the highly branched structure of glycogen, the molecule can be **hydrolysed** (broken down) rapidly to release glucose very quickly for use in respiration (see section B2.10). This is why glycogen is stored in muscles, to release energy very quickly to allow muscle contraction to occur.

> ### Key term
>
> **Hydrolysis:** literally 'splitting with water'; it is the reverse of a condensation reaction; most large biological molecules are broken down in hydrolysis reactions.

Cellulose, like starch and glucose, is insoluble. However, unlike starch and glycogen, which are food (energy) storage polysaccharides, **cellulose** is a structural polysaccharide. Cellulose is formed by condensation of β-glucose. Because of the way the —OH group is positioned on carbon-1, cellulose forms long straight chains that are held together by hydrogen bonds in bundles called **microfibrils** (Figure 14.13). This makes cellulose fibres very strong and well suited to their role in forming plant cell walls (see section B1.4).

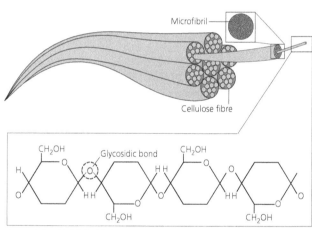

Long chain of 1,4 linked β-glucose residues. Hydrogen bonds link these chains together to form microfibrils.

▲ Figure 14.13 The structure of cellulose

Lipids

Cellulose, like starch and glucose, is insoluble. However, unlike polypeptides and polysaccharides, **lipids** are not polymers. Instead, lipids such as **triglycerides** and **phospholipids** are formed from fatty acids and glycerol.

Triglycerides are formed by the condensation of one molecule of **glycerol** and three molecules of **fatty acid** (Figure 14.14).

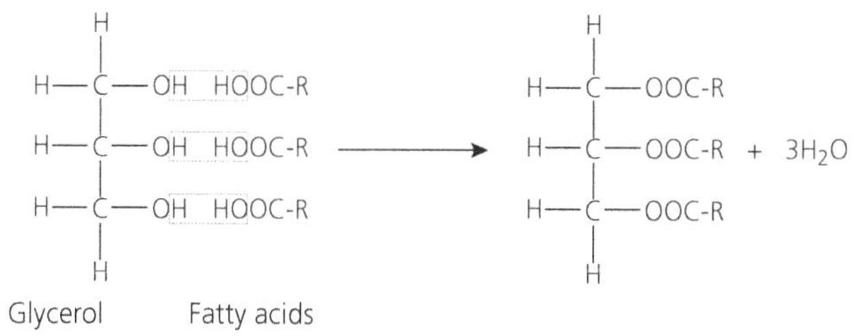

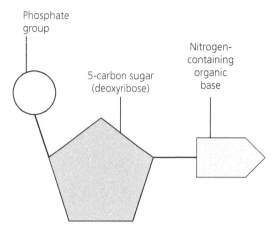

H—C—OH HOOC-R   H—C—OOC-R
H—C—OH HOOC-R   →   H—C—OOC-R + 3H₂O
H—C—OH HOOC-R   H—C—OOC-R

Glycerol   Fatty acids

▲ Figure 14.14 Formation of a triglyceride from glycerol and fatty acids; R represents a long carbon chain; the R groups can be the same or different

Phospholipids are formed when one of the fatty acids of a triglyceride is substituted by a phosphate-containing group. This is shown in a diagrammatic form in Figure 14.15.

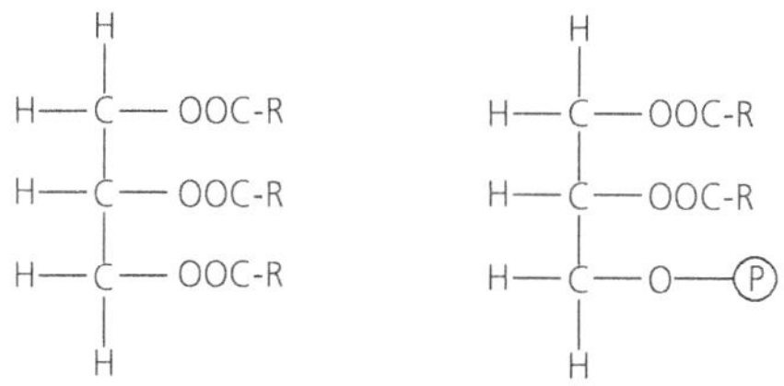

H—C— OOC-R   H—C—OOC-R
H—C— OOC-R   H—C—OOC-R
H—C— OOC-R   H—C—O—(P)

▲ Figure 14.15 General structure of a triglyceride (left) and a phospholipid (right). The P in a circle represents the phosphate group of the phospholipid

The long carbon chains of fatty acid molecules are **hydrophobic**, meaning they repel water. Glycerol molecules are **hydrophilic**, meaning that they attract water. When a triglyceride is formed, however, the molecule is hydrophobic.

Triglycerides are important as energy stores in **adipose tissue**. Adipose tissue also functions to provide thermal insulation under the skin and a physical cushioning around organs such as the kidneys.

A phospholipid is made up of two parts: a hydrophilic head (containing the polar phosphate group) and a hydrophobic tail. This molecular structure forms a bilayer that is important for all membrane functions (see section B1.11). Figure 11.14 in section B1.11 illustrates the structure of phospholipids within the bilayer, showing the importance of the hydrophilic and hydrophobic parts of the molecules.

Nucleic acids

Like polypeptides and polysaccharides, **nucleic acids** are polymers composed of **nucleotide monomers**. Nucleic acids can be extremely large molecules.

Each nucleotide in DNA is made up of **deoxyribose** – a **pentose** sugar (named because it contains five carbons), a phosphate and an organic base (Figure 14.16).

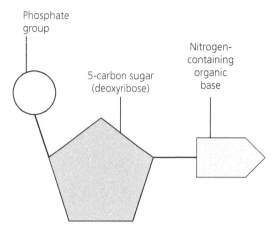

▲ Figure 14.16 A diagram of a DNA nucleotide

DNA is made up of two strands of nucleotides joined together by hydrogen bonds (Figure 14.17). DNA contains four types of organic base:
▷ guanine (G)
▷ cytosine (C)
▷ adenine (A)
▷ thymine (T).

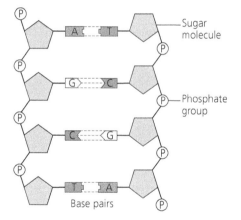

▲ Figure 14.17 Part of a DNA molecule showing the two strands held together by base pairing

Notice in Figure 14.17 how A forms hydrogen bonds with T, and G forms hydrogen bonds with C. This is known as **complementary base pairing** and is fundamental to how DNA replicates (see section B1.16).

The two strands of DNA wind around each other to form a double helix structure (Figure 14.18).

▲ Figure 14.18 The double helix structure of DNA

The double helix structure is very stable because of all the hydrogen bonds between the two strands, like rungs on a ladder. This is important, because DNA stores genetic information and passes it on to future generations (of cells and offspring) through the process of DNA replication, as explained in section B1.16.

The genetic information stored in DNA is used in the production of proteins, as we will see in the next section.

Test yourself

1 What are the **three** types of bond that help to maintain the tertiary structure of a protein?

2 What are the monomer units in the following biological polymers?
   a Proteins.
   b Starch.
   c Cellulose.
   d Nucleic acids.

3 Describe **two** functions of triglycerides.

4 Explain the difference in physical properties between triglycerides and phospholipids.

5 Explain the importance of hydrogen bonds in DNA.

Enzyme and protein structure

We have seen how the function of a protein depends on its tertiary structure, and that the tertiary structure of a protein is determined by its primary structure. We will now look at how proteins are made and how the tertiary structure of an enzyme determines how it functions as an enzyme to increase the rate of biological reactions.

B2.2 The role of DNA bases in the production of amino acid chains, which form proteins

At this point, it would be useful to review sections B1.13 and B1.14 on the role of DNA and RNA in inheritance (genetics).

We saw in section B2.1 that DNA nucleotides consist of a sugar molecule (deoxyribose) attached to a phosphate group and a nitrogen-containing base (Figure 14.16). The bases in DNA are guanine (G), cytosine (C), adenine (A) and thymine (T).

The **base sequence** along a single strand constitutes the **genetic code**. A sequence of three DNA bases is known as a **triplet** or a **codon**. Each codon codes for a specific amino acid or a start or stop codon. Start or stop codons signal the start or finish point of the polypeptide.

Key term

Base sequence: the order of the bases, attached to nucleotides, in a DNA molecule.

There are three key facts about the genetic code that you should remember:
▶ It is universal, meaning that it is the same in all organisms on Earth.
▶ It is non-overlapping (see below).
▶ It is **degenerate**, meaning that each of the 20 amino acids can be coded for by more than one codon.

The sequence of bases within a **gene** specifies the sequence of amino acids that are linked together to form a polypeptide chain. The sequence is read or decoded in groups of three bases – the codons. Table 14.1 shows how this is done in a way where the groups do not overlap.

A DNA base sequence	CCGCGATCGGACTTG
The sequence above contains these codons	CCG CGA TCG GAC TTG
But not these codons	CGC GCG GAT ATC CGG GGA ACT CTT

▲ Table 14.1 Codons within a DNA base sequence do not overlap

B2.3 How the process of protein synthesis occurs

DNA acts as a template that provides the instructions for synthesising (making) a polypeptide via the coding sequence of bases. The process of converting this information into a polypeptide chain consists of two parts:

▶ transcription
▶ translation.

We saw in section B1.16 how complementary base pairing enables the process of **semi-conservative replication** of DNA. Complementary base pairing is also involved in transcription and translation.

Transcription

In transcription, the section of DNA that contains the genetic information (the gene) is used as a template to make a complementary sequence of **messenger RNA** (mRNA) – we encountered RNA in sections B1.13 and B1.14. This process is illustrated in Figure 14.19.

Transcription involves the following steps:

▶ The DNA double helix is unwound to separate the two strands at the gene to be transcribed.
▶ RNA nucleotides bind to the exposed DNA bases on one of the strands. Note that the base U (uracil) replaces T (thymine) in RNA (that is, U pairs with A).
▶ The enzyme RNA polymerase joins the RNA nucleotides to make a complementary sequence of mRNA.
▶ The DNA rewinds to form the double helix again.
▶ The mRNA leaves the nucleus and meets a ribosome in the cytoplasm where the next step (translation) occurs.

Translation

Translation is the process in which the messenger RNA now acts as a template for assembly of the polypeptide chain, which occurs on a ribosome (see section B1.3).

This involves another type of RNA called **transfer RNA** (tRNA). There is a different type of tRNA molecule for each type of amino acid. One part of the tRNA molecule contains a sequence of three bases, known as an **anticodon**; the anticodon sequence will determine which amino acid is carried by the tRNA. Complementary base pairing between the anticodon on the tRNA and a codon on the mRNA ensures that the ribosome inserts the correct amino acid to the growing polypeptide chain. We can think of tRNA as bringing the correct amino acid to be incorporated into the growing polypeptide chain.

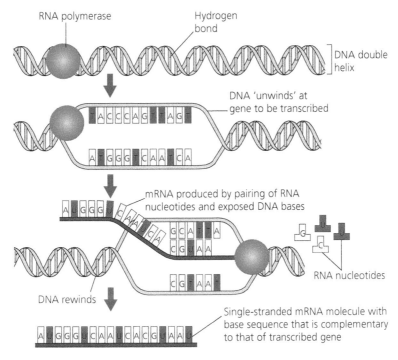

▲ Figure 14.19 The process of transcription showing how a sequence of DNA is used as a template to make a complementary sequence of mRNA

The process of translation is illustrated in Figure 14.20.

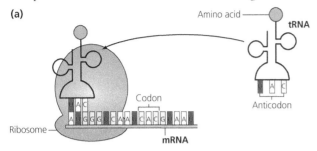

(a)

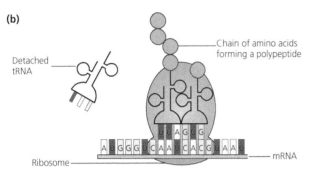

(b)

▲ Figure 14.20 The process of translation. (a) A ribosome attaches to the mRNA; a tRNA with a complementary anticodon binds to the codon on the mRNA; another tRNA will bind to the next codon and a peptide bond is formed between the amino acids. The 'empty' tRNA then leaves the ribosome. (b) The ribosome moves along the mRNA and the process repeats and so the polypeptide chain gets longer.

In this way, the sequence of amino acids in the polypeptide will correspond to the sequence of codons in the mRNA (which itself corresponded to the sequence of codons in the DNA).

B2.4 The properties of enzymes that are determined by their tertiary structure

We have seen how the structure of a protein is closely linked to its function – it might be better to say that the function is determined by the structure. This is certainly true of enzymes.

Enzymes are biological **catalysts** – they speed up reactions in the cell. Think about the formation of sucrose from glucose and fructose (Figure 14.11):

glucose + fructose → sucrose + water

This reaction is catalysed by the enzyme sucrose synthase; the **substrates** of the enzyme are glucose and fructose, and the product is sucrose (and water). The substrates bind to a part of the enzyme known as the **active site**.

Key terms

Substrate: the substance on which an enzyme acts to form the products.

Active site: the part of the enzyme where the substrate binds.

This binding of the substrates to the active site will depend on:
- the shape (specifically, the tertiary structure) of the active site
- bond formation between parts of the substrates and amino acid R groups in the active site; these could be hydrogen bonds or ionic bonds.

The shape of the active site is complementary to the shape of the substrate or substrates. This means that enzymes will be **specific** for a particular substrate (see Figure 14.21). This also means that, if there is a change in the shape of the protein (and, therefore, the active site), the substrates will no longer bind effectively.

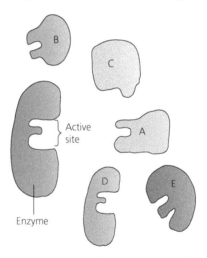

▲ Figure 14.21 The active site of this enzyme has a shape that is complementary to the shape of its substrate, molecule A. Other molecules do not have complementary shapes and so will not bind to the enzyme

In section B1.40, we saw how the rate of a chemical reaction increases as temperature increases. This is also true of enzyme-catalysed reactions – up to a point. Look at Figure 14.22 and you will see that the rate of reaction increases as the temperature increases, up to about 40 °C. This is known as the **optimum temperature** because the rate of reaction is highest.

Above 40 °C, the rate of reaction drops rapidly. This is because, as the temperature increases, the bonds that hold the tertiary structure of the protein weaken (increased temperature causes the atoms to vibrate more).

As a result, the shape of the protein, and therefore of the active site, changes; the substrates no longer fit correctly and the enzyme becomes less effective.

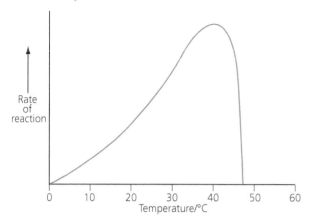

▲ Figure 14.22 The effect of temperature on the rate of an enzyme-catalysed reaction

Eventually, the bonds become so weak that the protein loses its structure completely – we say it has **denatured** (Figure 14.23). This is the same process that occurs when the soluble proteins in a raw egg are converted into insoluble proteins in hard-boiled egg, or when meat is cooked.

A similar process occurs when an enzyme is exposed to different pH values (degrees of acidity or alkalinity). Hydrogen bonds and ionic bonds are affected by pH and so, if the pH changes, the shape of the protein also changes and so the activity of the enzyme will change. This is illustrated in Figure 14.24.

There are two important things to note from Figure 14.24:
1 There is a relatively narrow range of pH over which an enzyme is active.
2 The optimum pH for different enzymes can vary quite widely.

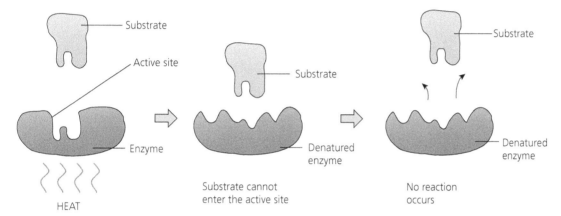

▲ Figure 14.23 The effect of heat on enzyme activity

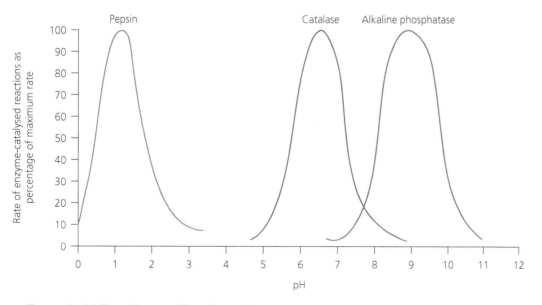

▲ Figure 14.24 The effect of pH on three different enzymes

B2.5 How enzymes' mechanism of action allows them to catalyse a wide range of intracellular reactions

We have seen that enzymes catalyse reactions in the cell (intracellular reactions) and how their activity is affected by temperature and pH. But how do they actually work?

An early model of enzyme action was the lock and key hypothesis. This suggested that a substrate is complementary in shape to the active site of the enzyme and so fits into the active site like a key fitting into a lock (Figure 14.25).

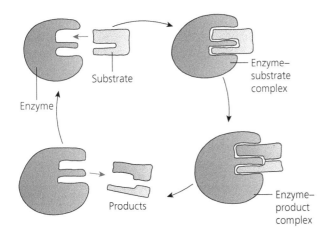

▲ Figure 14.25 The lock and key hypothesis; the substrate is an exact fit for the active site; once the reaction is over, the products leave the active site and the enzyme is ready to accept more substrate

The lock and key hypothesis explains why enzymes show **substrate specificity**. The active site is only complementary in shape to certain substrates but not others (Figure 14.26).

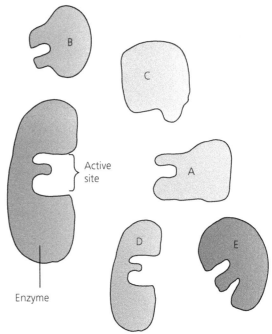

▲ Figure 14.26 The active site of this enzyme has a shape complementary to substrate molecule A, but not to any of the others

This explains why the enzyme **sucrase** will hydrolyse the disaccharide **sucrose**, but not the disaccharides **maltose** and **lactose** that have similar, but not identical, structures.

However, the lock and key hypothesis fails to explain one key aspect of enzyme action: how does the enzyme reduce the activation energy of the reaction? You will see from section B1.41 that reducing activation energy is the way in which all catalysts increase the rate of reactions.

The answer to this lies in a development of the lock and key hypothesis known as **induced fit**. This proposes that the active site of the enzyme changes shape and moulds around the substrate (Figure 14.27).

Induced fit helps us to understand an important feature of enzyme action. As the shape of the active site changes, R groups on amino acids in the active site come closer to parts of the substrate. This means that stronger bonds are formed with the substrate and that the substrate is not only held in place but is put under strain. This can weaken bonds in the substrate, making them more likely to react.

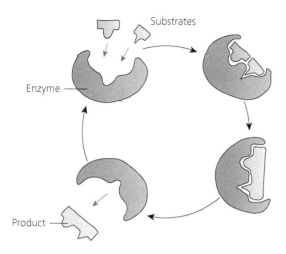

▲ Figure 14.27 Induced fit – notice how the active site changes shape as the substrate molecules bind

The effect of enzyme concentration and substrate concentration on the rate of the reaction

You may be aware that, if you increase the concentration of reactants in a chemical reaction, you will increase the rate of reaction. This will be covered in section B2.21 (page 273). This also applies to enzyme-catalysed reactions but, as with temperature, only up to a point. When a substrate binds to an enzyme, an **enzyme–substrate complex** (ESC) is formed. Once an ESC forms, the reaction can take place. Therefore, the rate of reaction will be proportional to the number of ESCs, and anything that increases the number of ESCs will increase the rate of reaction. There are two ways we can increase the number of ESCs:

▶ increase the concentration of substrate
▶ increase the concentration of enzyme.

However, once all the enzyme active sites are occupied, no more ESCs can be formed – unless we add more enzyme. If we do add more enzyme, eventually we will run out of substrate, and so increasing the enzyme concentration will not increase the rate of reaction. This is illustrated in Table 14.2.

If you look at the graphs in both Tables 14.2 and 14.3 you will see that the curves increase and then reach a plateau (they level off). At the point where the curve levels off, another factor has become limiting. In the case of Table 14.2 it is the substrate that has become the limiting factor, whereas in Table 14.3 it is the enzyme that has become the limiting factor.

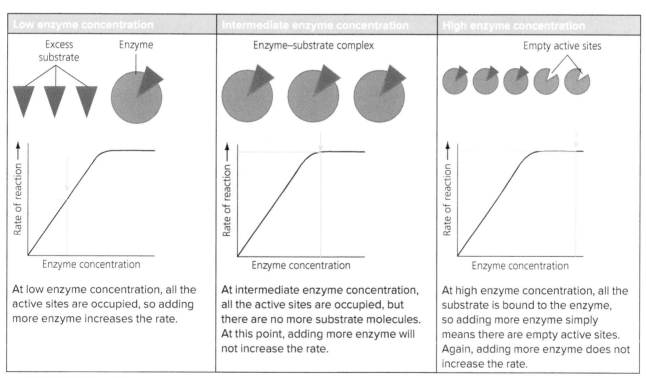

▲ Table 14.2 The effect of enzyme concentration on rate of reaction

Table 14.3 illustrates the effect of substrate concentration on the rate of reaction.

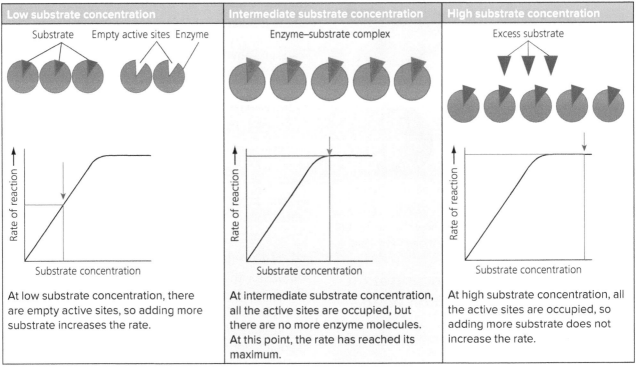

Low substrate concentration	Intermediate substrate concentration	High substrate concentration
At low substrate concentration, there are empty active sites, so adding more substrate increases the rate.	At intermediate substrate concentration, all the active sites are occupied, but there are no more enzyme molecules. At this point, the rate has reached its maximum.	At high substrate concentration, all the active sites are occupied, so adding more substrate does not increase the rate.

▲ Table 14.3 The effect of substrate concentration on rate of reaction

Reflect

Enzymes are being used more widely in industrial processes. What do you think the advantages might be? To make a process as efficient as possible, you need to consider the time taken (that is, the rate of reaction) because time is money in industry. Think about the factors that influence the rate of enzyme-catalysed reactions: temperature, substrate concentration and enzyme concentration. What combination of these do you think would give the most cost-effective process? What additional information would you require?

Test yourself

1 What are the **three** key facts about the genetic code?
2 What are the functions of mRNA and tRNA in protein synthesis?
3 Describe and explain the effect of increasing substrate concentration on the rate of an enzyme catalysed reaction.

Cell cycle

In eukaryotes (see B1.2 page 172), a fertilised egg is a single cell known as a **zygote**. This cell divides to form all the cells in the adult organism. The cells go through a cycle of division, growth and further division that is known as the **cell cycle**. This is shown in Figure 14.28.

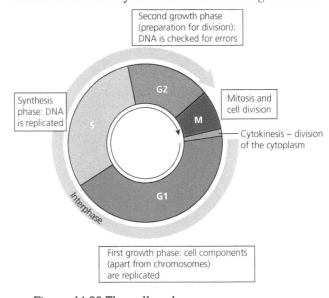

▲ Figure 14.28 The cell cycle

You can see from Figure 14.28 that the cell cycle consists of **interphase**, where cell components (mitochondria, ribosomes, endoplasmic reticulum, Golgi apparatus, and so on) are growing and replicating (G1 phase), DNA is replicated (S phase) and then checked for errors (G2 phase) in preparation for the next stage, mitosis (**nuclear division**, M phase) followed by cell division.

We will now focus on nuclear division and its importance. We saw how DNA replicates in Chapter B1 Core science concepts: Biology in the section on genetics, as well as the significance of this process in passing on the genetic information to future generations of cells.

DNA is contained in the **chromosomes**. Therefore, when the DNA replicates, the chromosomes replicate. Once the chromosomes have replicated, the process of nuclear division can begin.

> **Key term**
>
> **Chromosomes:** consist of DNA and proteins and carry the genetic information.

B2.6 The function of both mitosis and meiosis in nuclear division within cells

There are two types of nuclear division: **mitosis** and **meiosis**. The different features and functions of mitosis and meiosis are shown in Table 14.4.

Mitosis	Meiosis
Produces two daughter nuclei that have the same number of chromosomes (**diploid**) as the parent cell and each other. If the parent cell is diploid the daughter nuclei will also be diploid	Produces four daughter nuclei each with half the number of chromosomes (**haploid**) of the parent cell
Each of the daughter cells has an exact copy of the DNA of the parent cell, meaning the daughter cells are genetically identical	Produces cells that are not genetically identical
Produces **somatic cells** (normal body cells)	Produces **gametes** (sex cells)

▲ Table 14.4 The features and functions of mitosis and meiosis

> **Key terms**
>
> **Diploid:** a cell that contains two copies of each chromosome; in humans the **diploid number** is 46 chromosomes.
>
> **Haploid:** just one of each pair (half the number) of chromosomes; in humans the **haploid number** is 23 chromosomes.
>
> **Gametes:** the **haploid** sex cells (i.e. in animals, the sperm and egg) that join in the process of **fertilisation** to produce a new organism.

From Table 14.4 you will see that mitosis is important in growth of the organism as well as in producing new cells to replace damaged or worn-out cells (repair and replacement). Meiosis, on the other hand, plays an important role in bringing about variation in living organisms. We will see how that comes about in section B2.8.

B2.7 How the process of mitosis results in the formation of two genetically identical daughter cells

If you look at Figure 14.28 you will see that **interphase** occupies almost all the cell cycle, apart from mitosis. Interphase always preceeds mitosis as the DNA and organelles are replicated during interphase, so that there are now two identical copies of each chromosome. Interphase is also when organelles are replicated.

The characteristics of each of the stages of mitosis, including the behaviour of chromosomes and the cellular structure at each stage, are illustrated in Table 14.5. The left-hand column has diagrams of how the nucleus appears when viewed in the light microscope. During interphase, even when the DNA has replicated, the chromosomes are too long and thin to be seen with the light microscope.

Notice how we use the terms **pole** and **equator**. If you look at the diagrams below, you will see the reason for this – the cell looks a bit like a globe with the lines of longitude drawn on.

Prophase	Fibre of spindle	The chromosomes become visible and the nuclear envelope (see section B1.3) disappears. The chromosomes can be seen to consist of two **chromatids** joined by a **centromere**. Protein fibres form a spindle stretching from one side (pole) of the cell to the other.
Metaphase		The chromosomes arrange themselves at the centre of the cell (the equator) attached to the spindle fibres by the centromeres.
Anaphase		The centromeres holding sister chromatids together split. The spindle fibres contract, pulling one of each pair of chromatids to opposite poles of the cell. By the end of anaphase we no longer use the term chromatid and instead we can call them chromosomes.
Telophase		The two sets of chromosomes collect at the opposite poles of the cell. The nuclear envelope starts to reform around the two sets of chromosomes. The chromosomes become long and thin again and so cannot be seen clearly in the light microscope.

▲ Table 14.5 The characteristics of the stages of mitosis

Key terms

Chromatids: the two identical copies of a chromosome formed by DNA replication (sometimes called *sister chromatids*). The use of the term chromatid can cause confusion. We use chromatid as a shorter way of saying 'one of a pair of identical chromosomes'.

Centromere: joins the two chromatids together during the early stages of mitosis.

Strictly speaking, telophase is the end of mitosis. The cytoplasm then divides to form two daughter cells in a process known as **cytokinesis**. Each daughter cell will receive about half the organelles. Replication of organelles during the first growth phase of interphase (G1) restores the full number.

B2.8 How the process of meiosis, including phase 1 and phase 2, results in the formation of haploid gametes from diploid cells in the reproductive organs

Meiosis takes place in the reproductive organs to form haploid **gametes**. Because fertilisation involves the fusion of gametes, it is necessary for them to be haploid to maintain a constant number of chromosomes from one generation to the next. Meiosis is often referred to as a **reduction division** because the number of chromosomes is halved.

As in mitosis, the chromosomes replicate before the start of meiosis. For this reason, meiosis involves two phases or divisions (**meiosis I** and **meiosis II**), meaning that each diploid cell divides to produce four haploid gametes. The phases of meiosis are summarised in Figure 14.29.

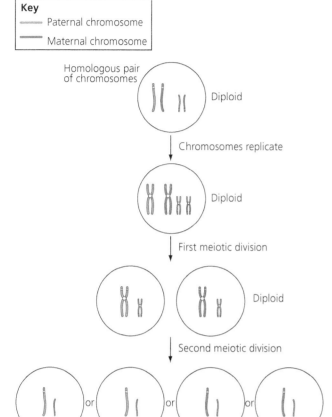

▲ Figure 14.29 The phases of meiosis; for simplicity, this cell contains two pairs of chromosomes

The first meiotic division (meiosis I) involves the separation of **homologous chromosomes**, so that each daughter cell receives only one of each pair of homologous chromosomes. The second meiotic division (meiosis II) is like mitosis in that the chromatids separate. This means that the daughter cells are haploid.

If you look carefully at Figure 14.29, you can see that the four haploid cells produced are not all different. In the first meiotic division, the diploid cells each contain one member of each homologous pair. However, it is entirely random whether these cells receive the **maternal** or **paternal** member – this is known as **independent assortment**. There are actually four possible combinations – you could cut out some paper chromosomes and work this out. For two pairs of chromosomes, there are 2^2 (that is, four) possible combinations. For a cell with n pairs of chromosomes, there will be 2^n possible combinations.

How many possible combinations will there be in human cells, with 23 pairs of chromosomes? What does this tell you about the amount of variation in the gametes produced?

In meiosis I the chromosome number is halved and the process of **crossing over** occurs. Crossing over involves the homologous pairs coming together to form a **bivalent**. They then exchange different segments of genetic material to form recombinant chromosomes (Figure 14.30). Crossing over, together with the independent assortment that takes place in meiosis I, creates huge genetic variation.

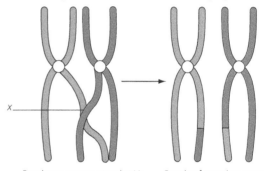

Breakage occurs at point X; fragments join 'wrong' chromosome   Result of crossing over

▲ Figure 14.30 A homologous pair of chromosomes forms a bivalent. Crossing over occurs when the chromatids break at the same point and DNA is exchanged between the chromatids of opposite members of the pair

Key terms

Homologous chromosomes: they carry the same genes but different alleles; in animals, one of each homologous pair is inherited from the female parent (mother) and one from the male parent (father); in humans there are 23 pairs of chromosomes.

Maternal chromosomes: the members of a homologous pair inherited from the mother (female parent).

Paternal chromosomes: the members of a homologous pair inherited from the father (male parent).

B2.9 The significance of the differences between mitosis and meiosis

Having studied the two types of nuclear division, we are now in a position to understand the significance of the differences between them and relate this to the importance of the two types of division.

Mitosis produces cells that are genetically identical to parent cells, so it produces new cells from the original that always have the same set of genetic information. However, although the cells produced by the process of mitosis are genetically identical, we know that there are many different types of specialised cells. This is the result of the process of differentiation that was described in section B1.5 and results in a huge variety of cells and tissues that perform the function they were intended to perform.

If cells are damaged or die, it is important that new cells produced have identical structure and function to the cells that have been lost; mitosis is, therefore, the process by which new cells replace damaged or dead ones.

Meiosis only occurs in reproductive cells to ensure that the cells produced have half (haploid) the number of chromosomes so that when gametes (for example, eggs and sperm) combine, the resulting zygote (fertilised egg) has the correct number of chromosomes (diploid).

The two phases of meiosis (rather than just one round of division in mitosis) result in genetic variation within daughter cells compared to the parent cells.

Test yourself

1   Describe **two** differences between the daughter cells produced by mitosis and meiosis.
2   Describe **two** differences between the processes of mitosis and meiosis.
3   Name the four phases of mitosis.
4   Name **two** ways in which meiosis increases genetic diversity.

Cellular respiration

Cellular **respiration** is the process that uses molecules such as carbohydrates and fats and converts the chemical energy stored in them into chemical energy in the form of ATP.

Key term

Respiration: the process by which energy is transferred from biological molecules such as glucose to **ATP** for use in energy-requiring processes in the cell.

Practice point

If you work in a field allied to medicine, you are likely to encounter another use of the term respiration, used to mean breathing. For example, respiration rate means the rate of breathing. In biology, we use ventilation (of the lungs) to describe the process of breathing.

Can you think of other terms that have different meanings in different areas of science? What about terms that have different meanings in science and in everyday life?

As scientists, we must always take care to use terms correctly and accurately to avoid misunderstanding.

B2.10 How respiration results in the breakdown of glucose to produce the energy-carrying molecule ATP

When we talk about energy, we must take care to describe energy *transfer* and not energy *production*. Chapter B1 Core science concepts: Physics described the various ways in which energy is transferred in physics and we follow that here. That means it is acceptable to talk about production of the energy-carrying molecule adenosine triphosphate (ATP), but not just 'production of energy'.

Respiration can be **aerobic** (when oxygen is present) or **anaerobic** (in the absence of oxygen). Here we will consider just aerobic respiration, which is the chemical breakdown of substrate molecules (for example, glucose) in cells to produce ATP when oxygen is present.

Aerobic respiration can be summarised by the following word equation:

glucose + oxygen → carbon dioxide + water (produces ATP)

This corresponds to the chemical equation:

$C_6H_{12}O_6 + 6O_2 \rightarrow 6CO_2 + 6H_2O$ + (produces ATP)

Do not be misled by these two equations. Energy has simply been transferred. Chemical energy is present in the glucose and respiration transfers some of that chemical energy into ATP.

This equation is similar to the one for the combustion of glucose, except when we burn glucose the energy is transferred in the form of heat rather than to ATP. Of course, body cells do not contain miniature furnaces burning glucose! Instead, the process of aerobic respiration involves a series of oxidation and reduction reactions. As is so often the case in biology, chemical conversions happen in a series of many small steps, allowing reactions to take place at 37 °C rather than the much higher temperatures that are often required in the chemistry laboratory.

B2.11 How ATP provides a source of energy for biological processes

ATP consists of an adenosine molecule bonded to three phosphate groups in a row (Figure 14.31).

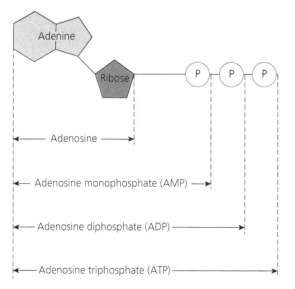

▲ Figure 14.31 The structure of ATP and ADP

The bond between the phosphate groups in ATP is easily hydrolysed to form ADP and inorganic phosphate (P_i), with energy released in this reaction:

ATP + water → ADP + P_i (energy released)

This reaction is catalysed by many different enzymes that all have the general name of **ATPase**.

Although it seems impossible, we produce our own body weight in ATP every day. We do not consist entirely of ATP by the end of the day however, because the ATP is broken down to transfer energy to other processes, as shown in Figure 14.32.

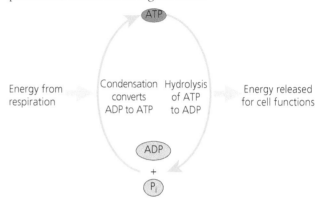

▲ Figure 14.32 The energy cycle showing how ATP transfers energy from respiration to processes that require energy

ATP is particularly well suited to its role in the transfer of energy:

▷ It is small and soluble, so can diffuse quickly throughout the cell.
▷ It is polar and so cannot diffuse across the phospholipid bilayer of the plasma membrane; this means it cannot leak out of the cell.
▷ It can be hydrolysed rapidly, so energy is transferred quickly.
▷ The amount of energy transferred when a single ATP molecule is hydrolysed is relatively small. This means that it is transferred in manageable amounts and little energy is wasted.

B2.12 The comparative amounts of energy produced by different respiratory substrates (lipids, proteins and carbohydrates)

In section B2.10 we saw how glucose was converted to carbon dioxide and water, with the energy contained in the glucose being transferred to ATP. Of course, our diet does not consist only of carbohydrates. Lipids and proteins can also be used as sources of energy – carbohydrates, lipids and proteins can all act as **respiratory substrates**.

Lipids are first hydrolysed to fatty acids and glycerol which enter the respiratory pathway.

Unlike lipids, **proteins** cannot be stored as an energy reserve in the body. Proteins in the diet are digested to amino acids and any surplus amino acids are **deaminated** (they have the amino group removed); the remaining part enters the respiratory pathway. Body proteins can be broken down to amino acids but are only used as respiratory substrates in extreme circumstances, such as starvation.

Table 14.6 shows the mean energy values of the three classes of respiratory substrates.

Respiratory substrate	Mean energy value (kJg^{-1})
Carbohydrate	15.8
Lipid	39.4
Protein	17.0

▲ Table 14.6 The mean energy values of carbohydrates, lipids and proteins

From this you can see that lipids release more than twice as much energy as the same mass of carbohydrates.

Reflect

Why are lipids a source of more energy per gram than carbohydrates?

Consider the equation we looked at above for the respiration of glucose:

$$C_6H_{12}O_6 + 6O_2 \rightarrow 6CO_2 + 6H_2O$$

Now compare this with the equation for the respiration of stearic acid, a typical fatty acid:

$$C_{17}H_{35}COOH + 26O_2 \rightarrow 18CO_2 + 18H_2O$$

We have left ATP out of the equations so that they can be balanced.

To compare them properly, we should probably multiply the top equation by three so that we have the same number of carbon atoms:

$$3C_6H_{12}O_6 + 18O_2 \rightarrow 18CO_2 + 18H_2O$$

Notice that the stearic acid requires 26 oxygens while the three glucose molecules only require 18 oxygens. Why does stearic acid require more oxygen? The reaction with oxygen releases energy (which is why burning produces heat). Does this explain why fats release more energy per gram than carbohydrates?

Test yourself

1 Write the word equation for respiration of glucose.
2 Explain **two** ways in which ATP is well adapted to its function.
3 Explain how respiration and combustion of glucose are different even though the chemical reaction is the same.
4 Explain why polar explorers and mountaineers carry chocolate (which is rich in triglycerides) rather than sugary sweets.

Pathogens

B2.13 The definition of a pathogen

Pathogens do not just cause disease in humans, or even just in animals. There are plant pathogens that cause disease in plants and even types of virus that infect and destroy bacteria.

B2.14 Examples of different types of pathogens and the diseases they can cause

Sections B1.24 and B1.25 looked at the nature of infection and examples of different types of pathogens and the diseases they cause. Here are a few more examples.

Bacteria

Escherichia coli (*E. coli*) causes gastrointestinal disorders. *E. coli* are common gut bacteria in humans and other animals, but some strains can cause food poisoning. This is why good hygiene is important when preparing and storing foodstuffs.

Fungi

Candida auris (*C. auris*) causes fever, bloodstream infections and sepsis. Fungal diseases can be difficult to treat because fungi, like all animals, are eukaryotes. Most antibiotics are not active against fungi and so more aggressive treatments are required, many of which can also damage human cells.

Prions

Prions are misfolded proteins that can cause normal proteins of the same type to misfold and thus cause disease. They are classed as pathogens as they can cause prion diseases, such as Creutzfeldt-Jakob disease (CJD) in humans or scrapie in sheep. These diseases involve damage to the nervous system and eventual death.

Protists (Protoctists)

Protists are another class of microorganism and some of them can cause disease. Several species of *Plasmodium* (the pathogenic parasite) are transmitted by the *Anopheles* mosquito. It is the *Plasmodium* that causes malaria, not the mosquito. It has proved difficult to develop an effective treatment or vaccine for malaria, so the disease is controlled mostly by controlling the mosquito.

Viruses

Viruses are acellular and can only replicate when they infect host cells. There are many viruses that cause human disease, including the SARS-CoV-2 virus that causes COVID-19 and the hepatitis A virus (HAV) that causes hepatitis A. Hepatitis A, also known as infectious jaundice, is a liver disease that is associated with poor sanitation and intravenous drug use. Severe cases can lead to liver failure requiring a liver transplant.

Project practice

Choose one of the following relevant to your workplace:
- ▶ How the cell can perform reactions at 37°C faster and more efficiently than a chemical laboratory that uses elevated temperature and pressure.
- ▶ How nuclear division maintains the genetic information in body cells but increases genetic diversity in the sex cells.
- ▶ Pathogens come in many shapes and sizes, some of them are not even alive, but all can cause disease – and not just in humans.

Research your chosen topic.

Prepare an infographic, wallchart or slide presentation that could be used for training and education purposes.

Make your presentation to the group and answer any questions they may have.

Write an evaluative reflection on your work.

Assessment practice

1 Describe the four levels of protein structure.

2 Starch and cellulose are both polymers of glucose.

   a Describe **one** similarity in their physical properties.

   b Explain their different structural properties and how they are adapted to their function.

3 Describe how the structure of DNA is well suited to its function.

4 A student investigated the hydrolysis of triglycerides in milk by the enzyme lipase. They used a pH meter to record the pH. The results are shown in the graph below.

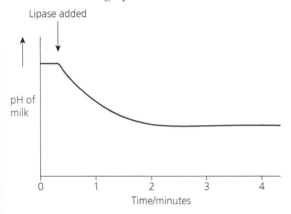

a Explain why the pH decreases when the lipase is added.

b Explain why the pH remained constant after two minutes.

c The experiment was carried out at 25 °C. Describe how the curve would be different if the experiment was repeated at 15 °C.

5 An enzyme has an optimum pH of 7. When the pH is reduced to 6, the activity of the enzyme decreases but increases again if the pH is returned to 7. If the pH is reduced to 2, the enzyme loses all activity and it does not regain activity if the pH is returned to 7. Explain this behaviour.

6 Some anticancer drugs target rapidly dividing cancer cells and prevent them from dividing. They do this by interfering with the formation of the spindle in mitosis. Explain how this type of drug will target cancer cells and prevent them from dividing.

7 The following table shows the base sequence of a short section of DNA.

Base sequence of DNA	A	C	G	A	T	G
Base sequence of mRNA						
Base sequence of anticodon on tRNA						

a How many amino acids does this sequence code for?

b Complete the table to show:

   i The base sequence of the mRNA.

   ii The base sequence of the anticodon on tRNA.

8 During a 100 m sprint, an athlete's leg muscles are contracting rapidly but oxygen cannot be delivered quickly enough by the blood for aerobic respiration to occur. However, the first stage of respiration can use glycogen stored in the muscles to provide sufficient energy for use in muscle contraction.

a What molecule transfers energy from glycolysis to muscle contraction?

b Explain the advantage of using glycogen during the sprint.

B2.15–B2.31: Further science concepts: Chemistry

In this chapter we will build on the content of Chapter B1 Core science concepts: Chemistry and look at three areas that are likely to be of practical use in many areas of chemistry:

▶ how to work with chemical equations
▶ how to understand the factors that affect the rate of reactions – this is particularly important in a production environment where time is money
▶ a range of analytical techniques.

Formulae and equations

Although there are conventions that you must follow when writing chemical equations, they do have some things in common with mathematical equations. For instance, it will be easier to understand if you think of the arrow in a chemical equation connecting the reactants and products as being like the equals sign in a mathematical equation.

Valency is a concept we should consider before looking at formulae and equations in more detail. We can think of valency as a measure of the combining power of elements, but it may be better to think of it in terms of the number of electrons in the outer shell, which determines the number of bonds an atom can make. In section B1.34 we saw that sodium has one electron in its outer shell and can lose that electron to form the Na^+ ion. This is true of all group 1 metals. At the same time, chlorine has seven electrons in its outer shell and so can gain one electron to form a Cl^- (chloride) ion. This is true of all group 7 non-metals. This means that Na and Cl combine in a 1:1 ratio to form NaCl. They each have a valency of 1.

On the other hand, group 2 metals lose two electrons to form ions, such as Ca^{2+} and Mg^{2+}; we say they have a valency of 2. Likewise, oxygen has six electrons in its outer shell and so gains two electrons to form an O^{2-} (oxide) ion and also has a valency of 2. This means that calcium and oxygen, both with a valency of 2, will also combine in a 1:1 ratio to form CaO (calcium oxide). Because sodium has a valency of 1, it combines with oxygen in a 2:1 ratio, so the formula for sodium oxide is Na_2O.

This also works if you think about the need for charges to balance. In NaCl, CaO and Na_2O there are equal numbers of positive and negative charges (note **charges** not **ions**). This is true in all ionic compounds.

We can apply similar logic to covalent molecules and compounds – see Table 15.1 for examples.

Atom or group	Valency (number of covalent bonds formed)	Element example	Compound or molecule examples
Hydrogen	1	Hydrogen (H)	H_2O, HCl, H_2
Group 7	1	Chlorine	$HCl\ Cl_2$, CH_3Cl
Group 6	2	Oxygen	H_2O, CO_2
Group 5	3	Nitrogen	NH_3
		Phosphorous	PCl_3
Group 4	4	Carbon	CH_4, CO_2

▲ Table 15.1 Examples of how valency affects the ratio of elements in covalent molecules and compounds

As with so many things in chemistry, there are exceptions to these rules. In some compounds chlorine can have a valency of 3 or even 6, but you will not encounter many such examples.

B2.15 How to balance a given equation

In maths, the two sides of an equation must be equal, and the same is true in chemistry. This is also a consequence of the law of **conservation of mass**. The total mass of all the reactants must equal the total mass of all the products. This means that the number of each type of atom on the left of an equation must equal the number of each type of atom on the right. This is the principle that we follow when we balance chemical equations.

Think about the following example, taken from section B1.38:

$$NaOH\ (aq) + HCl(aq) \rightarrow NaCl(aq) + H_2O(l)$$

We can account for all the atoms, as shown in Table 15.2.

Type of atom	Reactants (left side)	Products (right side)
Na (sodium)	1	1
O (oxygen)	1	1
H (hydrogen)	2	2
Cl (chlorine)	1	1

▲ Table 15.2 An example of the conservation of mass

Notice that we count the atoms individually, rather than thinking of what they are combined with. Chemical reactions simply break bonds and make new bonds to rearrange the atoms in the reactants and thereby produce the products. That is, therefore, exactly what the chemical equation does too.

Now we will look at some of the examples we first encountered in section B1.33 and see how we arrive at the balanced equations.

Group 1 metals with water and oxygen

We saw above that group 1 metals have a valency of 1, meaning that they form 1+ ions. Oxygen has a valency of 2. This explains why the formula for water is H_2O. From section B1.33, we know that sodium reacts with water to produce sodium hydroxide (NaOH) and hydrogen gas (H_2). This gives the **unbalanced** equation:

$$Na(s) + H_2O(l) \rightarrow NaOH(aq) + H_2(g)$$

If we look at this equation, we can see the following:

Type of atom	Reactants (left side)	Products (right side)
Na	1	1
O	1	1
H	2	3

▲ Table 15.3 Reactants and products are unbalanced

In cases like this, where we have an odd number on one side and an even number on the other, it usually helps to double one of the components to make an even number, although it may not yet be balanced

$$Na(s) + H_2O(l) \rightarrow 2NaOH(aq) + H_2(g)$$

This gives us the following:

Type of atom	Reactants (left side)	Products (right side)
Na	1	2
O	1	2
H	2	4

▲ Table 15.4 Reactants and products are still unbalanced

Clearly, we now need to double the amount of sodium and water:

$$2Na(s) + 2H_2O(l) \rightarrow 2NaOH(aq) + H_2(g)$$

If we do a final check, we will see that the equation is now balanced:

Type of atom	Reactants (left side)	Products (right side)
Na	2	2
O	2	2
H	4	4

▲ Table 15.5 Reactants and products are now balanced

We can use the same approach with a reaction between a group 1 metal, in this case, potassium, and oxygen to produce potassium oxide. Using valency, we can work out that potassium oxide must be K_2O, so we get the following **unbalanced** equation:

$$K(s) + O_2(g) \rightarrow K_2O(s)$$

Once again, because we have an uneven number of oxygen atoms on the right and an even number on the left, it helps to double on the right:

$$K(s) + O_2(g) \rightarrow 2K_2O(s)$$

Finally, we balance the potassium atoms and the balancing is complete:

$$4K(s) + O_2(g) \rightarrow 2K_2O(s)$$

Transition metals with oxygen and strong acids (hydrochloric, sulfuric and nitric acid)

Transition metals are more complicated, for two reasons:
▶ They are less reactive than group 1 metals, although some are more reactive than others.
▶ They can have multiple valency – they can form ions with different charges, such as Fe^{2+} and Fe^{3+}.

An example of multiple valency can be seen in the reaction between iron and oxygen. Iron oxides exist in several forms, the two most common being FeO (containing Fe^{2+}) and Fe_2O_3 (containing Fe^{3+}).

We can write the equations as:

$$Fe(s) + O_2(g) \rightarrow FeO(s)$$

and

$$Fe(s) + O_2(g) \rightarrow Fe_2O_3(s)$$

Balancing the equations is not difficult:

$$2Fe(s) + O_2 \rightarrow 2FeO(s)$$

and

$$4Fe(s) + 3O_2(g) \rightarrow 2Fe_2O_3(s)$$

Copper is more straightforward. If we heat a copper strip strongly in air it forms copper oxide. Although copper can form Cu^+ ions, most of the time it forms Cu^{2+} ions, so copper oxide is CuO:

$$Cu(s) + O_2(g) \rightarrow CuO(s)$$

And this is easy to balance:

$$2Cu(s) + O_2(g) \rightarrow 2CuO(s)$$

Other transition metals react only very slowly or not at all with oxygen. This is one reason why gold and silver are used in jewellery, and metals such as platinum, palladium and iridium are used as catalysts in catalytic converters – they provide a surface on which toxic gases can react to form less-harmful products without the catalyst reacting with oxygen in the air.

Iron will react with hydrochloric acid to produce iron chloride, and with sulfuric acid to produce iron sulfate. In both cases hydrogen is also produced and the iron forms Fe^{2+} ions, which makes the equations straightforward:

$$Fe(s) + 2HCl(aq) \rightarrow FeCl_2(aq) + H_2(g)$$

$$Fe(s) + H_2SO_4(aq) \rightarrow FeSO_4(aq) + H_2(g)$$

The sulfate ion has a 2– charge (SO_4^{2-}); as this remains unchanged in the reaction (we say it is a **spectator ion**) we can treat the SO_4 as a single unit without having to count the individual atoms if we want to save time.

The reaction with nitric acid is a little different because concentrated nitric acid is a stronger oxidising agent than either sulfuric, hydrochloric or dilute nitric acids. This means that the Fe is oxidised to Fe^{3+} rather than Fe^{2+}. Also, the nitrate (NO_3^-) in nitric acid is reduced to nitrogen monoxide (NO). This makes for quite a challenging equation to balance.

For that reason, we will look at a slightly simpler example: the reaction between copper and dilute nitric acid to produce copper nitrate and nitrogen monoxide:

$$Cu(s) + HNO_3(aq) \rightarrow Cu(NO_3)_2(aq) + H_2O(l) + NO(g)$$

Notice the formula for copper nitrate. The nitrate ion (NO_3^-) is shown in brackets with a subscript 2 **outside** the brackets. This shows that there are two NO_3^- ions for every Cu^{2+} ion – if you think about balancing the charges, that makes sense.

Unfortunately, there is no simple way of balancing this equation that does not require a greater understanding of redox (oxidation–reduction) reactions than is possible here. So, only a lot of trial and error is likely to get you to the actual balanced equation, which is:

$$3Cu(s) + 8HNO_3(aq) \rightarrow 3Cu(NO_3)_2(aq) + 4H_2O(l) + 2NO(g)$$

B2.16 How an empirical formula represents the simplest ratio of atoms of each element in a compound

When you make an organic compound, often the first thing you do is to send it for **elemental analysis**, sometimes called **CHN** analysis. This will tell you the chemical composition of the compound and the percentage by mass of each element. From this you can calculate the **empirical formula**.

Think about butane, which has the formula C_4H_{10}. This means there are four carbons and ten hydrogens. But the simplest ratio is not $4:10$, it is $2:5$ (not $1:2.5$, because the ratio must be whole numbers). This means that, for butane, the empirical formula is C_2H_5.

> ### Key terms
>
> **Empirical formula:** the smallest whole number ratio of the atoms present in a compound.
>
> **Molecular formula:** the exact number of each type of atom in a compound.

B2.17 How to use the empirical formula and relative molecular mass to work out the molecular formula of a compound

Having determined the empirical formula of our compound, the next thing we do is send it for analysis by mass spectrometry (MS) – this was covered in section B1.42. MS will tell us the relative molecular mass (M_r) of the compound. We can then use this information to arrive at the **molecular formula**.

The procedure we use is:

▶ Calculate the mass of the atoms in the empirical formula (add up all the relative atomic masses). For C_2H_5 this is $(2 \times 12) + (5 \times 1) = 29$.

▶ Next, divide the relative molecular mass (for butane the MS analysis will show that this is 58) by the mass of the atoms in the empirical formula to obtain the ratio of the molecular formula to the empirical formula. This gives us $58 \div 29 = 2$. Note that this ratio must always be a whole number, because the empirical formula must be the smallest **whole number** ratio of the atoms.

▶ Now, multiply the ratio by the empirical formula to arrive at the molecular formula. In this case it is $2 \times C_2H_5 = C_4H_{10}$, which is the molecular formula of butane.

B2.18 The definition of an isotope and relative isotopic mass

We first came across isotopes in section B1.61 when we were studying the application of radioactivity in the health and science sector. In that section we looked at radioisotopes that are unstable and undergo radioactive decay. Not all isotopes are unstable.

We saw in section B1.35 that atomic nuclei contain protons and neutrons. It is the number of protons that defines which element it is. For example, carbon has 6 protons while sodium has 11 protons. However, atoms of the same element can have different number of neutrons.

Isotopes are atoms of the same element with different masses due to a different number of neutrons. An example is carbon-12 (also written ^{12}C) and carbon-13 (^{13}C). Each atom of carbon-12 has 6 protons and 6 neutrons while each atom of carbon-13 has 6 protons and 7 neutrons.

Relative isotopic mass is the mass of an atom of any isotope relative to one-twelfth of the mass of a carbon-12 atom. Carbon-12 was chosen to be the standard due to the high abundance and stability of the element (think of graphite or diamond, both forms of elemental carbon).

Relative atomic mass (A_r) is the weighted average mass of all the atoms of an element relative to one-twelfth of the mass of a carbon-12 atom. For example, a sample of magnesium has the following composition:

Isotope	Relative isotopic mass	Abundance (%)
^{24}Mg	24	79.0
^{25}Mg	25	10.0
^{26}Mg	26	11.0

▲ Table 15.6 Relative isotopic mass and percentage abundance of isotopes of magnesium

A_r can be calculated by multiplying the percentage abundance of each isotope by the relative isotopic mass and dividing the total by 100:

$$A_r = \frac{\left(79 \times 24\right) + \left(10 \times 25\right) + \left(11 \times 26\right)}{100} = 24.3$$

You will not be expected to perform this type of calculation – we have included it here to illustrate how the A_r is calculated.

Relative molecular mass (M_r) is defined in a similar way; it is the average mass of a molecule relative to one-twelfth of the mass of a carbon-12 atom.

The term **relative formula mass** (**RFM**) is one that you might also see. The RFM is the average mass of a formula unit compared to one-twelfth of the mass

of a carbon-12 atom. The term RFM is applicable more generally because it seems odd to talk about the relative molecular mass of a compound such as sodium chloride that is not a molecule – it is an ionic compound. In the case of sodium chloride, the formula unit is NaCl, so the RFM of sodium chloride is 58.5. To calculate the M_r or RFM of any compound, we just add up the A_r of all the atoms in the formula as was shown in section B2.17.

B2.19 The link between balanced equations and the ratio of moles of a substance in a reaction

You can see, from section B2.15, that we sometimes go to great lengths to balance chemical equations. There are two very good reasons for doing this.

First, it is only when we have balanced the equation that we can be more confident that we have the correct equation with all the relevant reactants and products. That might not always be the case – it is possible to balance an incorrect equation, but it is more likely than not that we have it right if it balances.

The second reason is that the balanced equation shows us the proportions of reactants and products. Think about the following reaction between hydrogen and chlorine:

$$H_2(g) + Cl_2(g) \rightarrow 2HCl(g)$$

This shows that hydrogen and chlorine react together in a proportion of 1:1. This means that, if we mix one molecule of hydrogen and one molecule of chlorine, we will produce two molecules of hydrogen chloride gas.

Be careful, though. Look at the reaction between hydrogen and oxygen to produce water (actually, in this case it is in the form of steam):

$$2H_2(g) + O_2(g) \rightarrow 2H_2O(g)$$

From this equation, we see that two molecules of hydrogen will react with one molecule of oxygen but will produce two molecules of steam. In this case, two plus one does not equal three because the number of molecules has decreased (the atoms have been rearranged into new molecules).

Talking about molecules all the time is not particularly useful. Instead, we use the **mole** as our standard amount of substance in chemistry. The abbreviation for mole is **mol** so, in the equation above, 2 mol of hydrogen reacted with 1 mol of oxygen to produce 2 mol of water.

Practice point

One **mole** of any substance contains the same number of particles, which is the number of carbon atoms in 12 g of carbon-12. This is the Avogadro number = 6.02×10^{23}.

The mole becomes even more useful because 1 mol of carbon-12 has a mass of 12 g. If this seems like a huge coincidence, it is not – the definition of 1 mole is based on the number of atoms in 12 g of carbon-12. As we saw in section B2.18, carbon was chosen as the standard for all relative atomic mass measurements.

So, M_r and A_r are measured relative to the mass of carbon-12, and 1 mole of carbon-12 has a mass of 12 g. We can now use this information to convert between the number of moles, as shown in the chemical equation, and the mass of reactants or products.

If we look again at the reaction between hydrogen and oxygen, we can now work out that 4 g of hydrogen will react with 32 g of oxygen to produce 36 g of water. If you are not sure how we arrived at those figures, a periodic table will help.

You will find that you use the following equation quite often in many different chemistry calculations:

$$n = \frac{m}{M_r}$$

Where:

n = number of moles

m = mass

M_r = relative molecular mass/RFM

Put into words, this means that the number of moles of any substance is the mass divided by the M_r.

If you rearrange the equation, you get:

$$m = n \times M_r$$

Which, in words, says the mass of any substance is the number of moles multiplied by the M_r.

Another term you will encounter is **relative molar mass**. This is simply the M_r of the substance in grams, so water (H_2O) has an M_r of 18 and a relative molar mass of 18 g.

B2.20 The relationship between the number of moles of solute and the volume in dm^3 of solvent as a measure of concentration (mol/dm³)

By now, we should be able to work out the mass of reactants or products we expect in any reaction. But not all reactions involve solids, liquids or gases – sometimes we have solutions, where one substance (the **solute**) is dissolved in a liquid (the **solvent**), usually but not always water. This is particularly important in titrations, as we shall see in section B2.29.

We can express concentration in terms of mass per unit volume, that is, g/dm^3. This can be quite useful, particularly if we need to prepare a solution, because we can measure both mass and volume.

However, concentration is often more useful when it is expressed in moles per unit volume, or mol/dm^3. In section B1.44 we covered the principles of titration and saw that neutralisation occurs when we have equal moles of acid and base – this is how we determine the end point in a titration (discussed further in section B2.29).

From this, we can understand that $25\,cm^3$ of a $0.1\,mol/dm^3$ solution of HCl will neutralise $25\,cm^3$ of a $0.1\,mol/dm^3$ solution of NaOH because both solutions contain the same number of moles.

The equation that we use to calculate concentration is:

$$c = \frac{n}{v}$$

Where:

c = concentration (mol/dm^3)

n = number of moles (mol)

v = volume (dm^3)

You can rearrange this equation so that, if any one term is unknown, you can calculate it. We will put this to use in section B2.29.

When dealing with volumes, you will often need to convert between cm^3 and dm^3. Remember that $1\,dm^3 = 1000\,cm^3$ or $1\,cm^3 = 0.001\,dm^3$. There is more about conversion between units in section B1.63.

Practice points

Practice points

Application of this idea of concentration of solutions occurs everywhere in science and it is highly likely that you will need to prepare solutions of known concentration of different substances. Make sure you are familiar with the key equations we have covered in these two sections:

▶ the relationship between moles, mass and M_r
▶ the relationship between concentration, moles and volume.

If you ever forget the second of these, just think about the units for concentration: mol/dm^3 means moles divided by volume.

Practice points

Molarity or concentration?

You are likely to encounter the term **molarity** to mean concentration, although it is becoming less common. A molar solution, written as 1M, contains $1mol/dm^3$ of solute. As mentioned before, you are quite likely to see litre (L) used as an alternative for dm^3. This means a 1M solution is the same as 1 mol/L.

We have used mol/dm^3 as the units of concentration. However, you can express the concentration of a solution in g/dm^3.

You will also see concentrations expressed in terms of percentage, either by weight or volume:

▶ 10% (w/v) means a solution containing 10g solute per $100cm^3$ solvent.
▶ 10% (v/v) is used with mixtures of liquids and means $10cm^3$ of one liquid in a total of $100cm^3$ of solution or mixture.

Kinetic changes

Kinetics is the study of the rate of chemical reactions. Understanding kinetics and the factors that affect the rates of chemical reactions helps us to understand those reactions. In section B1.39 we touched on the principles of collision theory and then we looked at the effect of temperature (section B1.40) or a catalyst (section B1.41) on the rate of chemical reactions.

We are now going to go more deeply into this subject.

B2.21 A range of factors affecting the rates of chemical reactions

Surface area

Reactions between solids and liquids can only occur at the surface of the solid, where it is in contact with the liquid. One example is the reaction between magnesium and hydrochloric acid. Figure 15.1 on page 274 illustrates how increasing the surface area of the solid increases the contact between the magnesium atoms and the acid, thereby increasing the rate of reaction.

Temperature

We have seen how increased temperature increases the rate of reaction because it increases the kinetic energy of the particles. This increases the probability of collisions between reacting particles and also increases the proportion of particles that have sufficient energy to react. We will cover this in more detail in section B2.23.

Test yourself

1 Balance the following unbalanced equations:
   a $K + H_2O \rightarrow KOH + H_2$
   b $Co + H_2SO_4 \rightarrow CoSO_4 + H_2$
   c $V + O_2 \rightarrow V_2O_5$
   d The combustion of ethanol (C_2H_5OH) to produce carbon dioxide and water. You would not be expected to balance this equation, but it is a good exercise in applying your new skills!
2 A compound has an empirical formula of C_2H_4O and an M_r of 88. What is the molecular formula?

3 Bromine has two stable, naturally occurring isotopes: ^{79}Br and ^{81}Br. Calculate the number of protons and neutrons in each isotope.
4 Calculate the concentration in mol/dm^3 of each of the following solutions (use data from the periodic table):
   a 40g of sodium hydroxide (NaOH) dissolved in $1dm^3$ of water.
   b 1.65g of potassium iodide (KI) dissolved in $200cm^3$ of water.
   c A solution labelled $0.25g/cm^3$ of ammonium chloride (NH_4Cl).

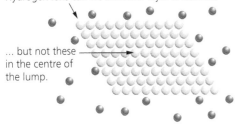

Hydrogen ions can hit the outer layer of atoms...

... but not these in the centre of the lump.

With the same number of atoms now split into lots of smaller bits, there are hardly any magnesium atoms that are inaccessible to the hydrogen ions.

▲ Figure 15.1 The rate of reaction increases when a large lump of magnesium (top) is broken into smaller pieces (bottom), which increases the surface area of the magnesium

Concentration and pressure

Increases in both concentration (for reactions in solution) and pressure (for reactions between gases) increase the probability of collision and, therefore, increase the rate of reaction.

B2.22 How to calculate the rate of reaction

Before we can study kinetics, we need to be able to calculate the rate of a reaction. This is simply the amount of reactant used or product produced per unit of time given by the equation:

$$\text{rate of reaction} = \frac{\text{amount of reactant or products}}{\text{time}}$$

For reactions that produce a gas, we can collect the gas produced in a gas syringe and measure the volume produced in a given time. Figure 15.2 shows the apparatus used to measure the rate of reaction between hydrochloric acid and calcium carbonate.

In this case, we are measuring the volume of gas produced in a given time, so the units of rate would be cm^3/min or cm^3/s.

If we collected $30\,cm^3$ of gas in two minutes, we calculate the rate of reaction as follows:

$$\text{rate} = \frac{30}{2} = 15\,cm^3/min$$

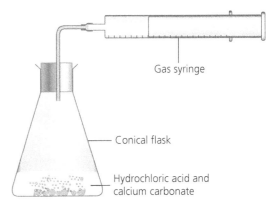

Gas syringe

Conical flask

Hydrochloric acid and calcium carbonate

▲ Figure 15.2 Measurement of the rate of reaction by collection of a gas produced

Reflect

What control variables would you need when using the apparatus shown in Figure 15.2?

How could you use this apparatus to investigate the effect of the following factors on the rate of reaction:

▷ concentration of hydrochloric acid
▷ temperature
▷ surface area of calcium carbonate?

Research

There are other ways to measure the rate of reaction:

▶ Use a colorimeter to measure the appearance of a coloured product or disappearance of a coloured reactant.
▶ Measure the loss of mass in a reaction that produces a gas.
▶ Measure the time taken for enough precipitate to form so that a cross drawn under the flask is no longer visible.
▶ The iodine clock reaction.

Use the internet to research these different methods. What do you think are the advantages and disadvantages? Which of these methods is subjective and which are objective?

You should have found that we can also use the change in mass or concentration of product per unit time as a way of measuring the rate of reaction. The following are examples of units used when measuring the rate of reaction:

▶ g/min or g/s (change in mass per unit time)
▶ $mol/dm^3/min$ or $mol/dm^3/s$ (change in concentration per unit time)

B2.23 The definition of activation energy

Figure 15.3 shows an enthalpy profile diagram of a reaction.

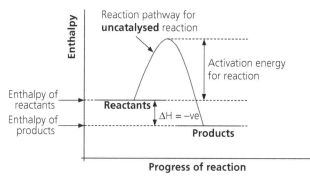

▲ Figure 15.3 Enthalpy profile diagram of an exothermic reaction

This diagram shows the **enthalpy** (heat energy) of the reactants and the products. The first thing to notice is that the products are at lower energy. This means that the **enthalpy change**, indicated by **ΔH**, is negative. The reactants lose heat as they are converted into products, and so the reaction gives out heat – we say it is **exothermic**. If the products were at higher energy than the reactants, they would gain heat as they are converted into products and the reaction would be **endothermic**.

> ### Key terms
>
> **Exothermic:** reactions that give out heat; ΔH is negative.
>
> **Endothermic:** reactions that take in heat; ΔH is positive.
>
> **Activation energy:** the minimum energy required to start a reaction.

Now, follow the reaction pathway from reactants to products in Figure 15.3 (for the uncatalysed reaction for now). If you remember what we learned in section B1.39 about collision theory, you will know that reactions involve breaking bonds and that this requires energy. This explains why the reaction pathway rises to a maximum – this represents the energy required to break the bonds. This is the **activation energy**.

If you keep following the reaction pathway you will see that it starts to fall until it reaches the products. This is where energy is released as new bonds are formed.

The enthalpy change, therefore, is the difference between the energy required to break the bonds in the reactants and the energy released when new bonds are made. The difference between these gives us the enthalpy change for the reaction, ΔH.

B2.24 The action of a catalyst, in terms of providing an alternative pathway with a lower activation energy

If you look at Figure 15.4, you will see another reaction pathway, this time for the **catalysed** reaction. In this case, the peak of the energy profile is lower. The catalyst has provided an **alternative pathway** for the reaction that has a lower activation energy (less energy is required to break bonds). The enthalpy change for the reaction is still the same though, because the difference in energy between the reactants and products is unchanged. The only difference is that the activation energy is lower. We will see in more detail how this increases the rate of the reaction in section B2.26.

You will sometimes hear the expression 'a catalyst reduces the activation energy'. Strictly speaking, this

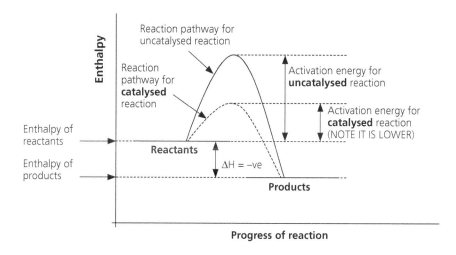

◀ Figure 15.4 Enthalpy profile diagram of an exothermic reaction showing catalysed and uncatalysed pathways

is incorrect (although widely used, including in this book). Figure 15.3 shows that the activation energy is unchanged – the catalyst has simply provided an alternative pathway that has a lower activation energy.

B2.25 The advantages of using a catalyst in industrial reactions

We have said already that, in industry, time is money. So, it will not be surprising that using a catalyst will increase the rate of reaction giving a faster turnaround time and so reduce costs.

We have seen that temperature increases the rate of chemical reactions. Another advantage of using a catalyst is that, by reducing the activation energy, we can carry out the reaction at a lower temperature. This reduces energy consumption and, therefore, reduces costs.

Catalysts can also be used as a way of selectively producing one product rather than another. For example, poly(ethene) produced using a catalyst will be the high density form (HDPE, see section B1.33, page 204). If a catalyst is not used, low density poly(ethene) is produced.

Another example is the use of zeolite catalysts in production of dimethylbenzenes. These exist as three possible isomers, but one is more valuable as it is used in production of polyesters. The zeolite catalyst can be treated so that yield of the more valuable isomer increases from 25% to 97%.

B2.26 How to use the Maxwell–Boltzmann distribution of molecular energies

A Maxwell–Boltzmann distribution of molecular energies is a plot of the number of gas molecules against their energy (at a fixed temperature). This is shown in Figure 15.5.

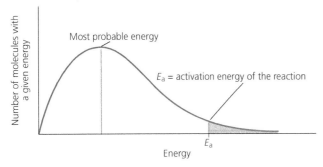

▲ Figure 15.5 The Maxwell–Boltzmann distribution of molecular energies

You will see that the activation energy, E_a, is marked on Figure 15.5. This is the minimum energy needed by molecules to react if they collide. The number of molecules with at least the minimum energy needed is represented by the grey shaded area.

When we increase the temperature, the energy of the particles increases. There will be more particles with higher energy, although the same number of particles overall. This means that the curve shifts to the right, although the peak is lower. This is illustrated in Figure 15.6.

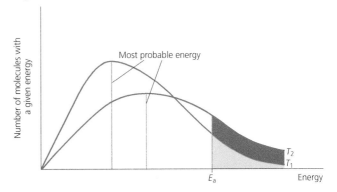

▲ Figure 15.6 Maxwell–Boltzmann distribution at two temperatures; T_2 is higher than T_1

If you look at the blue shaded area, this is the number of molecules with enough energy to react (energy greater than the activation energy) at some temperature (T_1). At the higher temperature, T_2, we must now add the number of molecules in the red shaded area, so the total number of molecules with enough energy to react is now a lot higher. This means there will be more reactions in a given time, and so the rate of reaction will have increased.

If we reduce the temperature of a reaction, then the number of molecules with energy greater than the activation energy will reduce, which explains why reducing temperature reduces the rate of reaction.

Now look at Figure 15.7, which shows the effect of adding a catalyst.

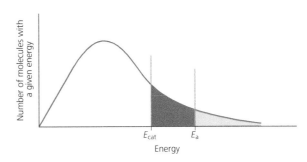

▲ Figure 15.7 Maxwell–Boltzmann distribution showing the effect of adding a catalyst

Remember that the catalyst provides an alternative reaction pathway that has a lower activation energy. This is represented by the energy marked E_{cat} on Figure 15.7. You can see that there has been a very large increase in the number of molecules with enough energy to react, so there is a large increase in the rate of reaction.

We saw in section B2.21 that increasing the pressure will increase the rate of reaction because it will increase the probability of molecules colliding. We also saw that increasing the temperature will increase the rate of reaction because it will increase the energy of the particles. This also increases the probability of molecules colliding.

We have now seen that increasing the temperature also increases the number of molecules with energy greater than the activation energy. This also increases the rate of reaction.

Which do you think will have a bigger influence on the rate of chemical reactions, temperature or pressure?

Test yourself

1 What would be the effect of the following on the rate of reaction?
   a Increasing the temperature.
   b Replacing magnesium powder with the same mass of magnesium pieces in a reaction with dilute sulfuric acid.
   c Increasing the pressure in a reaction between two gases.
2 In a reaction between an acid and a group 2 metal, 2.5cm³ of hydrogen gas was collected in five minutes. Calculate the rate of reaction in cm³/min.
3 What is meant by 'activation energy'?
4 As a general rule, increasing the temperature by 10°C will double the rate of reaction. Explain this observation in terms of the energy of the particles.

Analytical techniques

We covered a number of analytical techniques in Chapter B1 Core science concepts: Chemistry, sections B1.42 and B1.43. We are now going to look in more detail at some of the ones you are most likely to use.

B2.27 How chromatography can be used to separate substances due to their attraction to the mobile or stationary phase

We covered the steps involved in running a thin layer chromatography (TLC) plate, as well as the uses of the related method of column chromatography, in section B1.42. In that section, we saw that separation of different substances occurred because they had different solubilities in the mobile phase (solvent mixture). Changing the mobile phase would change the performance of the system.

We can think of solubility as the attraction or **affinity** of a substance for the solvent (mobile phase). But, there is also attraction between the substances being separated and the stationary phase, such as silica gel, alumina or other material. This is described as the **affinity** of the substance for the stationary phase.

Chromatography separates substances in a mixture due to their different affinities for both the mobile phase and stationary phase. This gives us greater opportunity to improve separation, because we can choose the most appropriate combination of mobile phase and stationary phase for any separation.

B2.28 How to calculate and use the R_f value to identify a substance

Having carried out TLC, we can measure the distance travelled by the substances (spots) because this will depend on the relative solubility in the mobile phase and affinity for the solid phase. The distance moved should, therefore, be characteristic of a given substance. Of course, the absolute distance moved will depend on the size of the TLC plate and the time we allow the TLC to run.

To overcome this, we measure the relative movement of the substance. This is known as the R_f value and we calculate it as the distance travelled by the substance (measure to the centre of the spot) divided by the distance travelled by the solvent:

$$R_f = \frac{\text{distance moved by spot}}{\text{distance moved by solvent}}$$

The R_f value should always be the same for a given substance as long as the TLC is run under the same conditions. This means that we can analyse a sample by TLC, calculate its R_f and look this up in a database of R_f values to help identify the substance.

This is an alternative to the method described in section B1.42 using standards run on the same plate.

B2.29 The stages of an acid–base titration, including the role of indicators in determining the end point

Section B1.44 covered the principles of titration, including the role of indicators in determining the **end point**.

The table shows the colour changes of the two indicators that you need to know about. This is illustrated in Figure 15.8. Figure 15.9 shows the apparatus used for a titration.

Indicator	Colour in lower pH range (acidic)	Colour in upper pH range (alkaline)
Phenolphthalein	Colourless	Pink
Methyl orange	Red	Yellow

▲ Table 15.7 The colours of Phenolphthalein and Methyl orange in acid and alkali

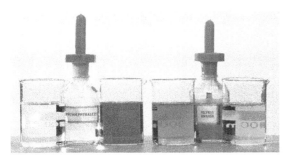

▲ Figure 15.8 Phenolphthalein (left) and methyl orange (right) in acid and alkali

In Figure 15.8, can you identify which beakers contain acid and which contain alkali? Which indicator do you think will give the clearest colour change at the end point of the titration?

We are now going to look at the practical aspects of this technique which is still very widely used in chemical analysis.

Titration is used to determine the number of moles of substance in a volume of solution of unknown concentration. To do this, we use a **standard solution** of known concentration. We will now look at the different stages of titration.

Health and safety

The substances used in titration are corrosive or harmful or both. You should always wear a lab coat and eye protection in the laboratory, but you should also wear gloves and take care not to splash any exposed areas of skin. You should also take care when filling the burette; it should be firmly clamped on the stand and a funnel used to prevent spillages.

When using a volumetric pipette make sure you use a safety filler. You should never pipette by mouth.

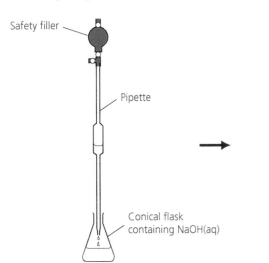

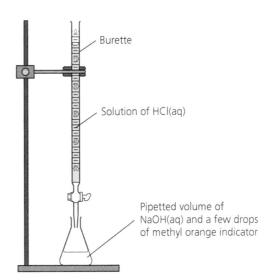

▲ Figure 15.9 The apparatus used for titration

Calculation of mass required to make a solution of a given concentration

Based on the information in section B2.20, you should be able to calculate the mass of substance you need to make a solution of a given concentration (the **standard solution**). This will normally be in mol/dm³. To calculate the number of grams required, you need the M_r of the substance you are going to weigh out (normally you will find this on the label of the bottle it came in, but you can also work it out by using A_r data from the periodic table).

Performing the titration

Having prepared the standard solution, we then have to perform the titration. For example, use a standard solution of 0.050 mol/dm³ hydrochloric acid to determine the concentration of 250 cm³ of sodium hydroxide solution of unknown concentration. The analyte is usually placed in the flask and the standard solution in the burette.

Stage 1: Approximation

1   Put 25 cm³ of the analyte (sodium hydroxide solution) into a 250 cm³ conical flask using a volumetric pipette.
2   Add the appropriate indicator (see below); if we use phenolphthalein the solution should be pink.
3   Put the flask on a white tile or sheet of paper so that the colour change of the indicator is easier to see.
4   Place the burette on its stand and fill with the hydrochloric acid.
5   Take an initial reading from the burette and note it in your laboratory notebook.
6   Run liquid from the burette into the flask, swirling the flask quite rapidly to mix the liquids.
7   Stop the flow as soon as the indicator changes colour (for phenolphthalein it will turn colourless).
8   Take another reading from the burette and note this. Calculate the volume delivered by the burette (final reading minus initial reading); this is the **titre**. This is used as a rough estimate of how much standard is needed.

Stage 2: Accurate titration

1   Empty the flask and rinse it. It doesn't matter if the flask is still wet when you add the next lot of analyte – titration measures the number of moles in the flask not the concentration.
2   Refill the burette if necessary and repeat the titration as in stage 1. Record the initial burette reading in your notebook.
3   At 2 cm³ before the approximate volume for the end point (worked out in stage 1), start to add the standard dropwise until the indicator just turns colour.

4   Record the final burette reading in your notebook and calculate the accurate titre. The rough titre should be greater than the accurate titre, but no more than 1 cm³ greater.
5   Repeat the procedure until you get **concordant** titres (within 0.1 cm³ of each other).
6   Calculate the mean titre – you can include any titres that are within 0.1 cm³ of each other when calculating the mean but not those outside this range.
7   You should now have a table of results in your laboratory notebook that looks like this:

Titration	Rough	1	2	3	4	5
Final burette reading/cm³	22.70	44.60	22.20	44.30		
Initial burette reading/cm³	0.00	22.70	0.00	22.20		
Titre/cm³	22.70	21.90	22.20	22.10		

▲ **Table 15.8** Titration results

Remember that we cannot read a burette to less than ±0.05 cm³ (half the smallest division).

There was a difference between titrations 1 and 2 of 0.30 cm³ (that is, they were not concordant) so titration 3 was carried out. This was concordant with titration 2, so we can stop. We can use the two concordant titres to calculate the average titre – in this case it is 22.15 cm³.

Although all the volumes are recorded to two decimal places, we might have a titration where the concordant titres were 21.10 cm³ and 21.15 cm³. This would give a mean of 21.125 cm³. We can either leave this as three decimal places, or round it to 21.13 cm³.

Calculation of concentration in mol/dm³

The following material is not required in the specification, but it is included here because it is something that you are likely to encounter in your working life. The whole point of titration is to determine the concentration of an unknown solution.

This is based on what we covered in section B2.20.
1   With an average titre (22.15 cm³) calculate the number of moles (n) of hydrochloric acid used to neutralise the sodium hydroxide (don't forget the volume must be converted to dm³ by dividing by 1000). It is better not to round any intermediate results:
n(HCl) = 0.050 × (22.15 / 1000) = 1.1075×10^{-3} mol
2   Use the chemical equation to work out how many moles of acid and base react:
NaOH + HCl → NaCl + H_2O
In this case 1 mol NaOH reacts with 1 mol HCl
Therefore, n(NaOH) = 1.1075×10^{-3} mol

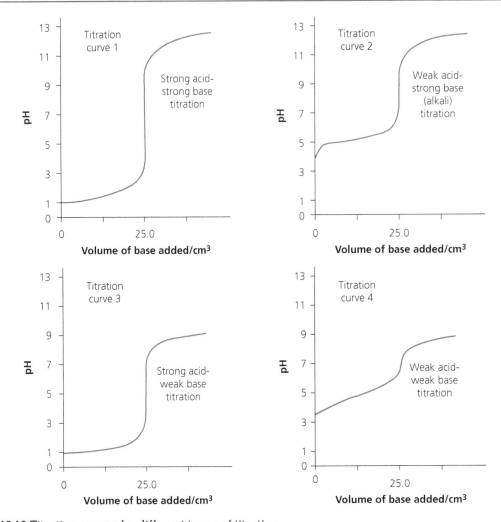

▲ Figure 15.10 Titration curves for different types of titration

3 This was contained in 25 cm³ (0.025 dm³) of solution, so the concentration of NaOH is given by:
conc = 1.1075 × 10⁻³ ÷ 0.025 = 0.0443 mol/dm³

The final step is to consider how many significant figures to use in our final answer (see section B1.64). In this case, the concentration of the standard solution was quoted to three significant figures, so this is what we should use, giving a final answer of 0.044 mol/dm³.

Choice of indicator

It will help you to understand how we choose an indicator if we look at how the pH changes in the course of a titration. These are known as titration curves and their shape depends on the type of titration (strong or weak acid or base) as shown in Figure 15.10.

We saw in section B1.44 that the end point of the titration is the point where the indicator changes colour. It should, ideally, coincide with the **equivalence point** of the titration (the point where we have equal moles of H⁺ ions and OH⁻ ions). In Figure 15.10 this is represented by the vertical region of the titration curve.

If you look at titration curve 1 in Figure 15.10, you will see that either of these indicators will change colour within the vertical region. So, when titrating strong acid and strong base, either indicator will work.

If you look at titration curve 2 in Figure 15.10, you will see that methyl orange is likely to have already turned yellow in the weak acid before adding any strong base (alkali). This is because the pH of a weak acid is relatively high. For titration of weak acid and strong base, we would use phenolphthalein.

The reverse is true for titration curve 3 (strong acid and weak base) and so we would use methyl orange for this titration.

Finally, if you look at titration curve 4 (weak acid and weak base) you will see that there is no clear vertical region. This means that there would be a gradual colour change and it would not be possible to determine a clear end point with either indicator. In fact, the only way you could titrate a weak acid and weak base would be to use a pH meter, plot the titration curve and estimate the equivalence point.

Back titration

This is a technique that you are quite likely to use, although it might not be called back titration; it might simply be referred to as a procedure including titration. It is often used in cases where you cannot use titration in an acid–base reaction. One example is where the acid or (more commonly) the base is insoluble, so you cannot prepare a solution for titration.

What you can do is react the insoluble base with a known amount of acid that is in excess, meaning all the base will react and some of the acid will remain. You can then titrate the unreacted acid with an alkali, such as NaOH, to determine the number of moles. This allows you to calculate the moles of acid that reacted, from which you can calculate the moles of the insoluble base.

This technique is used in applications such as:
▶ calculating the percentage of magnesium hydroxide in an indigestion tablet
▶ calculating the calcium carbonate content of a sample of limestone.

Can you think of any applications in your workplace?

B2.30 The applications of chromatography in industry

Chromatographic methods can be **qualitative**, where they simply detect the presence of a substance, or **quantitative**, where they also measure the amount of that substance. The principles of thin layer chromatography (TLC), gas chromatography (GC), high performance liquid chromatography (HPLC), ultra-high performance liquid chromatography (UPLC) and mass spectrometry (MS) were covered in section B1.42.

Forensic investigation

Chromatography is used widely in forensic science for the detection, separation and identification of a range of substances:
▶ TLC is used to analyse inks and dyes in banknotes, or fibres left at a crime scene.
▶ HPLC (page 215) is used to separate and identify drugs and other non-volatile substances that are not suitable for GC analysis (page 215). HPLC and its more recent development UPLC (page 216) are powerful techniques when combined with mass spectrometry (MS, page 216) to separate and then identify the components of a mixture.
▶ GC can be used in a similar way to HPLC, except that it works only with volatile compounds. GC can

also be combined with MS to identify substances in a wide range of applications, such as:
– detecting drugs of abuse or performance-enhancing drugs in urine samples
– investigating arson, poisoning or explosions to identify the compound(s) used.

Alcohol (ethanol) is routinely detected in human breath samples in connection with suspected drink-driving offences. The methods used include fuel cells (where the reaction of oxygen and ethanol generates an electric current) or infra-red absorption (where ethanol absorbs infra-red radiation allowing its concentration to be determined).

Detection and measurement of alcohol in human tissues, such as blood, liver or brain, is performed using a technique known as headspace GC. In this technique, the sample is placed in a chromatography vial and left to reach equilibrium; any volatile substances will then be present in the gas phase above the sample (headspace). This can be analysed by GC to detect the presence of alcohol or other volatile substances.

Water analysis

GC or GC-MS is widely used to detect pesticides in samples of river water and other environmental samples.

Another chromatographic method used in water analysis is **ion chromatography**. This is an extremely powerful tool that can detect very low levels of various cations, anions and even organic acids. These might be present as contaminants or impurities, or they might be what gives bottled mineral water its particular characteristics.

B2.31 The applications of chromatography and titration in industry

Quality control

Testing food products for consistency is an area where both chromatography and titration are used widely. Here are some examples of titration:
▶ The acidity of orange or other fruit juices can be determined by simple acid–base titration. If necessary, an acidity regulator can be added to the foodstuff to ensure consistency.
▶ Vitamin C content can be measured by iodine titration. Vitamin C (ascorbic acid) will reduce iodine to iodide. Once all the ascorbic acid has reacted, the excess iodine will then react with starch indicator to form a blue–black colour. This is an example of a redox titration.

- The **iodine number** of fats can be used to determine the number of C=C double bonds in a fat. Unsaturated fats have none of these, whereas polyunsaturated fats have several. The method involves reacting a known amount of iodine with the fat (iodine reacts with C=C double bonds) and then using titration with sodium thiosulfate to measure the amount of unreacted iodine. This is another redox titration.
- The **Kjeldahl method**, originally developed in the nineteenth century, is a method of determining the nitrogen content and is an important method for analysing proteins in foodstuffs. The method involves digestion of the protein and then conversion to ammonia, which can be measured by titration with acid. The method is now largely automated.

Here are some examples of the use of chromatography:
- Organic acids can be analysed by HPLC using ion chromatography columns. Applications include:
  - Determination of vitamin C content.
  - Monitoring contamination of apple juice with malic acid and sugar (this is a way of extending the volume of the apple juice by diluting with water). Malic acid occurs naturally in apples but all commercially available (synthetic) malic acid is contaminated with fumaric acid, so this acts as an indicator of the presence of synthetic malic acid.
- Use of GC to monitor the flavours and fragrances (these are all volatile compounds) added to foodstuffs to ensure batch-to-batch consistency.
- Identification and quantification of cannabinoid content in safe and legal cannabis-derived medicines.

Purity analysis

As well as analysis of final products (quality control), titration and chromatography are used widely in the testing of raw materials to ensure that they comply with the required specifications.

Karl Fischer titration is used to determine the water (moisture) content of raw materials.

Saponification value is determined by hydrolysis of oils (triglycerides) with potassium hydroxide in ethanol and then titrating the unreacted alkali with acid. High values indicate a rancid oil (one that has gone off and contains large amounts of free fatty acids) while low values can indicate contamination with mineral oil (which is a hydrocarbon).

HPLC and GC are both widely used to detect impurities in the chemical and pharmaceutical industries.

The amount of chloride in water can be calculated by titrating with silver nitrate solution using potassium chromate as an indicator. Chloride ions react with silver ions to produce a precipitate of silver chloride – this is a common qualitative test for the presence of chloride ions. In this case, it can be a quantitative test because, as soon as all the chloride ions have reacted with the silver ions, the silver ions will react with the chromate ions and produce a brownish–red precipitate; this indicates the end point of the titration.

Test yourself

1. Explain how TLC or column chromatography separates the substances in a mixture.
2. Describe how you would calculate the R_f of a spot on a TLC plate.
3. Explain the function of an indicator in an acid-base titration.
4. Give **two** uses of chromatography in:
   a. Forensic science.
   b. Quality control of food products.

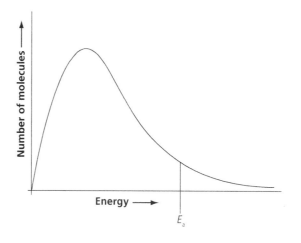

Project practice

You are a technician working in an analytical laboratory that deals with a wide range of ingredients and raw materials used in food production and processing. Your department is responsible for chemical analysis only (another group carries out microbiological analysis).

Research a range of tests that might be used for the following:
▶ *either* acidity regulators, such as citric acid and preservatives such as ascorbic acid and sodium benzoate
▶ *or* artificial flavourings and natural fragrances, such as plant extracts.

Prepare an evaluation of the different tests, based on:
▶ accuracy and reliability of the test methods
▶ ease of use of the equipment and procedures
▶ cost and availability of equipment and consumables.

Submit your evaluation in the form of a written report or scientific poster.

Discuss your findings with the group.

Write a reflective evaluation of your work.

Assessment practice

1 Nickel is a metal widely used in making alloys to produce coins. Balance the equations for the following reactions of nickel:
   a With sulfuric acid to produce nickel sulfate.
      $Ni(s) + H_2SO_4(aq) \rightarrow NiSO_4(aq) + H_2(g)$
   b When heated strongly in air.
      $Ni(s) + O_2(g) \rightarrow NiO(s)$.
   c. The partially balanced equation for the reaction of nickel with nitric acid is
      $3Ni + 8HNO_3 \rightarrow \_Ni(NO_3)_2 + \_NO + \_H_2O$.
      Fill in the blanks on the right hand side of this equation.

2 A sample of tungsten is being investigated to find out if it came from the same source as a batch of stolen tungsten. The stolen batch has an A_r of 183.90. The sample was analysed by mass spectrometry and the results are shown in the table below.

Isotope mass/charge ratio	Abundance (%)
182	26.50
183	14.35
184	30.67
186	28.48

The scientist performing the analysis concluded:
   – the sample of tungsten is an alloy of tungsten consisting of only four isotopes
   – the A_r for the sample is 183.90
   – the sample is from the same source as the stolen batch of tungsten.

Evaluate the scientist's conclusion. Your response should demonstrate reasoned judgements and/or conclusions.

3 A compound with the empirical formula C_2H_4O has a relative molecular mass of 88. Which of the following gives the molecular formula of this compound?
   A C_2H_4O
   B C_3H_7COOH
   C $C_4H_8O_2$
   D $C_5H_{12}O$

4 The curve in the graph below shows the Maxwell–Boltzmann distribution of molecular energies in a reaction at 300 °C.

a Make a copy of this curve and sketch the curves you would expect at:

　i　200 °C

　ii　400 °C.

b Explain the effect of these changes in temperature on the rate of reaction.

c Suggest what would happen to the height of the peak and shape of the curve if you halved the number of molecules in the reaction mixture at the same temperature.

5 A chemical engineer recommended the following conditions for a reaction between two gases:

Although the reaction is normally carried out at 500 °C, the use of a catalyst means the temperature can be reduced to 250 °C. The platinum catalyst should be a very thin layer on an unreactive support.

Evaluate these recommendations.

6 A scientist was analysing amino acids using TLC. The result of the analysis is shown below.

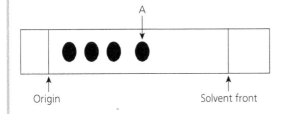

The table shows R_f values of several amino acids.

Amino acid	R_f
Arginine	0.16
Cysteine	0.37
Glutamic acid	0.31
Isoleucine	0.53
Methionine	0.51
Tyrosine	0.55

a Use the data in the table and a calculation to identify the amino acid labelled A.

b If the TLC had been run in the same solvent mixture but for half the time, what would be the effect on the R_f value of spot A?

7 Ammonia solution behaves as a weak base (alkali). A technician carried out an experiment to measure the concentration of ammonia in an industrial cleaning fluid.

a Outline the steps of the acid-base titration the technician would carry out.

b Identify **two** safety precautions that the technician should follow.

8 Give two examples of the use of chromatography and titration in industry.

Chapter B1 Core science concepts: Physics covered some of the fundamental concepts in physics. In this chapter we will look at gases (specifically the gas laws) and the properties of fluids, such as pressure and viscosity. As gases and liquids are both classed as fluids, the whole of this chapter is about fluids. The concepts covered here are likely to be of use in many areas of science where we encounter fluids.

Learning outcomes

The core knowledge outcomes that you must understand and learn:

Gas laws

B2.32 how the gas laws describe the behaviour of gases in particular conditions

B2.33 the use of the Kelvin temperature scale in describing the behaviour of gases in particular conditions

B2.34 the effect of compression when storing gases in cylinders

Pressure/fluid/viscosity

B2.35 the definitions of density, pressure, fluid, viscosity

B2.36 the properties of Newtonian and non-Newtonian fluids, as defined by Newton's law

B2.37 how depth affects hydrostatic pressure in a liquid

B2.38 the definitions of volumetric and mass flow rates

B2.39 the difference between steady and turbulent flow

B2.40 the coefficient of viscosity of a fluid.

In chemistry, our understanding of how temperature and pressure affect the rate of reactions is based on collision theory (see sections B1.39 and B1.40). Our understanding of the mole as an amount of substance, where 1 mole of any substance contains the same number of particles, is based on the proposal that equal volumes of gases at the same temperature and pressure will contain an equal number of molecules. All of this is based on three fundamental gas laws that were discovered between 1662 and 1802. These laws describe the behaviour of gases in response to changes in pressure, volume and temperature. The laws are **empirical**, meaning that they are based on observation and experiment, rather than on theory. Note that you will be expected to know and use the equations for the three gas laws. Elsewhere in this chapter, equations have been provided for those who find they help their understanding. However, you will not be required to know or use these other equations.

B2.32 How the gas laws describe the behaviour of gases in particular conditions

Boyle's law

Robert Boyle worked in Oxford, in rooms on the High Street, which is now part of University College, Oxford. As well as discovering the law that bears his name, Boyle's research student, Robert Hooke, created a microscope with which he made the first observation of cells, on which cell theory is based (see section B1.1). This one location has thus been the source of two fundamental strands of modern science.

Boyle and Hooke carried out a series of experiments in 1662. From these, Boyle realised that the pressure acting on a gas and the volume occupied by the gas were inversely proportional to each other. His results would have been similar to those in Figure 16.1.

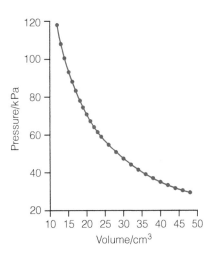

▲ Figure 16.1 Graph of Boyle's law showing the inverse proportion between pressure and volume

Boyle's law states that for a fixed mass of an ideal gas at constant temperature, the pressure of the gas is inversely proportional to its volume.

This can be expressed in the equation:

$$P \propto \frac{1}{V}$$

Where P is the pressure acting on a gas of volume, V.

A more useful version of this equation has the same mass of gas at the same temperature, but at different pressures (P_1 and P_2) and volumes (V_1 and V_2):

$$P_1 V = P_2 V_2$$

This form is useful because, if we know three of the terms, we can calculate the fourth, as in the following example:

A gas has a pressure of 300 kPa and a volume of 20 cm³. What would be the volume of the same gas at the same temperature but at a pressure of 400 kPa?

Rearrange the equation to give:

$$V_2 = \frac{P_1 V_1}{P_2}$$

Therefore:

$$V_2 = \frac{300 \times 20}{400} = 15 \, \text{cm}^3$$

Remember, this law only applies when the temperature is constant.

Charles's law

In 1802, the French chemist Joseph Gay-Lussac published what he called Charles's law. It was named after his friend Jacques Charles who was a hot air balloonist and had observed the behaviour of balloons at different temperatures. The experiments carried out by Gay-Lussac showed that the volume of a gas was proportional to temperature (Figure 16.2).

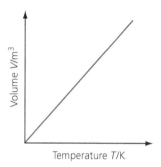

▲ Figure 16.2 Graph of Charles's law

The modern version of Charles's law states that at constant pressure, the volume of a fixed mass of an ideal gas is directly proportional to its absolute temperature (the temperature is expressed on the Kelvin scale, as described in the next section).

This can be expressed in the equation:

$$V \propto T$$

Where V is the volume and T is the temperature in Kelvin.

As with Boyle's law, the relationship can be expressed in terms of the same mass of gas at the same pressure, but different volumes and temperatures:

$$\frac{V_1}{T_1} = \frac{V_2}{T_2}$$

Which can be rearranged to give:

$$V_1 T_2 = V_2 T_1$$

Remember, this law only applies when the pressure is constant.

The pressure law

This law, also known as Amontons' law, was discovered in 1702, coming in between the other two laws. Amontons was trying to build air thermometers and realised that there was a linear relationship between the pressure and temperature of a gas as long as the mass and the volume of the gas were kept

constant. Amontons was not able to build accurate thermometers and so his ideas, when published, lacked quantitative data. With modern thermometers, we can obtain a graph like the one shown in Figure 16.3.

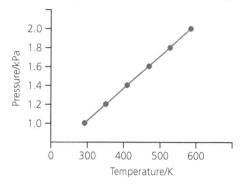

▲ Figure 16.3 Graph of the pressure-temperature law (Amontons' law)

In modern terms, the pressure law states that the pressure of a fixed mass and fixed volume of gas is directly proportional to the absolute temperature of the gas.

This can be expressed in the equation:

$$P \propto T$$

Which is the same as:

$$\frac{P}{T} = \text{constant}$$

From this, we get another more useful form:

$$\frac{P_1}{T_1} = \frac{P_2}{T_2}$$

Combining the three gas laws

If we combine these three gas laws, we get the following equation:

$$\frac{PV}{T} = \text{constant}$$

Which can be expressed in a more useful form as:

$$\frac{P_1 V_1}{T_1} = \frac{P_2 V_2}{T_2}$$

This gives us an equation that we can use to work out what will happen to a gas when a change is applied. For example, if we know the starting volume and pressure of a gas at a given temperature, and change the temperature and pressure, we can calculate the new volume.

We can summarise these laws in the following three sentences:

Boyle's Law: the pressure of the gas is inversely proportional to its volume at constant temperature.

Charles's Law: the volume of a fixed mass of a gas is directly proportional to its absolute temperature at constant pressure.

The Pressure Law: the pressure of a fixed mass and fixed volume of gas is directly proportional to the absolute temperature of the gas.

B2.33 The use of the Kelvin temperature scale in describing the behaviour of gases in particular conditions

Amontons realised that if he **extrapolated** his data to lower and lower temperatures, there would be a temperature at which the pressure of a gas dropped to zero. This is illustrated in Figure 16.4.

Key terms

Extrapolate: to extend a trend or line on a graph so that we can infer unknown lower or higher values.

Absolute zero: the temperature at which all molecular motion stops and the pressure of a gas falls to zero.

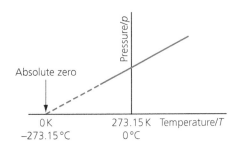

▲ Figure 16.4 Extrapolation of the pressure-temperature graph to absolute zero

Pressure in a gas is caused by the particles colliding with the walls of the container and so exerting a force. If the pressure drops to zero, this must mean that the particles in the gas have stopped moving. The temperature at which this happens is called **absolute zero**.

Amontons calculated absolute zero to be $-240\,°C$, which was fairly accurate considering the thermometer technology that was available to him. In 1848, Lord Kelvin used the concept of absolute zero to construct a temperature scale with absolute zero as the zero point in the scale. Kelvin had access to better thermometer technology and predicted absolute zero to be $-273\,°C$ or $0\,K$. Absolute zero is now defined as $-273.15\,°C$.

B2.34 The effect of compression when storing gases in cylinders

From looking at the gas laws in theory, we now move on to looking at some practical applications and consequences of those laws. We are most likely to encounter this when working with compressed gases in cylinders. Compressing a gas means that it occupies a much smaller volume, which is more convenient for storage. However, it does mean that the compressed gas will be at a much higher pressure, hence the need for strong and heavy cylinders to contain the gas.

The high pressure in a gas cylinder can constitute a serious potential hazard. If the cylinder suffers severe damage, either to a valve or to the cylinder itself, this could lead to an explosion. Even if the damage is not sufficient to cause an explosion, leakage of gas from a pressurised cylinder can be very dangerous – particularly if the gas is flammable or toxic.

We have seen, from the pressure-temperature law, that changes to temperature can affect the pressure. This is why gas cylinders should never be kept near sources of heat (including direct sunlight). It also works the other way – if the temperature drops, the pressure inside the cylinder will fall. This is not usually an issue, although you might find that the output from a gas cylinder drops if you use it in a cold room at +4 °C.

For this reason, gas cylinders must be stored at a determined temperature range. This information will usually be given on the cylinder, or the MSDS (see section A4.18). Large gas cylinders are also very heavy, meaning that they have to be secured to prevent accidents.

In Figure 16.5, notice the gauges to measure the pressure of the gas and the flow rate of the gas when the black regulator knob is opened. Pressure and flow rates are covered in the next section.

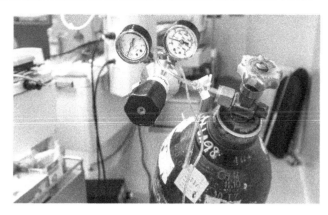

▲ Figure 16.5 Gas cylinder in use in a laboratory showing the regulator and pressure and flow gauge

Test yourself

1 Give the definitions of the following:
   a Boyle's law.
   b Charles's law.
   c The pressure law.
2 Explain why the pressure of a gas at absolute zero (0K) is zero.
3 Calculate the following:
   a 100°C in Kelvin.
   b 120K in °C.
   c A temperature increase of 27.5K in °C.
4 A new location for storage of compressed gas cylinders is required. One proposed site is in full sunlight for most of the day. Discuss the suitability of this location.

Pressure/fluid/viscosity

A **fluid** is a substance that can flow because its particles can move; a fluid has no fixed shape. By that definition, both liquids and gases are fluids. In the following sections we are going to look at two important features of fluids: **pressure** and **viscosity**.

Key terms

Pressure: the force per unit area exerted by a fluid on its container.

Viscosity: a measure of the resistance to flow of a fluid.

B2.35 The definitions of density, pressure, fluid and viscosity

Density

Density is a measure of how compact a substance is – how close together the particles are. The density of a given liquid will be uniform everywhere and it does not vary with the shape or the size of the liquid. The density of a gas can vary.

Density can be defined as the mass per unit volume and is given by the equation (you do not need to know this equation):

$$\rho = \frac{m}{V}$$

Where:

ρ = density

m = mass

V = volume

The symbol for density, ρ, is the Greek letter 'rho'. The units used are kg/m^3 although you will also see density expressed in kg/dm^3 or g/cm^3. The density of water is $1\,kg/dm^3$, which explains why 1 litre ($1\,dm^3$) of water has a mass of $1\,kg$.

A less dense material will float on top of a material that is more dense. Figure 16.6 shows three immiscible liquids forming three distinct layers; the least dense is at the top, the most dense at the bottom.

▲ Figure 16.6 Three immiscible liquids of different densities

The same behaviour can be seen with solids and liquids. A solid object of lower density will float on a liquid of higher density.

Pressure

The **pressure** of a fluid is the force exerted at right angles to any surface in contact with the fluid. The pressure is defined as the force per unit area and can be calculated using the equation (you do not need to know this equation):

$$P = \frac{F}{A}$$

Where:

F = the force in newtons (N)

A = the area of the surface (m^2)

The units of pressure given by this equation are Pascals (Pa); you will also see pressure expressed in kPa and MPa.

The pressure in a liquid will depend on the density; as the density increases the pressure will increase, as there will be more particles in a given volume and therefore more collisions with the walls of the container.

Viscosity

Viscosity is a measure of the resistance of a fluid to deformation. We can think of viscosity as a measure of the internal friction that arises between layers of particles in a fluid. In everyday terms, viscosity is a measure of the thickness of a liquid; syrup has a higher viscosity than water. This means that syrup (high viscosity) will have a lower flow rate than water (low viscosity) when poured or forced through a pipe.

As the layers of fluid flow over each other, they experience internal friction as a result of a force (such as stirring) being applied. This makes the different layers of liquid flow at different velocities. If you think about a fluid flowing through a pipe, there will be a difference in the velocity of the fluid across the diameter of the pipe. The velocity will be lowest next to the wall of the pipe because of friction between the fluid and the wall. The velocity will be highest in the middle. Internal friction means that this effect will be greater for high viscosity fluids than for low viscosity fluids.

B2.36 The properties of Newtonian and non-Newtonian fluids, as defined by Newton's law

Newton's law of viscosity states that, when a shear force is applied to a fluid, the velocity gradient is proportional to the shear stress between the layers in the fluid.

In a **Newtonian fluid**, the viscosity remains constant as the applied force (shear force) changes. This is because the shear force causes a velocity gradient that is independent of the size of the shear force. Water is an example of a Newtonian fluid. When you stir a glass of water, the viscosity does not change. The viscosity of a Newtonian fluid only changes with temperature or pressure.

In a **non-Newtonian fluid**, the viscosity does not remain constant as the applied force (shear force) changes. There are many examples of non-Newtonian fluids in everyday life:

- ketchup – when you shake the bottle, the ketchup is easier to pour because the viscosity decreases
- non-drip paint (also known as thixotropic) is thick (high viscosity) and so will not drip off the brush; however, when applied to a surface, the shear forces make the paint flow readily
- cornflour suspended in water looks milky when it is stirred slowly, but when stirred vigorously it feels like a very viscous liquid.

B2.37 How depth affects hydrostatic pressure in a liquid

As the depth of a liquid increases, the number of particles above that point also increases. The weight of these particles adds to the pressure at that point and so an increase in depth causes an increase in **hydrostatic pressure**.

Key term

Hydrostatic pressure: the pressure in a liquid.

This is expressed in the equation (you do not need to know this equation):

$$P = h\rho g$$

Where:

P = pressure (Pa)

h = height of column of liquid (the depth) (m)

ρ = density of the liquid (kg/m³)

g = gravitational field strength (N/kg); on Earth, this is 9.8 N/kg

B2.38 The definitions of volumetric and mass flow rates

We saw that, in chemistry and elsewhere, we can measure amounts of a substance by volume or mass. This also applies when we measure the flow of fluids.

Volumetric flow rate

The **volumetric flow rate** is defined as the volume of a fluid moving through a given area per unit of time. The equation is (you do not need to know this equation):

$$Q = vA$$

Where:

Q = the volumetric flow rate

v = the velocity of the flow (m/s)

A = the cross-sectional area of the surface (in m²) through which the flow moves

The units of volumetric flow rate are m³/s. You may sometimes see the symbol $\dot{V}$ (pronounced 'vee-dot') used instead of Q.

If a fluid is moving through a pipe (a cylinder) then the area we need is the cross-sectional area of the pipe.

A flow meter that is placed in the pipe will measure the volumetric flow rate.

Practice points

You can work out the units of any variable when you have an equation. Simply substitute the units into the equation. In this case, the units of velocity are m/s and the units of area are m², so we get:

units = m/s × m² = m³/s

This also works in reverse. From the units (m³/s) we can infer that this is volume flowing per unit time.

Mass flow rate

The **mass flow rate** is defined as the mass of a fluid moving through a given area per unit of time. The symbol for mass flow rate is $\dot{m}$ (pronounced 'm-dot') and the units are kg s.

It is often more practical to measure flow rates by volume. If you think about the density of a substance, it is the mass per unit volume. If you look back at the equation for density in B2.35, we can rearrange it to give (you do not need to know these equations):

$$m = \rho V$$

So, since mass equals density multiplied by volume, it should follow that mass flow rate equals volumetric flow rate multiplied by density:

$$\dot{m} = \rho Q = \rho v A$$

This means that we can calculate the mass flow rate either from:

- the volumetric flow rate multiplied by the density of the fluid, or
- the rate of flow multiplied by the cross-sectional area multiplied by the density of the fluid.

Can you think of occasions where it would be more practical to measure volumetric flow rate and convert it to mass flow rate? What would you need to measure in order to calculate the volumetric flow rate?

Research

You will encounter volumetric and mass flow rates in many areas of science:

- ▶ **Hydraulics** is the science of the mechanical properties and uses of liquids.
- ▶ **Pneumatics** is the corresponding study of gases.

Using your own area of interest or work placement, research the uses of liquids and gases. How do factors such as density, pressure and viscosity affect how they are used or handled? Can you find any examples of measurement of volumetric or mass flow rates? What type of instrument is used?

B2.39 The difference between steady and turbulent flow

Steady flow is when the rate of flow does not change over time. This means that all the particles are moving at a constant velocity at a certain point.

Laminar flow is when all parts of a fluid have the same velocity at a certain point. If we think of a fluid consisting of layers of particles, then all the layers are moving at the same rate in laminar flow. Laminar flow does not have to be steady; the rate of flow (velocity) can increase or decrease, but all the layers of particles will still have the same velocity.

Turbulent flow is when different parts of the fluid have a different velocity.

The difference between laminar flow and turbulent flow is illustrated in Figure 16.7.

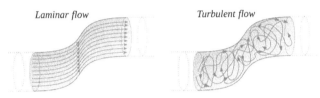

Laminar flow *Turbulent flow*

▲ Figure 16.7 Laminar flow and turbulent flow

In laminar flow, the viscosity of the fluid keeps the fluid moving smoothly. In turbulent flow, there is greater kinetic energy in parts of the fluid flow (they are moving at different speeds) and this overcomes the tendency of the fluid's viscosity to keep the fluid moving smoothly. For this reason, turbulent flow is more likely in low-viscosity fluids. If you look at Figure 16.7 you will see **vortices** where the flow is revolving around an axis rather than flowing in the general direction of travel. This creates **drag** caused by increased friction. As a result, the energy required to pump the fluid through the pipe will increase.

Here are some examples of laminar and turbulent flow:

- Water flowing out of a garden hose appears clear because the flow is laminar. However, if it is forced through a nozzle the flow becomes turbulent and the water mixes with air, making it appear milky.
- Aircraft are designed to maximise laminar flow over the wings to reduce air resistance. This increases fuel efficiency. The same is true of cars and other vehicles.
- When a river is quite deep and flowing over a smooth bed the flow will be laminar – the surface of the water appears smooth. However, if the river becomes shallower with an uneven bed (such as flowing over rocks) the flow will become turbulent and the surface will become more disturbed and appear white.
- When honey or syrup is poured from a jar it displays laminar flow. The viscosity of the fluid keeps the layers moving at the same velocity.

B2.40 The coefficient of viscosity of a fluid

We saw in section B2.35 that the viscosity of a fluid is a measure of its resistance to deformation due to resistance between the layers in the fluid. In section B2.36 we saw that, in a Newtonian fluid, applying a shear force will create shear stress between the layers in the fluid which leads to different layers of the fluid moving at different velocities.

If we now apply that logic to a fluid that is flowing, then the coefficient of viscosity is a measure of the resistance to flow of a fluid.

The SI unit of coefficient of viscosity is $N\,s/m^2$ (Newton-second per square metre). As the Pascal is equal to $1\,N/m^2$, we can also use $Pa\,s$ (Pascal seconds) as the unit.

Measuring the coefficient of viscosity is important in the food and other industries. For example:

- The viscosity of liquid chocolate will affect the density and texture (and hence taste) of the solid chocolate.

- Viscosity is important when fluids must be pumped as more viscous fluids require more powerful pumps.
- Measuring the coefficient of viscosity of a fluid raw material or intermediate can be used to check consistency from one batch to the next.

The table shows the coefficient of viscosity of some fluids.

Fluid	Temperature (°C)	Coefficient of viscosity (Pas x 10⁻³)
Water (liquid)	0	1.8
Water (liquid)	20	1.0
Water (liquid)	100	0.3
Water vapour	100	0.013
Ethanol	20	1.2
Engine oil	30	200
Glycerol	20	1500

The units of coefficients of viscosity are all $Pa\,s \times 10^{-3}$, so liquid water at 20°C has a coefficient of viscosity of $1.0 \times 10^{-3}\,Pa\,s$. We have not followed the rules of standard form (see page 239) to make comparison easier. Note also that water can be both liquid or solid at 0°C and liquid or gas at 100°C.

Compare the coefficients of viscosity of water at 0°C, 20°C and 100°C. What does this tell you about the effect of temperature on viscosity?

Compare the coefficients of viscosity of liquid water and water vapour at 100°C. What does that tell you about the difference in viscosity between a liquid and a gas?

If you took a sheet of glass and placed it at 45° and then used a pipette to drip ethanol, engine oil and glycerol on the top edge, which do you think would reach the bottom first? Why?

Test yourself

1   The table shows the densities of water and tetrachloroethene (a solvent used in dry cleaning).

Substance	Density (kg/m³)
Water	1000
Tetrachloroethene	1622

The two liquids are **immiscible**. This means that, if they are shaken together in a flask, on standing they will separate into two layers. Explain which will be the upper layer.

2   Explain why the hull of a submarine must be stronger than the hull of a car ferry of similar size.

3   Complete the following table to show whether the fluid is Newtonian or non-Newtonian.

Fluid behaviour	Newtonian/non-Newtonian
When a force is applied, the viscosity remains constant, but when the temperature is increased, the viscosity decreases.	
A quantity of paint is put in a tin. The paint does not reach the sides of the container. The lid is put on and the paint is shaken. When the lid is removed, the paint now fills the container evenly all the way to the sides.	
'Flubber' or 'slime' flows under low stress and can be poured, but when you squeeze or poke slime, it feels solid.	

4   A fluid is flowing through a pipe with a cross-sectional area of $0.25 \, \text{m}^2$. The fluid is flowing at a velocity of $3 \, \text{m/s}$. Calculate the volumetric flow rate.

Project practice

You are working as a technician in a school science laboratory. You have been asked to prepare a series of engaging experiments and/or demonstrations based around the gas laws that can be used to explain topics in biology, chemistry and physics, such as:

▶ ventilation of the lungs (breathing)
▶ the effect of concentration on the rate of reaction
▶ the effect of temperature on the volume or pressure of a gas.

Use the internet to research possible experiments or demonstrations that you could use.

Select possible experiments or demonstrations and justify your selection based on relevance, availability of apparatus and the ability to engage students.

Prepare a plan, including:
▶ apparatus and equipment list
▶ timescale for carrying out the experiments or demonstrations
▶ a risk assessment
▶ extension activities that could be undertaken by students.

Present your plan to the group and get feedback from the group.

Write a reflective evaluation of your work.

Assessment practice

1. A closed gas syringe contains a fixed mass of nitrogen gas at 25 °C. To what temperature, in degrees Celsius, does the gas in the syringe have to be raised so that its volume doubles at constant pressure?

2. An empty tin contains air at a temperature of 16 °C and a pressure of 140 kPa. If the pressure inside the tin reaches 240 kPa the lid will blow off.

   a. At what temperature will the lid blow off if the air in the tin is heated evenly with a Bunsen burner?

   b. The lid blows off more quickly if you put some water in the tin. Explain why.

   c. What safety precautions should you take if you carry out this experiment?

3. A truck tyre has a volume of 25 dm³ and the air inside is pumped to a pressure of 2.7 atm **above** atmospheric pressure (1 atm). The temperature remains constant. Calculate the volume the air would occupy at atmospheric pressure if it were released from the tyre.

4. The graphs below show the behaviour of a gas under different conditions.

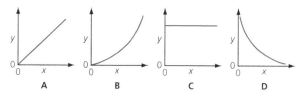

   a. Identify the graph that shows variation in pressure (y-axis) of a fixed mass of gas at constant volume plotted against the absolute temperature (x-axis).

   b. Identify the graph that shows variation in volume of a fixed mass of gas (y-axis) at constant temperature plotted against the pressure of the gas (x-axis).

   c. Identify the graph that shows the variation of the product ($p \times V$) of a fixed mass of gas (y-axis) at constant temperature plotted against the pressure (x-axis) of the gas.

5. A technician was investigating lightweight materials for use in automotive and aerospace applications. They had two magnesium alloy blocks of identical size and shape. One of the blocks floated on water while the other sank.

   a. The block that floated was made of a magnesium alloy foam containing hollow silicon carbide particles. The alloy foam had a density of 0.92 g/cm³. Explain how this block was able to float.

   b. Metal foams on their own are not usually very strong. The silicon carbide particles added to this foam could withstand a pressure of 170 MPa. Suggest how adding these particles would make this foam suitable for use in automotive and aerospace applications.

6. In a biological safety cabinet, laminar flow of air prevents entry of airborne micro-organisms into the cabinet. Explain why it is important to site biological safety cabinets away from open doors or windows.

7. A company was developing a new product. This required honey to be dispensed by pumping from a bulk tank, along a pipe and through a nozzle into jars on a packing line. A food scientist advised that the bulk tank, pipe and nozzle should all be kept warm to increase the speed at which the packing line could be run. Explain this advice.

8. The packing line was monitored in two ways:

   a. A flow meter in the pipe leading from the bulk tank.

   b. The increase in mass of the jars when they were filled was measured.

      Explain why method (a) gave a flow rate with units m³/s whilst method (b) gave a flow rate with units kg/s.

9. The packing line had previously been used with fruit juice, which has a much lower coefficient of viscosity than honey. The person monitoring the packing line noticed that the flow was much smoother with honey than it had been with fruit juice. Explain this observation.

The following questions go beyond the specification but are included to stretch and challenge your understanding – give them a try!

10. A gas cylinder has a volume of 5.0 dm³ (5.0 x 10⁻³ m³) and contains gas at a temperature of 5.0 °C and a pressure of 2.25 MPa. What would be the volume of the gas if it were used to inflate a balloon on a warm day (22 °C) at atmospheric pressure (0.1 MPa)?

11. An instrument designed for underwater use is marked 'tested to withstand pressures up to 400 kPa. The density of sea water is 1030 kg/m³. Use a calculation to determine whether the instrument can be used at a depth of 30 m.

Core skills

This chapter covers the core skills that you will need to demonstrate in the course of the employer-set project (ESP). You should find the information in here, together with many of the topics covered in previous chapters, very helpful in preparing for and carrying out the ESP.

However, it is more important than that. These core skills are ones that any employer will look for in an employee. They are sometimes described as **transferrable skills**, because you can apply them in any role and for any employer.

You will have accumulated a lot of knowledge and information during your course, and you will continue to do so throughout your professional life, but these core skills are the things that will enable you to build your career, wherever the future takes you.

Learning outcomes

CS1   Project management: to include independently producing a high-level project plan taking into account: timing of activities, resource and financial considerations, adherence to health and safety and the maintenance of quality outcomes

CS2   Researching: from independently identified sources including scientific literature and other appropriate sources, prior to the project commencement and referencing these sources appropriately

CS3   Working with others: for example, to ensure that any scientific techniques meet all safety, health and environmental requirements

CS4   Creativity and innovation: within a science context to improve practice processes and outcomes

CS5   Problem solving: within a science context and where appropriate making use of new technologies to solve problems

CS6   Communication: for example, providing results and recommendations in appropriate formats to clients and wider stakeholders which take into consideration 'business benefits' or show commercial awareness in a variety of formats including written reports and verbal presentations

CS7   Reflective evaluation: to be able to make improvements to own practice, for example having completed a task reviewing and suggesting improvements and considerations of lessons learnt for own professional development.

CS1 Project management

CS1.1 Independently produce a high-level project plan, written in a clear, unambiguous way and taking into account the document's purpose

The first step in the ESP will be to produce a project plan. You may not have to produce a project plan for a whole department and present it to senior management, at least not in the early stages of your career. However, you will certainly need to undertake projects of various sorts and you need to be able to plan and manage those projects.

Remember what we learned in sections A9.1 to A9.3 about experimental design – the importance of efficient management of time and resources; the need to define achievable outcomes and objectives; using critical path analysis to plan how the different parts of the experiment fit together and how long it will take – you will find the same is true about project management.

Project deliverables

A well-designed experiment has defined outcomes and clear objectives, and any well-designed project is the same. You need to define:
▶ a project scope statement that clearly and concisely outlines the intended outcomes
▶ a statement of the opportunities and benefits that may be presented by carrying out the project.

Project inputs

This covers all of the resources you will need to complete the project successfully, as well as taking account of who the project is designed for:
▶ People, for example, customers or clients; we saw in section A9.4 how their requirements can affect the scientific methodology, so they may also determine the scope of the project.
▶ Products and materials required, for example samples for testing, raw materials or consumables. We saw the importance of having adequate resources in section A9.1.
▶ Equipment: Is the necessary equipment available? If not, can it be acquired and how (bought, hired, borrowed)? How will that affect the project budget?

A timetable of activities, providing the appropriate level of detail to reflect the project's purpose

Section A9.3 (page 138) covered critical path analysis, and you might need to employ that technique in planning and managing the project:
▶ What is the total time required for the overall project? This may need to take account of those activities that can be undertaken in parallel (at the same time as each other) and those that are on the critical path.
▶ A breakdown of the time required for individual scientific activities.
▶ Important milestones. In some contexts, you might find that funding for a project is linked to the achievement of key milestones – in other words, it is only granted on the basis that these are achieved.
▶ Availability of resources, such as people, rooms or laboratories, and equipment.

A financial forecast, taking into account resource requirements

You are likely to use mathematical processes such as calculations, diagrams and data representations to support forecasting. You must be able to show what has to be done as well as how much it is likely to cost. A financial forecast is a prediction. You may not be able to guarantee that it will be 100% accurate, but you need to have confidence that you have accounted for all the costs that you can reasonably expect. You might also have to make an allowance for **contingencies**. These are unexpected costs that might arise, perhaps because experiments must be repeated, or additional equipment is needed. Make sure that you are confident using spreadsheet software as this usually incorporates many tools that are useful in financial forecasting.

Ethical considerations

We covered ethics in Chapter A7, where you will find more information about the importance of codes of practice and intellectual property rights. Considering these is an essential part of any project plan.

A completed risk assessment

Section A3.2 described how to carry out a risk assessment and you will find useful information to help with risk assessment in specific situations or environments in Chapter A4. An important aspect of risk assessment is to include details of how risks can be reduced or mitigated.

You should make sure the risk assessment is written in a style and level of detail that is appropriate to the document's purpose. For example, a project plan to be submitted to management for approval may only require a statement that a risk assessment has been performed and all risks have been or will be mitigated. On the other hand, a project plan that is intended as a working document for project management will require greater detail and may assume a greater knowledge of scientific language.

How quality outcomes will be maintained

Your organisation might have its own quality standards that must be followed. Alternatively, quality outcomes might be maintained through complying with relevant ISO standards. This has the advantage of being internationally recognised.

Practice points

Project management tools

Project management tools can be as simple as a large piece of paper and a selection of coloured pens or as complex as multi-user software products. The purpose is to help with planning and managing the project through all stages.

Software products designed for use with complex projects in large organisations – sometimes known as **enterprise solutions** – have a high price tag to match. However, there are several free versions of these tools.

Zoho Projects (**www.zoho.com**) is one example, but an internet search will reveal others. This is also a good way of getting access to advice on project management as the software companies that offer these tools will often provide advice and ideas on their websites.

Gantt charts are often used in project planning and management, and can be prepared using a spreadsheet. Microsoft Excel includes Gantt chart templates and provides advice on how to use a Gantt chart in project planning (Figure 17.1).

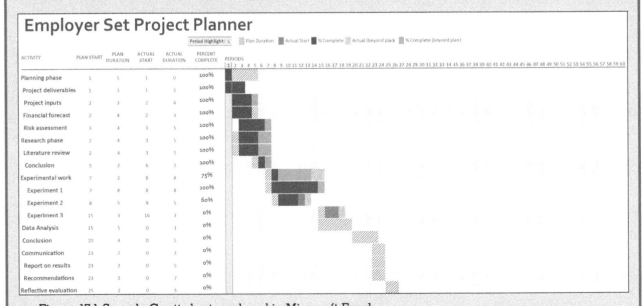

▲ Figure 17.1 Sample Gantt chart produced in Microsoft Excel

Another tool is the **Kanban board**. These can be physical boards with cards or sticky notes (Figure 17.2) or they can be software applications, such as **https://kanbanflow.com** or **https://trello.com**. An online search will provide you with a lot of advice and options.

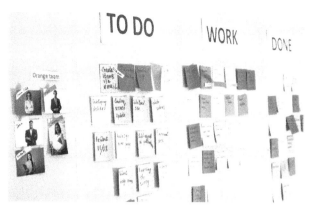

▲ Figure 17.2 A Kanban board using sticky notes to plan and track a project

Research

If you are interested in project management or think that it could be one direction in which you could develop your career, you could search for 'agile project management' or 'scrum methodology'.

Do you think any of these tools could help you? Can you see how ideas from other areas, such as software development, can be applied in scientific projects?

CS2 Researching

Sir Isaac Newton, who is recognised as one of the greatest mathematicians and physicists of all time, wrote: 'If I have seen further it is by standing on the shoulders of giants.' What he meant was that his achievements were made by building on the understanding gained by major thinkers who went before him.

In the same way, we need to base our own work on what has gone before and avoid unnecessary repetition of pre-existing work. For this reason, it is important to begin each project by researching the topic.

CS2.1 Conduct a review of independently selected scientific literature and other appropriate primary/secondary sources

We looked at the strengths and limitations of different data sources, including published literature, in section A5.4, and in section A9.6 we covered ways to access and critically evaluate scientific literature and other primary and secondary sources.

It is good practice to structure your background research in a logical way.

Practice points

Plagiarism is representing someone else's work as your own. Many educational establishments use anti-plagiarism software to check submitted work in case it has been copied (usually, but not always, from the internet). One way to avoid appearing to have plagiarised someone else's work is to reference it properly, sometimes known as citing or giving a citation. Citing previous work is widespread in science, particularly when you refer to a method or series of experiments carried out by someone else, that you are basing your own work on. Anyone reading your work can then go to the original publication if they want to get more information.

Research

Use of published literature has been covered in chapters A5 and A9. However, there may be a range of other sources of data or information that might be useful, including:
- Your organisation's own internal data, publications or reports
- Technical reviews and magazines – these might not all be covered by online databases, but they may be available in your organisation
- Newspaper articles.

Think about your employer. Are there any information sources that might be helpful to you? Do you know how to access them? Are you confident that you can identify appropriate sources?

Introduction

This will include the scope of the review and the criteria for the selection of different sources – what was included in the review, what was not, and why.

Main body

This is where you will make an evaluation of your sources. This was covered in section A9.6, and includes:

▶ the age and/or relevance of the literature
▶ the reliability of the sources, for example peer review, conflicts of interest, number of citations and impact factor
▶ the reliability of data, for example sample sizes and what collection method was used
▶ a logically ordered discussion of the themes, including how the literature relates to each other and to the project
▶ correct use of a recognised referencing system (for example Harvard, Vancouver or AMA in-text citations). Some organisations or educational institutions will have a preferred referencing system that you will need to follow, or you may be able to use the one that you are most familiar with.

Practice points

Managing your literature references can be a challenging task and formatting them into different referencing systems, if you need to, can be very time consuming. Reference-management software can make this process much easier. As well as incorporating tools to manage a database of literature references and incorporate selected ones into your reports, some also include tools that allow you to search online literature databases. Some of these can be expensive, but there are others that are free to use. Search for 'reference management software' online to see what is available.

Research

There are many different styles for references or citations, so you must make sure you use the correct one. Your college, other educational establishment or employer may have a preferred style. If your work is published in an academic journal you may find different styles are used.

Find out if your organisation uses a particular style for references. If it does not, choose one of the common styles such as Harvard or Vancouver as this will be easier to follow. Research your chosen style and see what style is used for a citation in the body of your work (sometimes called **in-text citations**) and for reference lists or bibliographies. Try to keep to that style consistently in your work, including the ESP.

Conclusion

Having researched and evaluated the literature, you then need to reach a conclusion. This will include:

▶ a summary of the key points, using appropriate technical terms
▶ any agreements and disagreements in the literature
▶ any gaps or potential future areas of study
▶ a full **bibliography** of sources (this is why reference-management software is so useful).

Key term

Bibliography: a list of all of the sources you have used in researching your work. It should include:
▶ the authors' names
▶ the titles of the publications (the book, journal, newspaper, and so on)
▶ the names and locations of the publishers (usually just for books)
▶ the date your copies were published.

For journal articles, the bibliography will also include:
▶ the title of the paper
▶ the title of the journal
▶ the year of publication, volume number and page numbers.

See below for more information about the different styles of academic referencing.

CS3 Working with others

We covered the basics of this in sections A2.5 and A2.6, and some ethical issues were covered in section A7.4.

CS3.1 Identify your own role in relation to the wider team

You need to know where you fit in to the wider organisation so that you can work most effectively and efficiently. Most companies and other organisations will have an organisation chart (see section A2.5 for more on this) and this can be very useful in helping identify your role. You also need to consider:

- the team structure, for example your position within the team and whether you have any direct reports
- methods of team working, such as using digital collaboration tools to meet with, share and collaborate with colleagues (see Practice points box for more information)
- the wider organisational structure, for example relationships between individual teams/departments; this is where consulting an organisation chart can be useful
- contact with external stakeholders/clients, for example directly or through third-parties
- establishing your own accountability for tasks and deliverables (objectives that have to be met, or things that have to be provided) – this was touched on in section A9.4 insofar as it relates to scientific methodology, but it has wider application
- establishing your own and others' area of expertise; this helps to identify what you can do yourself and where you will need help and support from others.

Practice points

Efficient team working is so critical to most organisations that it is not surprising that there are many tools available to make this easier and/or more efficient.

During the COVID-19 pandemic, it seemed as if everyone had discovered videoconferencing software such as Zoom and messaging apps such as WhatsApp as a way of keeping in touch with friends, family and colleagues. Other apps are available and offer varying degrees of functionality.

Microsoft Teams and Slack are two of the most widely used platforms for messaging, videoconferencing and collaborating in organisations of all sizes.

Some laboratory information management systems (LIMS, see section A6.3) can incorporate tools to support teamworking.

CS3.2 Meet your responsibilities when working in a wider team

When you work in a team, you have responsibilities to your colleagues as well as to the project. This means that you must ensure that the project is compliant with the following:

- health and safety requirements, for example if storing and handling hazardous substances; this was covered in depth in Chapters A3 and A4
- environmental requirements, for example when disposing of waste; this was covered in sections A4.2 and A4.3
- data protection regulations, for example when using information technology; this was covered in section A5.6
- SOPs specific to the lab in which they are working; the use and importance of SOPs was covered in sections A8.1 to A8.4
- project timescales, for example equipment and spaces are used in the allotted times
- keeping within an agreed budget, as discussed in sections A9.3 and A9.4.

Case study

Casey, Ryan and Umi are laboratory technicians working in a company providing chemical and microbiological analysis services. Because of the size of the company, they undertake a range of different roles as well as having their own specific responsibilities.

Casey is responsible for ordering laboratory reagents and consumables. This involves keeping records of all chemicals, particularly those that are hazardous. Casey uses the MSDSs supplied with each batch of chemicals and solvents to identify any potential hazards and can inform other members of the team if they need to be aware of any new hazards (for example, if a new chemical is being used or if updated information becomes available.)

Ryan operates the HPLC instruments and has been asked by the company's management to implement a new analytical procedure. This requires a solvent mixture that they have not used before. Ryan is informed by Casey that, according to the MSDS, one of the new solvents is harmful by inhalation and Casey feels that the HPLC equipment needs to be in a fume cupboard.

Ryan is not happy with this, because other chemicals that are used in the fume cupboard produce corrosive vapours and these could be damaging to the sensitive electronics in the HPLC equipment.

Umi is responsible for ensuring compliance with environmental regulations and suggests that it would be possible to use localised fume extraction (see Figure 4.8, page 55 for an example) as an alternative to a fume cupboard.

Umi consults the MSDSs that Casey has and sees that the harmful solvent requires special handling for disposal. As a result, Umi ensures that all the waste solvent from the HPLC equipment using the new solvent mixture is collected separately from other liquid waste so that it can be disposed of correctly.

Can you see any parallels with your own work? Do you know who to ask to obtain and information that you need? Which of your colleagues rely on you for providing the information they need to carry out their own work safely?

CS4 Creativity and innovation

Section A2.14 dealt with the importance and impact of innovation in the science sector and it should be obvious, by now, that innovation is central to how science progresses. Creativity might sound like something you would associate more with an artist or musician than a scientist. Nevertheless, creativity is an important part of innovation in science.

CS4.1 Make creative, innovative improvements to scientific practice, processes and outcomes by following an evaluation cycle

Being creative does not mean that we cannot be methodical or systematic. Creativity and innovation in science can flourish when they are part of a structured approach we call the evaluation cycle – this was mentioned in section A2.9 and is expanded on here.

Plan

We begin with a plan:
▶ Identify a potential area for improvement, taking into account:
  – who will benefit from the improvement
  – the desired outcome.
▶ Gather information to understand more about the need for the required improvement, for example:
  – talk to more experienced colleagues
  – collect data
  – research the literature.
▶ Generate ideas and screen them against the desired outcome. Which approach will achieve the best results?

Do

Having made a plan, we need to implement it:
▶ Use your knowledge of the context to find appropriate and approximate solutions; this will depend on the circumstances and your experience.
▶ Implement the improvement.
▶ Record the results. At this stage you need to organise and record data systematically before doing any scaling or processing that may be required; this was covered in Chapter A5.

Check

Did we achieve our objective? We must use data to analyse the results against the desired outcome. This is likely to involve statistical analysis as described in section A6.9.

Act

Once we have analysed the results, we can review the improvement and recommend the next steps.

Having done all this, we will most likely have to do it all again next time/on the next project. The evaluation cycle is part of our approach to creativity and innovation, but it is part of a process of continual improvement. We should always be looking for ways to improve, so we should see this as a regular part of our work.

Case study

What we have described here – the evaluation cycle – is methodical rather than creative, but innovation comes from a combination of creativity and method.

Sometimes creativity in science happens when we do not have the preconceptions that tell us that a particular approach will not work.

Dudley Herschbach was a professor of chemistry at Harvard University. He was studying the way in which molecules react when they collide. Herschbach came across a technique called molecular beams, which had been used in physics for many years. He had the idea to use crossed beams of different molecules such as chlorine and hydrogen to study how quickly the reactions occur when the molecules collide. Herschbach was told of the many reasons why the technique would not work, but he persevered. He and his co-workers spent several years collecting data and, as a result, gained important insights into the ways in which colliding molecules behave. Herschbach was awarded the Nobel Prize for Chemistry in 1986.

Herschbach was a well-established professor when he began the experiments that his critics said would not work. Do you think he would have been able to ignore those critics if he had been a young research student at the time?

It is unlikely that we are going to do ground-breaking research that will lead to a Nobel prize. However, the lesson of Herschbach is that we need to be creative and sometimes we need to have faith in our own ideas. If you are told that an idea you have is potentially dangerous, you should certainly pay attention. On the other hand, if you are told that it will not work because "that's not how we do things here" then maybe you should, as far as possible, persevere.

It is partly for this reason that teams working together are generally more creative and innovative than individuals working alone. When each person brings their own knowledge and experience – and sometimes a lack of knowledge or experience – the group can be collectively more creative.

CS5 Problem solving

If you look at the person specification (see section A2.4) for a wide range of jobs, you are likely to see that problem solving is a key skill that employers are looking for. As scientists, we are used to looking at problems and working out solutions; this is a skill that we should all try to develop further.

CS5.1 Solve a problem within a science context

By now, you know that we should take a methodical and systematic approach to problem solving and you need to demonstrate this approach in your ESP.

▶ Identify and clearly define the problem:
  – demonstrate a thorough understanding of the context of the problem.
▶ Decide on the change to be made, taking into account:
  – steps required to implement the change
  – success criteria for measuring the impact of the change (this is rather like clarifying the objective of any experiment we carry out).
▶ Implement the changes, using new technologies as appropriate for:
  – gathering data
  – recording results.
▶ Evaluate the impact and continue to monitor any changes:
  – make recommendations for further improvement.

You should be able to see that problem solving has a lot in common with the evaluation cycle that supports creativity and innovation.

Case study

Ahmed was planning a programme of experiments to produce a new compound. As part of the planning, Ahmed carried out a risk assessment. As a result, he found that a chemical reaction that was needed as part of the experiment would produce a toxic gas. Literature research showed that there was an alternative process that could be used that did not produce a toxic gas. However, it was not clear that the quality of the end product made by the alternative method would be acceptable.

Ahmed carried out a trial series of experiments on a small scale and analysed the end product for purity against the required specification. He discovered that the product did not meet the required specification and so he concluded that the method producing the toxic gas would have to be used instead of the lower-risk alternative.

Ahmed found a laboratory that had a fume cupboard. He checked with the safety officer that its performance was suitable for handling the toxic gas. As a result, Ahmed was able to carry out his experiments and produce the required quantity of the new compound. Testing showed that it met the required specification for purity.

▶ Do you think Ahmed would have been able to complete the work within the original budget?

▶ Would he have allowed for contingencies such as this?

▶ Why do you think he carried out the trial of the alternative method and did it on a small scale?

Use of new technologies

We saw, in section A5.5, how new technology can be used in recording and reporting information and data, for example:

▶ AI and machine learning
▶ Mobile technology and apps
▶ Cloud-based systems
▶ Digital information management systems
▶ Data visualisation tools.

Section A6.7 touched on the use of specialist software in analysis of large data sets.

All these techniques can contribute to the problem-solving process. For example, digital information management systems and data visualisation tools can help us understand the problem. AI and machine learning may help to offer a range of solutions and even recommend a course of action. Finally, we can use a range of IT systems to gather data and record results as a part of the implementation phase.

Reflect

If you search the internet for 'problem solving in science' you will discover many articles about problem solving in the context of learning and education – some of this might be helpful for you.

However, if you search for 'use of new technology in scientific problem solving' you are likely to come across articles about how science can be used to solve problems that the world faces, such as climate change, diseases, malnutrition, sustainable energy and water supply.

Thinking about your own work, can you see how learning to solve problems that you face in your work might make you better prepared to contribute to solving some of these larger scale problems?

CS6 Communication

Good communication is essential in the workplace, and you will have encountered many examples of the importance of good communication, both in this book and in your own workplace. In the context of the ESP, we should focus on how to communicate results and recommendations. This brings together several of the topics covered in Chapters A2, A5, A6 and A9.

Practice points

If you look back at section A9.4, you will see one example of why communication is important. Most projects involve teams and have customers or clients, either external or internal (other departments in a larger organisation, for example). Communication between and within teams is essential to provide an excellent product or service. These are the wider stakeholders that are essential to us providing service to the customer.

In this section, we will talk mostly about customers or clients, but the principles apply equally to these wider stakeholders.

CS6.1 Provide results and recommendations (written and verbal) to customers/clients

We should communicate in a clear and unambiguous way. This means tailoring the language we use and the technical information we include to the audience.

Before we can communicate our results and recommendations, we need to be confident in our conclusions. This will mean carrying out any necessary statistical analysis using the methods outlined in section A6.9.

We should think about and select the most appropriate way of presenting data. This could involve the use of images and other tools (for example, visualisations or infographics) to clarify complex information. Section A5.2 discussed different ways of presenting data and how this can be adapted for different audiences. Flow charts (see section A6.8) might also be useful, whilst section A9.5 covered ways in which we should provide results and recommendations in appropriate formats to customers or clients.

Communication is not all about speaking. It is essential that we actively listen to the client's contributions and ask them questions to test their understanding. Having done that, we must respond to the client's questions. Once again, we must use a tone and **register** (degree of formality) that reflects the audience. This could be quite different if we are dealing with other technicians or scientists in the client organisation compared with dealing with senior management, who may not be science specialists. In a similar way, we may be much less formal when discussing work with our colleagues than we would be when presenting that work at a conference or when writing a report or scientific paper.

Practice points

Written reports

We have covered the presentation of results and recommendations in various places. One important way of doing this is still the written scientific report and report writing is one of the transferrable skills valued by employers.

You should normally follow a standard structure:
▶ Title
▶ Abstract or summary
▶ Introduction (possibly including a summary of the brief)
▶ Methods used, with appropriate amount of technical detail
▶ Results (see section A5.2 and A9.5 for advice on presentation)
▶ Discussion, conclusion and/or recommendations.

It can also be helpful to include literature or other reference sources when appropriate.

When making oral (spoken) presentations, it is important to speak clearly and confidently – it helps to have practised beforehand for important presentations. Once again, we need to get the tone and register right.

We should also make sure that we can answer any questions relating to the brief or to the research undertaken. Preparation is important, because we will probably have to provide supporting documentation, possibly in various formats.

When making recommendations, it is important to highlight the commercial or business benefits for the customer or client, otherwise they may not be convinced to invest in or fund your work. This might involve the use of calculations, diagrams and data to support your arguments.

Case study

Henry works for an engineering services company that provides product development and manufacturing support services to a number of different manufacturers of components for use in the aerospace and medical device industries.

Henry's group was asked to investigate problems with a ceramic coating used on turbine blades in aircraft engines. A meeting with the client showed that the turbine blades were failing as a result of the ceramic coating becoming worn away more quickly than expected in the life of the component.

Initial work suggested that the problem might be due to the method used to produce the surface coating known as Physical Vapour Deposition (PVD).

The group investigated the problem over several months, which included meetings with the client to provide regular progress reports. At the conclusion of this work, Henry wrote a final report to the client recommending use of a modification to the PVD method based on advances in nanomaterials production methods. Henry's report focused on key advantages to the client:
▶ Lower production costs
▶ Lower failure rate in the finished components, meaning lower cost of replacement, repair and compensation.

In addition, Henry realised that this same improved method could be of benefit to other clients making components for use in medical devices and gave a copy of his report to the Business Development Director who was able to use this to generate new business and find new customers.

What lessons can you take from this about the importance of good communication and team working?

We can learn from good salespeople when thinking about communication with customers or clients. It is tempting, when selling a product, to focus on the features and benefits of the product. However, a good salesperson will start by finding out what the customer needs. They then look at ways that their product can meet those needs. In other words, they are selling a solution, not a product.

When you are planning your ESP, how can you ensure that you find out what the customer or end-user really needs? How can you tailor your recommendations so that you are demonstrating a solution to their problem?

Most of the project practice features at the end of each chapter include the presentation of your results or recommendations to the group and a group discussion. This gives you useful practice in communication skills.

CS7 Reflective evaluation

You will have seen that most of the project practice features at the end of each chapter include reflective evaluation as the final step. Again, it is important to do this – not just as practice for the ESP, but because it is a useful exercise at the end of every project. Remember: continuous improvement!

CS7.1 Evaluate the project's processes and outcomes

In your reflective evaluation, you should focus on the following points.

> Section A9.11 has some useful information on the evaluation of scientific methodology.

Experimental design

Was the project designed efficiently enough, that is, to yield the maximum results from the minimum repeated experiments? Did you follow the recommendations in Chapter A9 about the importance of experimental design?

The accuracy and reliability of the results

Did you use appropriate technical terms? Was the sample size sufficient to ensure the results are reliable (section A9.7)? Was the equipment appropriate to ensure accuracy? It might be useful to consult Chapter A10 regarding this.

Reproducibility

Can the results be replicated easily? Did you incorporate replication into the project where possible?

Suitability of equipment

Was the chosen equipment appropriate for the experiment? This is a wider question than just whether it was accurate. It could also include:
▶ Was it appropriate for the scale of the work?
▶ Did it require manual data collection? If so, could this be improved by some form of automation?

▶ If the project involved the analysis of large numbers of samples, could the equipment cope with the throughput? Again, would automation be possible and feasible in this situation?
▶ Was the equipment easy to use? Did it require a lot of setting up or calibration? Are there alternatives that could be used?

Suitability of methods

Were the chosen methods the most suitable? Good experimental design and planning should ensure that they were. Sometimes with hindsight, however, we realise that we could have done an experiment differently or taken a different approach.

Your own actions during the project

What did you do well and how can you improve? Be honest with yourself.

The quality of the data

This includes how the data was processed and scaled. If you based your conclusions on small samples and then applied this on a larger scale, was that justified?

Did you obtain sufficient data to draw valid conclusions? Sometimes it helps to use different methodologies, if necessary or possible, to improve the validity of your conclusions.

Fulfilment of objectives

Were the project's objectives met? Having carried out the work, think about whether the objectives were appropriate. Were they realistic, or were they too ambitious? Could you have aimed to achieve more? This does not have to mean criticism; it is simply another way we can learn from experience.

Recommendations for improvement

Having reflected on all of the above, you need to make some recommendations for how the experiment could be more effective. Again, recommending ways to improve is not an admission of failure. It shows that you have thought clearly about the project and have gained an insight into how to achieve more next time. This is how science progresses.

Assessment

Assessment overview

This qualification in assessed in the following ways:
- Core component:
  - examination papers A and B
  - employer-set project (ESP).
- Occupational specialism component:
  - synoptic assessment.

As well as the ESP, covered in the Core skills chapter, you will also be assessed through two written examinations:

Paper A covers the topics in specification sections A1–A10, covered in Chapters A1–A10. This is split into four sections:
- Section A – Working within the science sector
  - A1 Working within the health and science sector
  - A2 The science sector
  - A8 Good scientific and clinical practice
- Section B - Ethics, data and managing personal information in the science sector
  - A5 - Managing information and data within the health and science sector
  - A6 - Data handling and processing
  - A7 - Ethics
- Section C - Health and safety in the science sector
  - A3 - Health, safety and environmental regulations in the health and science sector
  - A4 - Application of safety, health and environmental practices in the workplace
- Section D - Scientific methodology, equipment and techniques
  - A9 - Scientific methodology
  - A10 - Experimental equipment and techniques.

Paper B covers the topics in specification sections B1 and B2 (Core and Further science concepts in Biology, Chemistry and Physics). This is also split into four sections:
- Section A – B1 Biology (Chapter 11 B1 Biology)
- Section B – B1 Chemistry (Chapter 12 B1 Chemistry)
- Section C – B1 Physics (Chapter 13 B1 Physics)
- Section D – B2 Further Scientific Concepts (Chapter 14 B2 Biology, Chapter 15 B2 Chemistry and Chapter 16 B2 Physics)

Each exam lasts 2.5 hours, so you will need to practise writing and concentrating for quite long periods.

At the end of each chapter of this book you will find some assessment practice questions that will help you check your understanding of the specific content for that topic. This chapter looks at the types of question you will encounter and how to approach them to give yourself the best chance of success.

Assessment objectives

Core Papers A and B will assess to what extent you have achieved the following assessment objectives (AOs):
- **AO1 Demonstrate knowledge and understanding of contexts, concepts, theories and principles in science**.
  Some of this will just be recalling what you have learned, but there will be other questions that test your understanding of familiar contexts or situations.
- **AO2 Apply knowledge and understanding of contexts, concepts, theories and principles in science to different situations and contexts**.
  Do not be put off by an unfamiliar context or situation. Having taken this course and studied for it, you should have the knowledge and understanding required to answer the question, otherwise it would not be included in the paper.
- **AO3 Analyse and evaluate information and issues related to contexts, concepts, theories and principles in science to make informed judgements, draw conclusions and address individual needs.**
  In 'evaluate' or 'assess' questions, you will almost certainly have to show knowledge and understanding (AO1), as well as being able to apply that knowledge (AO2), but make sure that you do actually analyse and/or evaluate the information

given and use it to reach a conclusion or judgement, as there will be marks reserved for candidates who do demonstrate these skills.

Command words

Before we think about the different types of questions, we need to think about the words used in questions. They are often known as **command words**. They are important because they indicate how you should go about answering, what you will be assessed for and the kind of skills involved. The different types of command words are usually linked to a particular AO, so we have grouped command words by the AO that they are testing. Here are some key examples:

AO1

These require you to recall and provide some information, so you are just showing your knowledge.

▶ **State, name** or **give** usually require a short statement or even a single word. It might be a definition or just a piece of information or knowledge. However, you might see a question that says: 'Give **two** reasons why …', in which case your answer needs to be more of an explanation.

▶ **Identify** requires you to name something or characterise it in some way.

▶ **Choose** means you should select from a range of alternatives or options. Make sure you pay attention to numbers, particularly when they are in bold as in the following examples:
  – Choose **one** of the following options …
  – Choose **two** ways to carry out …
  Having to choose more than one option usually means you could score more than one mark, too.

▶ **Complete** usually refers to tables where you have to fill in the blanks.

AO2

This group of command words requires you to apply your knowledge, sometimes to an unfamiliar situation, or to do something with the information you are given to demonstrate your understanding.

▶ **Calculate** should be self-explanatory. However, with calculation questions it is important that you use the correct number of decimal places or significant figures in your answer. The question may specify this but, if not, follow the advice in section B1.64.

▶ **Determine** usually means that you have been given some data or information that you need to use to get an answer. It may not involve a calculation.

▶ **Balance** questions require you to balance a chemical equation (see section B2.15).

▶ **Compare** means you must describe the similarities and/or differences between things. Make sure that you do not just write about one.

▶ **Describe** is asking you to recall some facts or information, or to say what a graph or set of data shows. For example:
  – Describe **two** differences between DNA and RNA.
  – Describe what the graph shows about the relationship between temperature and the rate of reaction.
  Be careful when you are asked to describe something – do not try to explain it. If you do, you will just be wasting valuable time. Also, in giving an explanation you might not give a proper description and so lose marks.

▶ **Explain** is asking you to make something clear or give the reasons why something happens or has happened. Your answer will often be '*statement* because *explanation*'. For example:
  – Explain the relationship between the primary structure and tertiary structure of a protein.
  – Explain why alpha radiation is not very penetrating.

AO3

This group of command words is expecting you to demonstrate your higher-level skills, such as making judgements and/or reaching conclusions.

▶ **Assess** is asking you to make a judgement. This might be 'did the results of the experiment support the hypothesis?' or 'was this statement correct?' or 'was this supported by the evidence?' For this qualification, the question will usually include the statement that 'your response (answer) should demonstrate reasoned judgements and/or conclusions'.

▶ **Evaluate** is quite like assess, in that you are being asked to make a judgement. However, in order to evaluate you need to use any information given in the question as well as your own knowledge and understanding to consider evidence for *and* against when making your judgement.

▶ **Suggest** means that you probably have not learned the answer to this question. However, you should be able to apply your knowledge and understanding to a new or unfamiliar situation to find a suitable potential answer.

Question types

Both papers contain a mix of different question types that require different approaches to answering them.

Multiple choice

In Science T Level exams, the multiple choice questions are spread throughout the paper.

These are usually quite short questions (so there is not too much to read) with four possible answers. Only one answer is correct but you will find that most of the other options are quite plausible – do not let them distract you from the correct answer. In some cases, the options are all correct statements in general, but only one of them will be correct in relation to the question.

Here is an example that illustrates a slightly different type of multiple choice question. In this case, you have to choose the correct combination of options. Nevertheless, the way you go about working out the correct answer is the same:

1 The following is a selection of health and safety practices when carrying out experiments:
  w Carry out a risk assessment before you start.
  x Always carry out the experiment in a fume cupboard.
  y Wear safety glasses or goggles.
  z Always wear a dust mask.
  Which of these should you do when carrying out an experiment for the first time?
  A w only
  B w, x, y and z
  C z only
  D w and y only

Here is one way to approach this:
▶ Looking at the different practices, you should know by now that you must always do a risk assessment before any new experiment (practice **w**), so we can rule out **C** (because it does not include practice **w**).
▶ Practice **x** (use of a fume cupboard) seems reasonable, but only when an experiment might produce a harmful or toxic gas, not all the time. That means we can exclude option **B**.
▶ Practice **y** (wear safety glasses or goggles) is also something that we should always do when working in a laboratory, so we can rule out **A** (because it only includes practice **w** but not practice **y**).
▶ That leaves us with option **D**, and we have already worked out that practices **w** and **y** are ones that we should always follow.

What about practice **z**? We ruled out option **C** because it did not include practice **w**, but we can double check by thinking whether we must always wear a dust mask. In general, we should not use more protective clothing or equipment than we need because it could cause other problems. For this reason, we would only wear a dust mask if the experiments created dust. If you wear a mask when it is not needed, you might increase the risk of touching your face with contaminated hands or gloves.

In this case, we used a process of elimination to help us by narrowing down the options. Of course, you might look at the question and know immediately that practices **w** and **y** are the only ones that you should always do and go straight to option **D**, the correct answer.

Short-answer questions

These are usually for just one, two or three marks, although you might find occasions where you have a list of four or five short answers within a question. Try to be concise and pay attention to the command word, usually **identify**, **state**, **name** or **give**, but you will also sometimes find **describe** or **explain** questions that only require a short answer.

Extended writing

These will usually be **assess** or **evaluate** questions and require you to write at greater length. It helps to think a little more before you write – to gather your thoughts and make sure you maintain a coherent and logical flow. Questions are normally worth between four and six marks, and you will most likely receive one mark for each valid point, so make sure you make at least the appropriate number of points in your answer. These need to be different points, however; you will only get one mark for saying the same thing in two different ways.

Some questions are worth 9 or 12 marks for the answer itself, plus 3 marks for **quality of written communication (QWC)**. These questions are **band marked**. Such questions do not have one mark for each point you make. Instead, your answer will be judged as a whole against criteria listed in the mark scheme. These questions require you to address all three of the **assessment objectives** (see above). So, make sure that you include knowledge, an explanation, and conclusions or judgements.

Quality of written communication

The three QWC marks are awarded based on how well you express yourself and how well you structure your answer. You must write in clear and grammatically correct formal English so that your meaning is clear. You must also use a wide range of relevant technical terms effectively.

These questions are worth 12–15 marks in total, so you should practise them so that you are comfortable answering them. It is also good preparation for your future professional life – you are effectively writing a short report, and report writing is a skill that employers value highly.

Calculations

Make sure that you show all your working in calculation questions. Some questions will explicitly remind you to do this, but you should try to make a practice of always showing your working. The reason for this is twofold:

▶ You are more likely to spot a mistake or, if you think your final answer is wrong, it makes it easier to see where you went wrong.

▶ There may be marks for intermediate steps in the calculation, or you could be given credit for 'an error carried forward', in other words where you make a mistake at the start but the rest of your working is correct.

Remember, therefore, to check through each stage of your answer.

Employer-set project

The ESP is an important part of the qualification. It gives you the chance to apply the core knowledge and, particularly, to demonstrate the core skills (see the Core skills chapter). You will develop and produce a substantial piece of work following an employer-set brief.

There will be a choice of ESPs, one for each of the occupational specialisms:
▶ laboratory sciences
▶ food sciences
▶ metrology sciences.

The instructions for each ESP will be broken down into a series of tasks, usually six, along the following lines:
1 Research a strategy.
2 Plan a project.
3 Gather and/or analyse data.
4 Present outcomes and conclusions.
5 Group discussion.
6 Reflective evaluation.

AOs for the ESP

The ESP has a different set of AOs to the core exams. You will need to demonstrate several of these at each stage in the ESP.
▶ **AO1** Plan your approach to meeting the project brief.
▶ **AO2** Apply core knowledge and skills to the development of a scientific project.
▶ **AO3** Select relevant techniques and resources to meet the brief.
▶ **AO4** Use English, maths and digital skills as appropriate.
▶ **AO5** Realise a project outcome and review how well the outcome meets the brief.

If you look back at the project practice features at the end of each chapter, you should be able to see how these address some or all of the tasks you will undertake in the ESP.

Synoptic assessment

Synoptic assessment simply means questions or tasks that cover different parts of the specification. For this T Level, this will be used to assess you on the content for your chosen occupational specialism. As you worked through this book you will have seen cross references to different chapters, which gives you an idea about some of the connections that you might have to make. Synoptic assessment might present you with a context that means you have to use knowledge of different areas of the specification. In such cases, try not to be too narrow in your thinking.

It will help you to answer this type of question if you are always on the lookout for connections between the topics you cover. Synoptic assessment is an opportunity for you to demonstrate the breadth as well as the depth of your knowledge and understanding.

This book covers the Core component, assessed through the two exams and ESP detailed above. The three Occupational Specialisms are beyond the scope of this book, but we have included some information here about the assessment.

You will be assessed through a number of Synoptic Assignments specific to your chosen Occupational Specialism. These will test your knowledge and understanding of the Core components covered in this book as well as knowledge and skills gained from your Occupational Specialism. There will also be at least one assessment of practical skills.

Occupational Specialism: Laboratory Sciences

You will be assessed against three performance outcomes:

- **Performance outcome 1:** Perform a range of appropriate scientific techniques to collect experimental data in a laboratory setting, complying with regulations and requirements
- **Performance outcome 2:** Plan, review, implement and suggest improvements to scientific tasks relevant to a laboratory setting
- **Performance outcome 3:** Identify and resolve issues with scientific equipment or data errors.

Occupational Specialism: Food Sciences

You will be assessed against four performance outcomes:

- **Performance outcome 1:** Perform appropriate activities to support the food supply chain, complying with regulatory requirements

- **Performance outcome 2:** Develop new food and food-related products to support the food supply chain
- **Performance outcome 3:** Identify and resolve issues in the food supply chain
- **Performance outcome 4:** Collect, analyse and interpret food production data.

Occupational Specialism: Metrology Sciences

You will be assessed against four performance outcomes:

- **Performance outcome 1:** Plan appropriate scientific measurement for any measurand to comply with regulatory requirements
- **Performance outcome 2:** Perform scientific measurement tasks using the most appropriate measurement for a measurand to ensure accuracy
- **Performance outcome 3:** Collect, analyse and interpret data from measurement tasks
- **Performance outcome 4:** Identify and resolve issues with measurement tools and equipment.

Glossary

Absolute zero The temperature at which all molecular motion stops and the pressure of a gas falls to zero.

Accident A separate, identifiable, unintended incident, which causes physical injury. This specifically includes acts of violence to people at work.

Accuracy Measurements that are close to the true value.

Acid A proton (H^+ ion) donor. An acidic solution contains H^+ ions.

Activation energy The minimum energy required to start a reaction.

Active site The part of the enzyme where the substrate binds.

Activity The rate at which a radioactive source decays.

Adsorbent Often used to describe the stationary phase in chromatography because substances become adsorbed to it during separation.

Adsorption When a substance (e.g. a gas, liquid or solute) binds to or attaches to another, usually solid.

Alkali A water-soluble base, such as sodium hydroxide. An alkaline solution contains hydroxide (OH^-) ions.

Allele A variant of a gene.

Amino acid A molecule with both an amino group and a carboxyl group. Amino acids are the small molecules (monomers) from which all proteins are made.

Ammeter Measures the flow of current; always connected in series.

Amplitude The maximum displacement of any point from the equilibrium position

Analyte The solution of unknown concentration in a titration.

Antibody A blood protein that is produced in response to a specific antigen. An antibody binds specifically to an antigen in a similar way to an enzyme binding specifically to its substrate.

Antigen A substance that is recognised by the immune system as self (the body's own cells) or non-self (foreign cells and pathogens) and stimulates an immune response. Antigens are found on pathogens but also on the surfaces of all body cells.

Aseptic Free from contamination with micro-organisms that might be pathogenic or cause food spoilage.

Atomic number Refers to the number of protons in the nucleus.

Attenuated A strain of a pathogenic organism that has been modified or weakened so that it does not cause disease.

Base A proton (H^+ ion) acceptor. Examples include hydroxides as well as ammonia and amines.

Base sequence The order of the bases, attached to nucleotides, in a DNA molecule.

Bias Anything that prejudices or influences the results of an experiment or observational study in one direction.

Bibliography A list of all of the sources you have used in researching your work.

Biohazard A micro-organism, cell culture or human endoparasite that may cause infection, allergy, toxicity or otherwise create a hazard to human health.

Biological agent See Biohazard

Burette A long glass tube that has a tap at the bottom and is marked in $0.1\,cm^3$ divisions. It is used to deliver an accurate volume of liquid (the titre) to reach the end point.

Calibration The process of comparing measurements, usually against a reference standard.

Categorical data Is divided into groups or categories, such as male and female, ethnic group, city or country of residence.

Cell-signalling The process by which cells communicate with each other, usually by release of chemicals such as histamine, cytokines and interleukins.

Centromere Joins the two chromatids together during the early stages of mitosis.

Charge A fundamental property of many subatomic particles. Electrons, by convention, have a negative charge.

Chromatids The two identical copies of a chromosome formed by DNA replication (sometimes called sister chromatids). The use of the term chromatid can cause confusion. We use chromatid as a shorter way of saying 'one of a pair of identical chromosomes'.

Chromatography The separation of the components of a mixture dissolved in a liquid or gas (the mobile phase) carrying it through a structure holding the stationary phase.

Chromosomes Consist of DNA and proteins and carry the genetic information.

Cilia (singular **cilium**) Are hair-like structures found on the plasma membrane of some types of cell, particularly in the lungs.

Condensation reaction A reaction between two small molecules to produce a larger molecule and water; most large biological molecules are formed by condensation reactions.

Consumables Items that are used and then disposed of. They are mostly single-use but might be re-used in some circumstances.

Continuous data Is numerical and can be measured. It is possible to have any intermediate value, for example, height, mass, length.

Corrosion The process where metals react with substances in the air to form oxides, carbonates, hydroxides or other compounds.

Cosmic radiation (also called cosmic rays) Originates from the sun as well as from outside the solar system. It is a mixture of high energy protons (hydrogen nuclei), alpha particles (helium nuclei) and beta particles (electrons).

Coulomb (C) The unit of charge.

Covalent bond Forms when atoms share one or more pairs of electrons; the electrons are attracted to the nuclei of both atoms, and this holds the two atoms together. Covalent bonds are very strong.

Cytoplasm Is the fluid component of the cell, enclosed by the cell membrane and surrounding the organelles.

Delocalised electrons 'Free' electrons that are not associated with any single atom.

Dependent variable (often denoted by y) A variable whose value depends on that of another variable

Diffusion Is the movement of a substance from a high concentration to a low concentration.

Diploid A cell that contains two copies of each chromosome; in humans the diploid number is 46 chromosomes.

Discrete data Is numerical and can be counted. For example, number of patients (you cannot have half a patient). This is sometimes referred to as integer (only whole numbers).

Disulfide bond Covalent bond between the sulfur atoms of the R groups of two cysteine amino acids.

Electromagnet Produced when a current flows through a coil of wire.

Electromagnetic spectrum Describes all the different types of electromagnetic waves.

Electromagnetic waves Include gamma rays, X-rays, visible light, microwaves and radio waves. Their energy is carried by oscillating electric and magnetic fields.

Electromotive force (emf) Like potential difference, except that it refers to power supplies such as cells (batteries), generators or mains power supplies. These transfer other forms of energy, such as light energy or kinetic energy, into electrical energy.

Eluate The mobile phase, containing dissolved substances, as it emerges from a column.

Eluent The solvent (mobile phase) used to wash substances out of a column.

Elution To wash out. In column chromatography this means 'washing out' a substance that has become adsorbed to the column (stationary phase).

Empirical formula The smallest whole number ratio of the atoms present in a compound.

Employment tribunals Responsible for hearing claims from people who think an employer has treated them unlawfully, for example, through unfair dismissal or discrimination.

Endoparasite A parasite that lives inside its host and obtains its nutrition directly from the host.

End point The point in a titration where the indicator changes colour.

Endothermic Reactions that take in heat; ΔH is positive.

Equivalence point The point of neutralisation where the number of moles of acid and base are equal. This should ideally be the same as the end point.

Exothermic Reactions that give out heat; ΔH is negative.

Extrapolate To extend a graph so that we can infer unknown lower or higher values.

False negative A negative result that is not true.

False positive A positive result that is not true.

Flagella (singular **flagellum**) Are similar in structure to cilia but are much longer and are involved in propulsion of the cell.

Fluid mosaic model Describes the structure of the plasma membrane and how its components are arranged.

Frequency In Hz, the number of complete waves that pass a given point in one second.

Gametes The haploid sex cells (i.e. in animals, the sperm and egg) that join in the process of fertilisation to produce a new organism.

Gaussian distribution See Normal distribution

Gene A sequence of bases in DNA that codes for (contains the information to make) a polypeptide.

Generator effect When a potential difference (voltage) is induced in a wire that experiences a change in magnetic field.

Genetics The study of how single genes, or a small group of genes, function and how they affect the appearance and functioning of the organism.

Genome The entire genetic material of an organism. This includes DNA that does not code for proteins as well as the coding DNA (genes).

Genomics The study of how all the genes in an organism interact, as well as the role of non-coding sequences of DNA.

Grievance Any concern, problem or complaint you may have at work. If you take this up with your employer, it is called 'raising a grievance'.

Group Refers to the columns in the periodic table. Elements in each group have the same number of outer shell electrons.

Half-life The time taken for half the unstable nuclei in a sample to decay.

Haploid Just one of each pair (half the number) of chromosomes; in humans the haploid number is 23 chromosomes.

Hazard Something that has the potential to cause harm.

Homologous chromosomes They carry the same genes but different alleles; in animals, one of each homologous pair is inherited from the female parent (mother) and one from the male parent (father); in humans there are 23 pairs of chromosomes.

Hydrogen bond The attraction between slight positive and negative charges when hydrogen atoms are attached to oxygen or nitrogen atoms.

Hydrolysis Literally 'splitting with water'; it is the reverse of a condensation reaction; most large biological molecules are broken down in hydrolysis reactions.

Hydrostatic pressure The pressure in a liquid.

Independent variable (often denoted by x) A variable whose value does not depend on that of another variable.

Indicator A substance that changes from one colour to another or from coloured to colourless depending on whether it is in acidic or basic solution.

Induced magnet An object that can become a magnet when it is placed in a magnetic field.

Inflammation A local response to injury and infection.

Innate immunity The non-specific mechanisms, present from birth, that protect against a wide range of pathogens. These mechanisms, including inflammation, phagocytosis and antimicrobial proteins, work quickly, but are not always very effective.

Intellectual property (IP) Creations of the mind such as inventions, literary or artistic works, designs, symbols or names used in commerce.

Ionic bond The attraction between a positively charged R group of one amino acid and the negatively charged R group of another amino acid.

Ionisation The formation of charged particles from neutral molecules or atoms by adding or removing electrons.

Ionising radiation Any form of radiation that interacts with matter, resulting in ionisation of that matter.

Ions Atoms that have lost electrons (positive ions) or gained electrons (negative ions).

Irradiation The process of exposing an object to radiation. It does not make the object radioactive.

Isotopes Atoms of the same element (they have the same number of protons) but with different number of neutrons.

Laws (legislation) Passed by Parliament. They state the rights and entitlements of individuals and provide legal rules that have to be followed.

Leading question A question that pushes the respondent to reply in a particular way; it almost contains its own answer.

Levels A way of grading a qualification or set of skills and the corresponding occupations.

Lines of magnetic flux Indicate the direction and strength of a magnetic field.

Lymphocytes Small white blood cells. B lymphocytes, or B cells, are responsible for antibody production. Different types of T lymphocytes, or T cells, play different roles in the immune response.

Magnet A material or object that produces a magnetic field.

Magnetic field A region where magnetic materials experience a force.

Magnetic materials Such as iron, steel, nickel and cobalt will experience an attractive force when placed in a magnetic field.

Magnetism The force experienced by some types of metals in the earth's magnetic field or in a magnetic field of a magnet. It is also defined as the attractive or repulsive force produced by a moving electric charge.

Magnification How much bigger the image is than the actual object we are viewing. It should not be confused with resolution.

Materials Include items such as ingredients or components used in the manufacture of a product.

Maternal chromosomes The members of a homologous pair inherited from the mother (female parent).

Membrane All membranes consist of a phospholipid bilayer together with proteins and other components.

Mole An amount of substance. This helps us to work out the reacting proportions in any reaction. We can also use the mole to work out reacting masses or volumes. The abbreviation is mol. This is the Avogadro number = 6.02×10^{23}.

Molecular formula The exact number of each type of atom in a compound.

Motor effect When a current-carrying wire is placed between magnetic poles. The magnetic field around the wire interacts with the magnetic field it is placed in. This causes the wire and magnet to exert a force on each other and can cause the wire to move.

Mutation A change in the sequence of bases in DNA. This can occur in a number of ways. When a mutation occurs within a coding region of DNA a new allele can be formed.

Non-pathogenic An infectious organism that does not cause disease.

Normal (Gaussian) distribution A symmetrical distribution of quantities about the mean value; it is also known as a bell curve because of its shape.

Null hypothesis A statistical tool that states that any differences between the observed and expected results or between mean values are due to chance; statistical tests are used to decide whether to accept the null hypothesis (that the differences were due to chance) or reject it (deciding that the results were not due to chance and the differences are statistically significant).

Orbitals Are where electrons are located. Each orbital can be empty or can contain one or two electrons.

Organelles Specialised structures within plant and animal cells that have specific functions. Some types of organelle are also found within bacterial cells.

Organism An individual plant, animal or single-celled lifeform.

Parallel circuits Where the components are each connected separately to the positive and negative terminals of the power supply.

Paternal chromosomes The members of a homologous pair inherited from the father (male parent).

Pathogen A micro-organism that causes illness or disease by damaging host tissues and/or by producing toxins.

Peptide A compound containing two or more amino acids joined together by peptide bonds. A dipeptide contains two amino acids bonded together.

Permanent magnet Produces its own magnetic field.

Phagocytes Produced in the bone marrow and circulate in the blood. Some leave the blood and are present in the tissues.

Phagocytosis The process of a phagocyte engulfing a pathogen or other foreign material.

Phospholipid A large molecule formed from a glycerol molecule covalently bound to two fatty acid molecules and a phosphate group.

Phospholipid bilayer A double layer of phospholipids with the hydrophobic tails arranged towards the middle and the hydrophilic head groups on the outside. It forms the basis of all biological membranes.

Photosynthesis The process that plants use to make complex organic compounds such as sugars from carbon dioxide and water using light energy.

Plant Any equipment used in the workplace, e.g. laboratory equipment.

Plasma membrane Sometimes called the cell-surface membrane, it is the membrane that surrounds all types of cell; animal, plant and bacterial.

Polymer A long molecule made from many small molecules called monomers.

Polypeptide A polymer of amino acids joined together by peptide bonds.

Population In statistics, a population is the collection of all objects or measurements; all the people in the UK represent a population.

Precise Measurements that are close to each other, but they may be inaccurate.

Pressure The force per unit area exerted by a fluid on its container.

Prospective These studies take a group of people and observe them over a period of time. This could involve looking for correlations between factors such as diet or exercise and development of cardiovascular disease. The advantage is that the data collection methods can be tailored to the question being asked. The disadvantage is that these can take many years to complete.

Protein A polypeptide with a recognisable three-dimensional structure. It may contain more than one polypeptide chain.

Qualitative data Is descriptive, for example, a patient's medical history.

Quantitative data Is numerical, for example, the results from a laboratory experiment.

Quantitative structure–activity relationship (QSAR) The use of statistical methods to refine the approach of correlating chemical structure with biological activity.

Radioactive decay The random process that occurs when an unstable nucleus loses energy by giving out alpha or beta particles or gamma radiation.

Radiocarbon dating The process that uses a radioactive isotope of carbon, carbon-14, to determine the age of an object containing organic material.

Radioisotopes Unstable isotopes that undergo radioactive decay emitting radioactivity as they do so.

Radiopharmaceuticals Therapeutic drugs (medicines) that incorporate radioisotopes. Some radiopharmaceuticals are used to treat disease whereas diagnostic radiopharmaceuticals are used to help in diagnosis of disease.

Read across A technique that attempts to fill gaps in experimental data about the properties, toxicity or potential environmental harm of substances. It does this by looking for patterns or chemical similarities in substances to predict the properties of substances that have not been fully tested.

Reference standard Something of known size, mass, concentration, etc. that we can use to calibrate equipment or methods.

Reportable injuries The following injuries are reportable under RIDDOR when they result from a work-related accident:

- the death of any person
- specified injuries to workers
- injuries to workers which result in them being unable to work for more than seven days
- injuries to non-workers which result in them being taken directly to hospital for treatment, or specified injuries to non-workers which occur on the premises.

Resolution The smallest change in a quantity being measured that gives a perceptible change in the reading. In microscopy, resolution is the ability to distinguish two points as being separate.

Respiration The process by which energy is transferred from biological molecules such as glucose to ATP for use in energy-requiring processes in the cell.

Retrospective These studies look backwards at data from a group of people over many years. This often involves examining published data classifying people according to risk factors or medical outcomes. Although these can give results more quickly than prospective studies, the disadvantage of retrospective studies is that there is little control over data collection. This type of study typically looks at published data from many different sources that might involve different methods of data collection or analysis.

Risk How likely a hazard is to cause harm.

Sample A subset of a population that is selected for a particular investigation; the size of the sample and how it is chosen can affect the outcome of the investigation.

Semi-conservative replication When DNA replicates two new double helix molecules are formed, but each one consists of one of the original strands and one newly synthesised strand.

Series Circuits where the components are connected in line, end to end between the positive and negative terminals of the power supply.

Standard solution The solution of known concentration in a titration.

Statistical significance This means that differences between data sets are not due to chance (random variation).

Stem cells Undifferentiated (non-specialised) cells that can give rise to one or more types of differentiated (specialised) cell.

STI or sexually transmitted infection Caused by a pathogen that is passed from person to person during sexual contact.

Substrate The substance on which an enzyme acts to form the products.

Titre The volume of standard solution needed to neutralise the analyte (i.e. to reach the end point of the titration).

Viscosity A measure of the resistance to flow of a fluid.

Wavelength The distance between the same point in successive cycles. For example, the distance from the peak of one wave to the peak of the next wave. The standard unit of wavelength is the metre, m. However, electromagnetic waves can have wavelengths from 10^4 m to 10^{-15} m, so you will often see wavelengths expressed in units such as nanometres, nm. ($1\,\text{nm} = 1 \times 10^{-9}\,\text{m}$).

Work-related An accident in the workplace does not always mean that the accident is work-related – the work activity itself must contribute to the accident. An accident is 'work-related' if any of the following played a significant role: the way the work was carried out; any machinery, plant, substances or equipment used for the work; the condition of the site or premises where the accident happened.

Index